INFORMATION AND BEHAVIOR
Volume 1

Editorial Board

INFORMATION AND BEHAVIOR
Volume 1

Edited by

Brent D. Ruben

Transaction Books
New Brunswick (U.S.A.) and Oxford (U.K.)

ISSN: 0740-5502
ISBN: 0-88738-007-7 (cloth)
Printed in the United States of America

Contents

List of Figures

List of Tables

Acknowledgments

Initiating a series such as *Information and Behavior* is a substantial task, one which is possible only with the energy and support of a great many individuals. I am particularly grateful for the dedicated services of Associate Editors Jim Anderson and Dick Hixson and Assistant Editor Gordon Miller, who have devoted countless hours to manuscript review, editing, and index preparation. To the members of the prestigious Editorial Board, whose encouragement and editorial assistance have been essential at every point in the project, I am also most appreciative. Special thanks to Jim Anderson, Bob Cathcart, Frank Dance, Young Kim, Elizabeth Loftus, Dick Hixson, Dan Nimmo, Ev Rogers, Herb Schiller, Fred Smith, Bud Wheeless, Rolf Wigand, and Fred Williams who, in addition to their other editorial responsibilities, graciously agreed to submit manuscripts for this first volume.

Thanks to Irving Louis Horowitz and Scott Bramson of Transaction, from whom I have come to expect only the best in colleagiality and candor. Valued editorial assistance on *Information and Behavior* was provided by Dalia Buzin of Transaction and by Barbara Kwasnik, Jacki McGuiness, Joan Chabrak, and Marsha Bergman of Rutgers. To each I express my sincere appreciation.

A special note of appreciation is expressed to Rutgers University, and in particular to Provost, Kenneth Wheeler; Associate Provost, Jean Parrish; Faculty of Professional Studies Dean, Bill Stuart; and School of Communication, Information and Library Studies Dean, Dick Budd, for support and encouragement which have made the series possible.

Finally, a word of thanks to Jann, Robbi, and Marc for their thoughtful tolerance of yet another important project.

Guidelines for Manuscript Preparation and Submission

Manuscripts are encouraged which address topics of information, information processes, and technology, including: the information concept, applications and impact of information and information-processing technology, communication and information policy; information networks; cognitive information processing; information systems; information organization; information storage and retrieval; information management; economics of information; scientific communication; information forms; information media; artificial intelligence; and individual, social, and cultural functions of communication and information.

Articles may be philosophical, qualitative, quantitative, or policy-oriented in perspective. Preference will be given to manuscripts that address issues in a manner that will be of interest to a broad interdisciplinary audience.

Manuscripts to be considered for publication in Volume 3 should be submitted in triplicate, following the American Psychological Association style, by October 31, 1985, to Brent D. Ruben, Professor and Director, Ph.D. Program, School of Communication, Information and Library Studies, 4 Huntington Street, Rutgers University, New Brunswick, New Jersey, 08903. A style summary sheet is available from the editor.

Foreword

Alfred G. Smith

Here is the first volume of still another series opening up one more channel for the diffusion of communication and information studies. These new series keep coming like babies in China. What brings on this ceaseless propagation of publications?

Maybe it is brought by a stork winging in with *Information and Behavior* dangling in diapers under the big bird's beak. Maybe the birth of *Information and Behavior* is unavoidable, consummating the great fertility of information research. Or maybe that fertility of communication scholarship is itself like the stork, a notion in want of evidence. Whatever the case may be, the question still remains: Why add another series to those already in our horn of plenty? Well, that depends on its novel distinction.

All noteworthy innovations are protests against the status quo. Every new achievement, if it is worth at least a sidelong glance, says the old order is wrong or just not good enough. It says the old boat leaks or drifts off course. The old program does not fit the changing times. So every new curriculum is a revolt against the prevailing system of education, and each new publication cries out, even if softly, against the prevailing achievement of its field. Whether or not the birth of this new publication is at all significant depends on the criticism it embodies.

Let me present two of these criticisms as I see them here, two fundamental and interrelated flaws of our publications today. One of these is a fault of the overall scholarly system, the other a fault of the individual research contributions. One looks at the forest, the other at the trees. Both are aberrations that I think *Information and Behavior* should be against. They also indicate what *Information and Behavior* can be for. In this way let me propose some new distinctions for this new series, and by implication for higher learning in general.

The publishers, editors, and scholars have neither authorized nor empowered me to deliver this pontifical service. I am self-ordained, blessing and sprinkling this infant with mystic waters from my own font.

The Publication Potlatch

Think of the university library and its stacks and stacks of books on information technology, information transfer, and information services; on international, interpersonal, and mass communication; on print and electronic journalism; on radio, television, and film; on public relations, advertising; and on and on, books as boundless as the seas. Think also of that library's periodical room and its hundreds of journals in information studies and communication, with many issues of each journal, scores of articles in each issue, journal after journal, year after year. All of this together is only the exposed and visible portion of a great whale.

Think of the toil and sweat behind it all, the years of study by each author, the months of research, the weeks at the desk, and even the weekends that families had to stay at home because of that paper in the typewriter. And for each paper published there were many rejected, work that is even more out of sight. This reminds us also of the work of editors, publishers, and the rest of the impresarios. Altogether there is enough labor here to build all the pyramids of Egypt and Disneyland besides.

What have we erected instead? Has our output been worth the input? How much has science been enlarged, or society bettered, or our spirits lifted? Not much for all the effort. For the system as a whole, the costs of producing this mountain of printed matter are far greater than the benefits received. Perhaps the one page that makes a real difference is worth the thousands of others, but this allocation of intellectual resources is an awfully wasteful business.

Even the visible portion of that whale is seldom seen. The number of people who read research publications is very small. Only a handful of professional scholars and scholarly professionals read any one of these books and journals. (This excludes textbooks, but they rarely enrich the field with original facts, findings, or inspirations.) We know that any one article in any psychology journal is read by less than 2 percent of the psychologists. The percentage in information studies and communication is probably much the same. There is no Say's law here: the supply of publications does not create its own demand. Instead, God willing, the mass of the readership corresponds to the weight of the contributions.

But these works should be taken at more than face value. Their significance is more than literal and apparent. If we suppose them to be primarily additions to human knowledge, then we misunderstand their basic mission. These works are ceremonial communications.

Shelf upon shelf of learned publications form a solemn academic procession, chanting a litany of paradigms and parameters, and invoking names like Shannon and McLuhan as if turning a prayer wheel. For most of us, authorship in this parade is our declaration of faith in the communion of scholarship. Each printed page is our pledge of allegiance to the fellowship of research. These publications are the ritual trappings of membership in a tribe.

They are like a potlatch. Totem poles and potlatches were the distinctive creations of the Indians of British Columbia and southern Alaska. Members of the Haida, Kwakiutl, and kindred tribes erected totem poles as huge monuments to their ancestry; they gave potlatches to celebrate their own social position. Both were implicit ways of self-promotion and advertising.

In these tribes anyone who assumed some special position or mark of note had to certify that distinction with some public theatrics. The Haida youth taking his seat in the men's council, or the Kwakiutl family claiming the right to a crest, gave a potlatch. This was a sumptuous feast that publicly displayed the host's wealth and social position. The climax of this feast was a dramatic performance of ultimate squandering and wastefulness. The host burned his blankets up, kicked his canoes in, and wiped his assets out. This haughty show of sheer extravagance proved and confirmed his council seat or crest or whatever his special distinction.

Among some of the academic tribes of North America we potlatch by publishing. Like the destroyed canoes that carry no one anywhere, and the burned up blankets that keep no one warm, our writings are rarely read. So much of our scholarship is merely for show, it is a barbaric waste, but that does not matter. The fundamental goal is the symbolic payment of dues.

This is the supreme excellence, the most admirable and praiseworthy performance by people of special distinction, and well beyond what is expected of ordinary members of the tribe. Most of them do not take part in the publication potlatch. Most academics even at major institutions have never published a book; many have not published an article in years; and only a few members of longer standing are vigorously occupied in any original scholarship. There is no academic imperative to publish and flourish or else.

Yet many scholars are convinced they have to follow this ritual to get a job or keep one, or to get a raise in rank or pay. That's the way it is, and supposedly, so it shall be forever. These scholars publish more for worldly profit than for innocent wisdom. They may not make love for money, but their academic passion is for sale.

It makes little difference to our store of knowledge whether we publish as a ritual exhibit or not at all. There is little innovation either way. Scholars in communication and information studies who produce six articles a year and a book every six years, who traipse from one conference to the next and potlatch in every hole and corner, may be displaying a clannish conformity that actually contradicts the creative skepticism of science.

The profusion of our publications comes largely from the redundancy of scholarly ritual. We need more than such ritual if we want *Information and Behavior* to be more than just another pretty journal. Particularly now when journals that are printed on paper are closer to last rites than to baptism, now when we have just begun to store and retrieve information electronically, we need to establish some special distinction for *Information and Behavior.*

Distinctions can be superficial and false, like Goldwater's glasses or Groucho's moustache. The series could, for example, vary its size and frequency. If one volume after another came out at unvarying intervals, and each volume had the same number of pages, we might have another redundancy machine like the ritually rotated prayer wheel. Instead *Information and Behavior's* distinguishing marks could be to follow the irregularity of innovations in scholarship by bringing out a fat volume one year, no volumes for perhaps seventeen years, and then a slim volume two years in a row. On the same order, the series could spend most of its budget, hundreds and hundreds of dollars, on a worldwide advertising campaign soliciting nonceremonial contributions to its treasury of knowledge. But such remedies are shallow and silly.

The heart of the matter is whether this new publication is necessary. If it is not, then no nostrums like these can ever make it so. What we need, instead, are other priorities.

Before considering them, however, I must consider my second criticism of scholarship in communication and information studies today.

The Refined Red Herring

The first choice of modern whalers is to hunt the great blue whale, the largest animal the world has ever known, up to 100 feet long and 150 tons plump. Remarkably, these great whales live on tiny things, plankton floating in the brine, little plants and fishes barely visible to the naked eye. That is also true of the great body of information studies and communication publications. It, too, feeds on minute morsels. Most of its individual studies are very small fry.

Go back to the university library, pick out a sample of our journals, and take stock of the individual contributions. They often put me in mind of "Adaptive Territoriality aboard Public Vehicles," and "Newspaper Readership among Celibate Vegetarians." The one contribution discusses nonverbal communication by teen-agers on a school bus in West Virginia, and the other contribution turns up some people who abstain three ways at once. Such little fishies are too often the staple of our current publications.

Now take stock, as well, of our vital needs and challenges, such as the conquest of cancer by medicine, the production of nuclear fusion by physics, and the understanding of human identity and diversity by anthropology. In communication and information studies we face problems just as critical, grim questions of access and privacy, of freedom and equality, of truth and ballyhoo. The hope of the human race everywhere depends on advances in science and research, but our studies are seldom so serious. Most often, in fact, they are of little consequence, scraps and fragments of fluff.

Again the foremost mischief maker is ritual. Now, added to the rites of tribal membership, we have the rites, alas, of research itself. They must bear much of the blame for so many of our small fry studies. For the most part they are guilty because their devoted routines divert our attention.

Research methods are the means of inquiry rather than its goals. When means become ends in themselves, they turn into rituals. They demand conformity. They become authoritarian. And with this they retard growth. Yet as a rule, methodology is the favorite yardstick that editorial boards use to evaluate the papers they receive. We are most likely to welcome with open arms the contribution that obeys the whip hand of statistics, the research design that is waterproof and shock resistant, and the manuscript that complies with the style sheets in outward appearance. We swear by the liturgy and form of research.

We generally give less credit to the content of a study: its vision of great issues, its implications, and its likely applications. We tend to avoid the significance of an idea and look primarily at the mechanics of studying it. We feel awkward groping with meaning, and are afraid of wrestling with ghosts in the dark. We feel more adequate wrestling with ritual routines.

If Plato submitted *The Republic* to us, we would soon put him in his place. "What kind of methodology do we have here? Are there any statistical controls? Where are the allowances for extraneous variable conditions? Oh, I see. Well, I'm afraid, Professor Plato, that this is not the way to go. This really will not do."

Our studies are smaller fish than *The Republic*. This makes them even more vulnerable when research methods become an evasion of our real quarry. When our ideas of "adaptive territoriality" and "celibate vegetarians" are slight and frothy to begin with, and our methodology refines them even further, these ideas often wither away into extinction.

When we first beget an idea to be studied, it generally compares two different things, such as cancer and diet, or fusion and the laws of conservation. Then we polish the idea into a researchable hypothesis.

Suppose we think of comparing young men from the country with those from the city, and we think that the ones from the country will make better soldiers. Such a possibility may be of some interest to an army recruiter, a sociologist, or an advocate of rustic virtues.

Now we have to lick this idea into shape by defining its terms operationally. All right, let us say that a young man is a male seventeen to twenty-two years old, that a city has at least 5,000 people, and that the test of soldierly proficiency will be a forced march of thirty miles, full pack and all, after which we count the blisters on the young men's feet: the fewer the better. We might even pretest for reliability and validity.

In this way we finally form the true-to-type hypothesis: males seventeen to twenty-two years old from population centers of more than 5,000 people get more blisters on their feet after a forced march of thirty miles than matched males from smaller population centers. The initial idea is now methodologically refined and ritually correct. With faithful compliance to the routines of scientific research, we have cast our net for little fishies, and landed one.

But stifle those cheers of rejoicing. What on earth does our hypothesis actually say? Not much. Is it even vaguely meaningful? Not really. And who cares that 5,000 people divided by thirty miles equals 2.67 blisters? Not a blessed soul. The hypothesis is claptrap: high-falutin hooey.

Our business is built on research design, and it is carried on with every analytic tool that will hold a sharp edge. We need all the scientific methods we have in stock, and any others we can order. But we do not need a hundred pound harpoon to catch little fishies.

We often think so much of our research instruments that the actual problem under investigation eludes us. The hypothesis forms, the questionnaire designs, the multivariate techniques, and all the other delicate scalpels and forceps of the trade become, in fact, red herrings. They lead us away from our real quarry. They get us to chase the means and routines of research rather than the substance. As a result, so many of our individual contributions to the vast literature of communication are puny.

Information and Behavior needs something more than refined hypotheses. It needs significant ones. It needs more than refined means of research. It needs refined ends. And so we come to some alternative priorities.

On Purpose

Individual studies are like French restaurants. We can rate them: three stars or four. *The Origin of Species* is like Bo Derek: a "10." Many of the contributions that win a Nobel Prize are a "9." The most significant communication and information studies are perhaps an"8," given the relative enlightenment of this field so far.

Each of us can take his own pick of studies that are top of the line. That itself is more than difficult. It means we have to compare, for example, *The Gutenberg Galaxy* with *Cybernetics,* which are like parsnips and moonbeams—hardly comparable. It means we will rate these studies tomorrow as we rate them today—hardly likely. The fact is, our ratings have hardly a leg to stand on. Yet we need them. We have to rank the little fishies. As students and teachers we have to decide what we want to study most. As editors and practitioners we have to decide what to use and what to hold back. Ratings are unavoidable yet variable, and not above suspicion.

A scholarly series that publishes only a "10" or "9" is not necessary. There are too few Darwins and Einsteins in our field. Accordingly, let *Information and Behavior* publish the "8."

Then the crucial question is: What makes an "8" an "8"? And why is a "2" a "2"? What is the test for a distinguished contribution?

I think scholars generally agree that "I think" pieces rate "1" or "2." These tinpot contributions make personal pronouncements without evidence or specifications. Ritual pieces rate a little higher, perhaps "3" to "5." At the low end of that range these contributions file, shuffle, and refile bits of raw data. They are the work of file clerks. At the high end of that range the pieces are still means-oriented, but the means are more sophisticated. They all follow rituals that are not so much the outward expression of inward meanings as they are self-accomplishing acts of devotion.

At "6" and "7" the contributions become more interesting. The "7" hopes to come upon clues to the way things work. The "6" hopes for clues that things do not work that way. Both steer a course toward scholarly goals, hoping to come upon their islands of enlightenment. Such islands are windfalls of meaningful data. Darwin, you see, came upon his islands in the Galapagos, where the variety of finches displayed the whole process of natural selection for him. Just as lucky was Malinowski, the anthropologist who came upon the Trobriand Islands. They displayed all the functionalist traits of reciprocity and integration that made Malinowski famous. Islands of enlightenment can perhaps be found anywhere: a school bus in the wilds of West Virginia, or a factory in Hawthorne, Illinois.

Then there is the "8." It does not believe in islands of enlightenment. It does not hope to receive cues, like manna from heaven, or to stumble on windfalls of meaningful data. The "8" is not a receiver; it is a sender. It produces evidence and generates significance. The Galapagos did not put Darwin on the map. Darwin put the Galapagos on the map. The Trobriands did not make Malinowski famous. He made the Trobriands famous. The "8" is a sender. It has something to say.

This brings us to the heart of the matter: What does the "8" send? I think its most important message is its development of its own goals. This emphasis on goals is its novel distinction.

The "1" and "2" think-pieces study *nothing*; the "3" to "5" ritual pieces study *how*; the "6" and "7" scholarly pieces study *what*; but the "8s" study *what for*. (The "9" and "10" sockdolager pieces study some other world light years away.)

Ritual pieces take goals for granted because the performance of ritual is generally its own fulfillment. A ritual study of newspaper readership may aim to refine the methods of readership studies, or even to increase that readership itself, but it seldom examines the goals of either one. Any ritual study, however, even of school buses and foot blisters, can become an "8" if it develops its purpose.

The "6" and "7" scholarly pieces also take goals for granted. These pieces come upon communication and information as if they were islands in uncharted waters, ready to be discovered and explored. They come upon readership, kinemes, and feedback as if they were berries growing on a bush. Their reason then for studying these islands and berries is merely the mountain climber's reason: because they are there. But that can just as well be the reason for studying the lint in my pockets: the origins of that lint, its variability, and its distribution in time and space. The existence of lint is not a sufficient reason for devoting time, effort, and resources to studying it. Gregor Mendel did not study peas, and Ivan Pavlov did not study dogs, just because they were there.

An "8" does not receive its reasons and goals from out there somewhere; it is a sender. Darwin and Malinowski did not land on their islands stumbling ashore with empty hands outstretched asking for a contribution. They brought their own contribution with them, ideas that were perhaps only questions, faint, nebulous, and uncertain, but their own offering nevertheless that gave their studies purpose and significance.

The distinguished contribution starts with its own goals and then develops them. Most often it starts with goals that are not consistent with one another and that are not all equal. Shannon began with goals of efficiency and freedom in transmitting any information electronically. Pool began with practically the same goals politically. But both saw that efficiency and freedom are often inconsistent: one compromises the other. Indeed, once the issue of purpose and goals is faced, it mushrooms. We face an everlasting process of winnow and sift, change and test, to transform a collection of goals into a well-ordered and organized system.

That system invariably goes beyond the individual study. It implies some view of communication, some view of scholarship, and some view of society. What are their goals and purposes? A distinguished contribution may hold that television, for example, is escapist, activist, or whatsoever. It may hold scholarship to be plebeian, patrician, or classless. It may hold society to promote private enterprise and individual initiative. The distinguished contribution develops its own purposes and considers those of its neighborhood and hinterlands.

That generally means it goes on to challenge or to justify the status quo. The "8" goes on to social criticism and reconstruction. It has moral implications. The distinguished contribution relates *information* and *behavior*: information as data and behavior as conduct. It relates scholarship and morality.

In all, the huge waste of potlatch scholarship and the ritual evasion of significant research show how we are left, after all our labors of learning, with few substantial residues. But distinguished contributions can rise above this petty puttering. Great whales and little fishies, we can develop our own goals, and thereby let *Information and Behavior* open up new vistas in the high country of scholarship! Then we may hear the echo of Dorothy's awestruck words, "Toto, I don't think we're in Kansas anymore."

Introduction

Brent D. Ruben

The history of discourse has seen the advancement of any number of schemes for explaining human behavior. Some frameworks have emphasized the role of physical forces; others physiological, psychological, or spiritual forces; and still others social or political forces. Implicit in each is a mechanism through which the substance of these forces is created, conveyed, attended to, and acted upon. That mechanism, which is seldom singled out for attention in its own right, is *information.*

At the most basic biological level, information and information processing is fundamental to life. It is a primary means through which living organisms adapt to their physical and social environments (e.g. Miller 1965; Lawson 1963; Ruben 1972; Ruben 1984; Thayer 1968). Information and communication processes play an indispensable role in physiological self-regulation (e.g. Cannon 1932; Selye 1956; Young 1960), mating and reproduction (e.g. Dawkins 1976; Lawson 1963; Tinbergen 1965; Shorey 1976), parent-offspring relations and socialization (e.g. Etkin 1964; Frings and Frings 1964; Sebeok 1968; Thielcke 1970). Locomotion and the establishment and maintenance of territory (e.g. Shorey 1976; Thielcke 1970; Frisch 1950; Wilson 1971), among other basic life activities, also require information processing.

Additionally, information and information processes are fundamental in higher-level human activity associated with perception and cognition (e.g. Boulding 1956; Bruner 1973; Goffman 1974; Miller 1969; Loftus and Loftus 1976; Rapoport 1973; Schroder, Driver, and Streufert 1967; Shriffrin, Castellan, Lindman, and Pisoni 1975; Zubik 1969). The same can be said for personality development (e.g. Allport 1955; Johnson 1946; Kelly 1963; Ruesch and Bateson 1951), and other forms of complex individual behavior.

At an interpersonal, social, and cultural level of analysis, information and information processes play an equally vital role. They are essential in the establishment and maintenance of relationships, groups, organizations, and societies, and in the cultures and normative realities, rules, and roles of each (e.g. Berger and Luckman 1966; Blumer 1969; Budd and Ruben 1979; Cherry 1971; Deutsch 1966; Duncan 1962; Gumpert and Cathcart 1982; Goffman 1971; Hall 1959; Kuhn 1970; Machlup 1962; McLuhan 1964; McQuail 1969; Miller and Steinberg 1975; Ruben 1979; Smith 1966; Wright 1959). Furthermore, information has been an important element in the more general framework of system theory (e.g. Bertalanffy 1968; Ruben and Kim 1975), cybernetics (Ashby 1964; Wiener 1950), even electrical engineering (Shannon and Weaver 1949; Cannon and Luecke 1980), and computer science (Simon 1969).

Information is a powerful, generic concept, one which has traditionally played a central role in theoretical perspectives throughout the behavioral and social sciences. In recent years, *information* has also become a very popular concept, owing in large part to the "Post-industrial Revolution" and its technological trappings. Beyond playing a fundamental role in theories of behavior, information and information processing technology has come to occupy a significant place in the work of computer science, artificial intelligence, and administration. This is also the case in the plethora of applications in professions such as librarianship, journalism, medicine, and law, where a concern with information products and services is paramount.

In each of these contexts the issues of concern, like the concept itself, transcend traditional disciplinary boundaries. There are questions related to regulation and control, the impact of new technology on office and home life, new concepts of literacy and learning, the merging of information-processing and transportation technologies, freedom and privacy, the relationship between available information and use, the economics of information, and many others.

The concerns of the Age are not exclusively in the domain of the scholar or the professional. They are not solely issues to be explored within the disciplines of information studies or communication, psychology or political science, computer science or engineering. Each is a matter of interest in its own right, and also in relation to one another, and to broader questions concerning the nature of information and behavior.

This, then, is the backdrop against which the *Information and Behavior* series has been created. Accordingly, the series will endeavor to provide a forum for the exploration of central issues of the information age—in terms of theory, empirical research, and policy. At the same time we will also pursue several broader, more fundamental goals: to explore the nature of information; to clarify its significance within various disciplines and contexts; and to seek an understanding of the generic role and functions of information, information technology, and information processes in the broad range of human affairs.

References

Allport, G.W. *Becoming.* New Haven: Yale University Press, 1955.

Ashby, W.R. *An introduction to cybernetics.* London: Chapman & Hall, 1964.

Berger, P.L. and T. Luckmann. *The social construction of reality.* Garden City, N.Y.: Doubleday, 1966.

Bertalanffy, L. von. General system theory. New York: Braziller, 1968.

Blumer, H. *Symbolic interactionism.* Englewood Cliffs, N.J.: Prentice-Hall, 1969.

Boulding, K. *The image.* Ann Arbor: University of Michigan Press, 1956.

Bruner, J.S. *Beyond the information given: Studies in the psychology of knowing.* New York: Norton, 1973.

Budd, R.W. and B.D. Ruben. *Beyond media: New approaches to mass communication.* New York: Hayden, 1979.

Cannon, D.L. and G. Luecke. *Understanding communications systems.* Dallas: Texas Instruments Learning Center, 1980.

Cannon, W.B. *The wisdom of the body.* New York: Norton, 1932.

Cherry, C. *World communication.* New York: Wiley, 1971.

Dawkins, R. *The selfish gene.* New York: Oxford University Press, 1976.

Deutsch, K.W. *The nerves of government.* New York: Free Press, 1966.

Duncan, H.D. *Communication and social order.* London: Oxford University Press, 1962.

Etkin, W. *Social behavior from fish to man.* Chicago: University of Chicago Press, 1964.

Frings, H. and M. Frings. *Animal communication.* Norman: University of Oklahoma Press, 1964.

Frisch, K. von. *Bees: Their vision, chemcial senses and language.* Ithaca, N.Y.: Cornell University Press, 1950.

Goffman, E. *Interaction ritual.* New York: Harper & Row, 1971.

———. *Frame analysis: An essay on the organization of experience.* Cambridge, Mass.: Harvard University Press, 1974.

Gumpert, G. and R. Cathcart. *Intermedia.* New York: Oxford University Press, 1982.

Hall, E.T. *The silent language.* Garden City, N.Y.: Doubleday, 1959.

Johnson, W. *People in quandaries.* New York: Harper, 1946.

Kelly, G.A. *A theory of personality.* New York: Norton, 1963.

Kuhn, T.S. *The structure of scientific revolutions.* Chicago: University of Chicago Press, 1970.

Lawson, C.A. Language, communication, and biological organization. In L. von Bertalanffy and A. Rapoport (eds.), *General systems: Volume VIII.* Ann Arbor, Mich: Society for General Systems Research, 1963.

Loftus, G.R. and E.F. Loftus. *Human memory: The processing of information.* Hilldale, N.J.: Erlbaum, 1976.

Machlup, F. *The production and distribution of knowledge in the United States.* Princeton, N.J.: Princeton University Press, 1962.

McLuhan, M. *Understanding media: The extensions of man.* New York: McGraw-Hill, 1964.

McQuail, D. *Towards a sociology of mass communication.* London: Collier-Macmillan, 1969.

Miller, G.A. *The psychology of communication.* Baltimore, Md.: Penguin, 1969.

Miller, G.R. and M. Steinberg. *Between people: A new analysis of interpersonal communication.* Chicago: Science Research Associates, 1975.

Miller, J.G. Living systems. *Behavioral Science*, 1965, 10, 193-237.

Rapoport, A. Man, the symbol-user. In L. Thayer (ed.), *Communication: Ethical and moral issues.* New York: Gordon & Breach, 1973.

Ruben, B.D. General system theory: An approach to human communication. In R. Budd & B.D. Ruben (eds.), *Approaches to human communication.* New York: Spartan, 1972, pp. 120-144.

———. General system theory. In R. Budd and B.D. Ruben (eds.), *Interdisciplinary approaches to human communication.* Rochelle Park, N.J.: Hayden 1979, pp. 95-118.

———. *Communication and human behavior.* New York: Macmillan, 1984.

——— and J.Y. Kim, (eds.), *General systems theory and human communication.* Rochelle Park, N.J.: Hayden, 1975.

Ruesch, J. and G. Bateson. *Communication: The social matrix of psychiatry.* New York: Norton, 1951.

Schroder, H.M., M.J. Driver, and S. Streufert. *Human information processing.* New York: Holt, Rinehart & Winston, 1967.

Sebeok, T.A. *Animal communication: Techniques of study and results of research.* Bloomington: Indiana University Press, 1968.

Selye, H. *The stress of life.* New York: McGraw-Hill, 1956.

Shannon, C.E. and W. Weaver. *The mathematical theory of communication.* Urbana: University of Illinois Press, 1949.

Shriffrin, R.M., N.J. Castellan, H.R. Lindman, and D.B. Pisoni. (eds.), *Cognitive theory: Volume 1.* Hillsdale, N.J.: Erlbaum, 1975.

Shorey, H.H. *Animal communication by pheromones.* New York: Academic Press, 1976.

Simon, H.A. *The sciences of the artificial.* Cambridge, Mass.: MIT Press, 1969.

Smith, A.G., ed. *Communication and culture.* New York: Holt, 1966.

Thayer, L. *Communication and communication systems.* Homewood, Ill.: Irwin, 1968.

Thielcke, G.A *Bird sounds.* Ann Arbor, Mich.: University of Michigan Press, 1970.

Tinbergen, N. *Social behaviour in animals with special reference to vertebrates.* London: Methuen, 1965.

Wiener, N. *Cybernetics: Or control and communication in the animal and the machine.* Cambridge, Mass.: MIT Press, 1950.

Wilson, E.O. *The insect societies.* Cambridge, Mass.: Belknap-Harvard University Press, 1971.

Wright, C.R. *Mass communication: A sociological perspective.* New York: Random House, 1959.

Young, J.Z. *Doubt and certainty in science: A biologist's reflection on the brain.* New York: Oxford University Press, 1960.

Zubik, J.P., ed. *Sensory deprivation: Fifteen years of research.* New York: Appleton-Century-Crofts, 1969.

PART I

INFORMATION AND COMMUNICATION: THEORETICAL ISSUES

1

The Coming of the Information Age: Information, Technology, and the Study of Behavior

Brent D. Ruben

The trappings of the information age are everywhere about us. The period ushers in many complex issues, at a time when the academic community has limited generic theories, concepts, frameworks, and institutions available to address these important concerns. The purpose of this introductory article is to look broadly at defining characteristics of the information age, to identify issues of this age that seem to have generic and lasting significance, and to suggest the need for new disciplinary alignments and cross-disciplinary scholarship to meaningfully address the dialogue of this age.

When historians, social critics, and playwrights reflect upon the present era in years hence, there is little question that information and information technology will figure centrally in their characterization of life during the period. It has become increasingly difficult to watch television or to leaf through the pages of a newspaper or magazine without being reminded that we are living in what may appropriately be called, *The Information Age*.

As Bell (1976), Dizard (1982), Edelstein, Bowes and Harsel (1978), Hammer (1976), Naisbitt (1982), Oettinger (1977, 1980), and a number of other writers have persuasively argued, the information age is not simply an intellectual abstraction, but a pragmatic reality. In 1967, information services and products accounted for some 46 percent of the nation's GNP (Porat 1977), and of the 20 million new jobs created in the 1970s, nearly 90 percent could be classified as information, knowledge, or service jobs (Birch 1981).

The consequences of the information age are even more apparent in the 1980s. In the United States, Japan, Sweden, England, and a number of other countries at least half of the society's labor force is engaged in information-related work (Salvaggio 1983). Naisbitt (1982, 22) notes:

> Today's information companies have emerged as some of the nation's largest. AT&T grossed $58 billion in 1981, far surpassing the GNP of many nations. Other information companies include IBM, ITT, Xerox, RCA, all the banks and insurance companies, the broadcasters, publishers, and computer companies. Almost all of the people in these companies and industries spend their time processing information in one way or another and generating value that is purchased in domestic and global markets.

Whether we are living through a "revolution," as some have suggested, or merely the less glamorous but no less significant consequences of a more natural evolution of electronic technology and its use, probably matters little. The significant point is that something of a dramatic nature is occurring. Much is being said and written about the period, so much in fact that it seems to become increasingly difficult to differentiate that which is central to an understanding of the times from that which is peripheral, issues which have lasting significance from those which are fleeting, implications which are generalizable from those which are of limited usefulness.

It is the purpose of this introductory article to look broadly at the information age: (1) to identify and discuss its defining characteristics; (2) to enumerate issues of this age which seem likely to be of general and lasting significance; (3) to consider the appropriateness of present concepts and disciplinary structure for the study of these issues; and (4) to suggest the value of the concept *information* as a means for integrating cross-disciplinary scholarship.

Pervasiveness and Convergence

Technology

In the information age, *technology* (usually termed *new* or *high* technology) occupies center stage. Evidence of the importance of technology in the culture of the age abounds. Witness the proportion of the square footage of our department stores and shopping malls allocated to communication and information hardware—television, stereo systems, photographic equipment, telephones, electronic watches, VCRs, computers, and a host of other technologies. Additional space is devoted to software for these technologies and to the new magazines and books about the new devices. Finally, there are what future archaeological teams may well construe as the community shrines of the early 1980s, where the young in particular tithed and paid homage to Space Invaders, Pac Man, and other electronic deities.

The culture of the information age is very much reflected in our language, as well. "Mouse," "CPM," "BASIC," "graphic equalizers," "joystick," and "HBO" have all but replaced "MPG," "five speed," and "front-wheel drive" as the social currency of our time. The "walkman" headphone cassette players and their more obstrusive counterpart, the "portable" FM cassette "boxes" certainly rival the sports car or the big-finned Cadillac as a portable symbol of status on the streets of our cities.

The new technologies have made an appearance in nearly all facets of contemporary social and professional activity. One area where their consequences are particularly apparent is entertainment, where cable, telecommunications, and various video recording and playback devices have greatly broadened the number of available leisure outlets. Portable video and audio devices provide increasingly greater flexibility as to when and where, as well as how, we wish to be entertained.

In the workplace, remote paging devices, word processors, electronic cash registers and bookkeeping systems are being widely adopted even by a number of smaller businesses. In larger organizations sophisticated management information systems and electronic mail and filing provide decision makers access to more data, from a larger number of sources, and with greater speed and ease of access. And, through computerized networks, geographically dispersed operating divisions and personnel are linked to one another.

In publishing and print journalism, the new technologies have encouraged fundamental changes in the collection and processing of information, as well as in the dissemination of the finished product. Writers and editors have access to a variety of databases that augment information gathered through interviews and other means, and articles are written, revised, edited, and arranged on-screen. The resulting products may be printed in a traditional manner or transmitted via satellite or cable to remote printing sites, as is the case with newpapers such as *USA Today* and *The Wall Street Journal*. Or, as is the case with videotex, the textual and visual information—in its original or specially modified form—can be transmitted directly to the consumer or to an intermediary computer database such as Compuserve or the Dow Jones News-Retrieval Service.

For libraries, the significance of the new technologies and electronic publishing is also substantial. The concept of the library as a place where people come to review and check out printed documents is giving way to a vision of a generic institution that aims to serve the information needs of its clients. This perspective is, in large measure, made possible by communication and information technologies that eliminate the necessity of physical contact between an individual and a book as a means of making information available.

These same basic information and communication technologies are also in use in medicine, augmenting data collection and patient record-keeping as well as traditional methods of diagnosis, patient care, and treatment. In the legal profession judicial decisions, regulations, and statutes can be easily stored, indexed, and retrieved using interactive computers, databases and telecommunications systems. Word processing facilitates the "personalized" preparation of contracts and wills in a fraction of the time that was required when each legal document had to be created from scratch.

In each of these applications the pattern is the same. At the core of the information age is the progressive convergence of media that were once distinct. Print, broadcast, and common carrier technologies are increasingly indistinguishable from one another because of changes in the devices themselves and changes in the uses to which the technologies are put.

Watches are now also television screens and calculators. Radios are equipped with built in telephones and tape players. Television, originally a medium primarily for the mass distribution of standardized fare, has also become a device for use with interactive video games, cable systems, and a display for alpha-numeric as well as graph computer output. Telephones, once designed for one-to-one conversation, are now used not only in this way but also in conjunction with computers for alphanumeric data transmission. Descendants of the typewriter, used for print correspondence and report preparation, now combine with the telephone and television monitor to provide the means for obtaining a current quote on a particular stock, reading restaurant reviews and making reservations, gathering information from an encyclopedia, and "chatting" with others who have similar equipment via online computer bulletin boards.

Thus, the information age is marked by the wholesale availability and application of electronic technologies in a wide range of personal and professional contexts. While some of these technologies are new, many others have resulted from the convergence and transformation of existing technological forms and uses.

Converging Concepts of Information

In the information age, *information* plays a major role alongside technology. The script has most certainly been committed to memory by all but the most isolated of our time: "Information is everywhere" (since the arrival of the information age). "Information is good" (some is good, more is better). "Information is expensive" (but new technologies are making it less expensive all the time). "Information is an advantageous resource which is unevenly distributed among individual decision-makers, businesses and countries" (which is the desirable and intended state of affairs if you have it, unequitable and discriminatory if you don't).

What is most striking about this litany is not the specific points of view that are generally held—though many of these are indeed worthy of discussion—so much as the pervasiveness of the concept in the culture of the time. Traditionally the preoccupation of a relatively small, homogeneous, and well-defined community of information scientists and communication scholars and practitioners, *information* has swiftly and unmistakably become a focus of interest among a broadening group of academics, professionals and lay persons from a number of fields.

Information has become an increasingly important marker of our age and our culture. Perhaps reflective of this, the term is used to refer to an ever-growing domain of products and services which were previously referred to with distinctive terms. The telephone business has become the information business; electrical and phone hook-ups are now information systems; "news at the top of the hour" has become "information at the top of the hour"; statistics and data are now information. Libraries are described more and more in terms of information providing and less as respositories of knowledge. Data processing and administrative functions are described in terms of information resource management and information policy.

The popular and convergent use of *information* seems to represent something beyond the mere cosmetics of doublespeak, of a "garbage collector" turned "sanitary engineer" or a "strike" turned "work stoppage." There is, it would seem, a broadly-based though not well-articulated recognition that there are generic similarities among a growing range of products and services traditionally described and regarded as distinct.

Issues of the Information Age

Because *technology* and *information* permeate such a broad range of personal and professional endeavors, any issue of concern to some is, by definition, an issue of concern to many. While some are specific to particular technologies or context, many issues transcend any particular new device or application. Among these are issues relating to: 1) Increases in the volume of available data; 2) the shift from linear to interactive media; 3) a broadening concept of literacy; 4) a merging of information-processing and transportation technology; 5) regulation of new technologies; 6) freedom and privacy; 7) the relationship between available information and use; and 8) variety and constraint in media and society.

Increase in the Volume of Data

With each new communication and information device comes an increase in the volume of potential information available in the environment, and with that a number of issues of concern. Pelton (1983) makes the point dramatically when he points out that during one year a network of six advanced communication satellites could transmit the amount of data estimated to be equivalent to the collective lifetime thoughts and speech of all New Yorkers.

> Twenty-first-century satellite platforms that have communication-carrying capabilities of 100 billion bits per second are already well-defined concepts. Two such platforms would be capable of handling in a day the entire U.S. population's thoughts and writings for a decade. (Pelton 1983, 61)

Notwithstanding the widely-held view that if some information is good, more must be better, it is clear that the technological capacity for increased data production and transmission is a mixed blessing, giving us more and more information, perhaps more than we can meaningfully use. Too little or too much information can be equally dysfunctional, if for quite different reasons (Pelton 1982; Schroder, Driver, and Streuffert 1967; Work 1982; Zubik 1969). At one extreme are the difficulties resulting from sensory deprivation (Zubik 1969), and in less extreme circumstances, "uninformed" decisions based on too little data. At the other are the host of personal and social problems, some of which result from the fact that each year humans are able to absorb a smaller percentage of the information available in the environment (Toffler 1970), and others thought to be consequences of "overload" (Artandi 1979; Schroder, Driver, and Streuffert 1967). Researchers also note that even our short-term memory limits severely the extent to which we are able to use large quantities of information at any one time (e.g. Loftus and Loftus 1976).

The issues here are numerous and complex: What is the relationship between varying levels of information availability and various individual and social outcomes? Is it possible to develop methods to determine what is the "right" amount of information for the purposes and decisions at hand, on an a priori basis? How can we apply knowledge relative to human information processing, attention, information reception, information selectivity, source selection, and use? Can methods be developed that will assist with the personal and social management of information?

Shift from Linear Media with Standardized Fare to Interactive Media with Individualized Fare

A second set of issues arises as a consequence of the shift away from traditional, linear, one-way media—such as newspapers, magazines, or network television—which disseminate standardized fare on a standardized time schedule to passive receivers. The shift is toward interactive media, so termed because of the increasing control over media selection afforded the user. Even with simple devices such as record and audio cassette recorders, an increasingly active, controlling role is provided to the audience member. With personal computers, video cassette recorders, videotext, and interactive databases there is a further progression from a sender-determined system to one in which the receiver has substantial options.

Using current technologies anyone who owns a personal computer with communication capability could create private databases or do his or her own publishing. In the future, a videotex subscriber could design customized entertainment and news programming. If, for instance, an individual enjoys sports and adventure programming but has no real interest in politics or international affairs, the personalized media prepared for him or her can be tailored to automatically exclude all but the selected items.

In effect, the new technological devices of the age are progressively shifting the traditional gatekeeping functions—of the television programmer, news editor, or book publisher—from the provider to the user, and at the same time from the professional to the layperson. Many important issues are involved. Information producers and providers have often been criticized for what is sometimes seen as the extreme power they have in controlling the media and media content, from advertisements to movies, children's textbooks to the evening news. It may be argued that interactive media minimize the tyranny of the elite minority by giving over control of media fare to the user (e.g. Masuda 1980; Pool 1983). Furthermore, with less standardization of information there should be greater support for pluralism within society. If an individual speaks Spanish as a first language, or is a religious liberal, or a Republican, he or she would be able to select programming which is consistent with his or her ideological or linguistic preferences and exclude inconsistent content.

From another perspective, of course, increased user control raises concerns about a lessening of quality control and a loss of the authenticating and legitimizing functions served by skilled gatekeepers (Horowitz and Curtis 1982). Furthermore, to the extent that interactive media result in less standardized fare, one may question the ability of the media to transmit cultural heritage in a society composed of individuals sharing an increasingly uncommon database.

Broadening Concept of Literacy

The convergence of communication and information-processing media, coupled with the increase in data available to the individual, raise other issues with personal, social, and especially, occupational implications (Bell 1983; Ruben and Kwasnik 1985; Work 1982). As we move into an age where information and information-handling technology play an increasingly central role in our lives, new competencies may well be required of the individual in order to fully participate, if not to survive, in an environment flooded with data and information-creating, storage, and transmitting devices.

Traditionally, professions such as journalism, librarianship, medicine, law, or business were quite distinct, each with its own content domain, and its largely unique set of operational competencies. Increasingly, the operational competencies of the journalist, the librarian, the organizational communication specialist, and the lawyer overlap as a consequence of the use of common information-handling technologies and processes, suggesting the need for a broadened notion of literacy.

Reading and writing (and to a lesser extent, speech), long regarded as the cornerstones of literacy, may well come to be seen as instances of far more generic information-processing competencies. Even today's innovative concept of computer literacy, stresses primarily limited technical skills. Viewed from the perspective of a broader, more generic concept of information literacy—one which includes information organization, management, processing, transmission, and use—today's cutting-edge concepts of computer literacy might be viewed as tapping only the most superficial and transitory of the necessary skills and competencies. The issues of what the new dimensions and competencies will be required for literacy in an information age, and where, when, and how these skills will be acquired, seem likely to become matters of growing concern.

Merging of Transportation and Communication Technology

There has always been an intriguing relationship between transportation and communication, with both serving a number of similar functions. In many respects, communication and information technologies were developed to obviate the need for travel. Instead of delivering a message in person, one could send it on horseback, on a ship, or by wire.

The movement of data can be an efficient, economical, and energy-saving alternative to moving things or people. The new technologies permit such activities as at-home banking, remote access to library resources and databases, teleconferencing, political campaigning, news reporting, and other activities which customarily require physical movement (e.g. Dordick 1981; Williams 1982b; Ciano and Carne 1982; Feather and Mayur 1982; Harkness and Standel 1982; Neustadt 1982; Ruben 1984; Turock 1983; Cherry 1971).

The new technologies for information processing and communication, then, should require the expenditure of less time and energy, leaving more time for leisure and other activities. There could also be less obligatory face-to-face interaction of the sort required for shopping or going to the library. Yet without the routine of interpersonal and social communication might there not be a weakening of social interdependency, a reduction in chance meetings through which many valued relationships begin, and a decrease in the sense of community? Alternatively, might not some of these functions be served as well through membership in an interactive network (Chesebro 1985; Kerr and Hiltz 1982)?

Converging Concepts of Office and Home

For most members of our society, *work* is a place travelled to and from, and is therefore distinct from *home*. The new technologies have the potential to change all that for the many citizens involved in information-related employment (e.g. Ciano and Carne 1982; Kraut 1984; Olson 1984; Ruben, Holz, and Hanson 1982; Williams 1982b).

For some of these individuals, at least, the notion of the away-from-home office, filled with colleagues who carry out the affairs of business largely through face-to-face interaction, could go the way of the general store. Why expend unnecessary capital on rent, transportation, and equipment to bring individuals together, when they can have access to necessary records and to one another through interactive terminals in their homes. The vision here is not unfamiliar to many in sales today, who can complete much of their necessary "paperwork" from home, freeing them to spend more time securing new clients.

From the standpoint of efficiency, cost-effectiveness, and formalized information transfer, the prospect of the home-office is encouraging. Yet here too there are a number of issues to be considered: What are the functions of the office? Are they compatible with the functions of home? To what extent do the necessary activities of the office require the physical presence of other people? Can the many at-work communication activities associated with the fulfillment of social or personal needs be accommodated in other ways?

At a recent national conference sponsored by Bell Communications Research on "Technology and the Transformation of White Collar Work," it was noted that the mere presence of technology for home-work does not assure it will be adopted and used. Technological innovations, in addition to being available, must satisfy needs of users and must be compatible with existing patterns of behavior and institutions. On these grounds, Kraut (1984) questions the likelihood of the widespread use of home-work technologies, noting that the office as institution has survived a number of major technological innovations. He suggests further that the stability of the office and the presence of co-workers is essential for job satisfaction, socialization (the enforcement of social norms), informal communication, collaboration, and time management (separation of work and family roles.) No doubt one of the important issues of the information age will be to give full consideration to the extent to which the new technologies have promise for serving these functions in alternative ways, and to identify the kinds of jobs, people, and technologies that may be best served by such innovations.

Regulation of the New Technologies

Another set of issues given greater urgency by recent developments has to do with regulation. As the late Ithiel de Sola Pool (1983, 6-7) argued forcefully in *Technologies of Freedom*, "The electronic transformation of the media occurs not in a vacuum, but in a specific historical and legal context." It is also clear that the historical and legal context in which the developments take place have a number of potentially important behavioral implications.

With regard to these legal and regulatory traditions, three types of communication systems can be distinguished—print, common carrier, and broadcasting—each with distinctive and characteristic policy (Pool 1983). For print media, the valued First Amendment concepts of freedom of public and religious speech have generally served as guiding principles. The courts have acted to prohibit the prior restraint, censorship, and licensing of publications. The operation of the mail, telephone, and telegraph systems, on the other hand, has been shaped by a very different philosophy and regulatory framework. Each of these has been regarded as a *common carrier*—channels to which all should have access.

Still a third regulatory pattern has emerged for broadcasting, where the intent has been to ensure the fair and appropriate use of what were, at least originally, a limited number of usable radio frequencies. In an effort to achieve this end, an extensive governmental regulatory mechanism has been instituted, including the licensing of station operators based upon an assessment as to whether they are operating in the public interest.

As Pool (1983) and others (e.g. Read 1983) note, an historical analysis of these three rather distinct philosophical and regulatory traditions leads to a basic question relative to newly emerging communication technologies: Will the new electronic media be thought about and handled in the protected, First Amendment tradition of speech and print, or in the more regulated tradition of the common carriers and, in particular, broadcasting? According to Read (1983, 95):

The conclusion to be drawn . . . is simply this: The rationale for a divergent, two-track legal approach for mass media has eroded. Once seemingly clear distinctions between print and broadcasting are no longer clear; "blur" is fast becoming an appropriate word. The question then is whether in an information society, both media should be placed under the print standard or under the broadcast standard? Or, perhaps, a standard yet to be developed.

The issues here are complex with far-reaching implications as Pool (1983, 10) eloquently argues:

> The move to electronic communication may be a turning point that history will remember. Just as in the seventeenth and eighteenth century Great Britain and America, a few tracts and acts set precedents for print by which we live today, so what we think and do today may frame the information system for a substantial period of time in the future.

There is little doubt that issues of regulation will figure centrally in the dialogue of the period.

Freedom and Privacy in the Information Age

Closely related to the questions about regulation are issues pertaining to freedom and privacy in the information age. Some writers have suggested interactive technologies of the information age will promote freedom to the extent that such technologies are not regulated by the government (e.g. Pool 1983) or controlled by big business (Rose 1983; Schiller 1983, 1985). Implicit in this view is the position that technological decentralization and deregulation are inherently positive, and that centralization and regulation are to blame for present inadequacies, will inevitably minimize freedom, and therefore, are inherently negative (e.g. Pool 1983).

This perspective raises a number of provocative and certainly debatable issues regarding the nature and origins of regulation and the impact of regulation and technology on human behavior. In some senses, of course, the premise that regulation inhibits freedom is patently obvious; more regulation is less freedom—freedom from regulation, at the least.

On the other hand, there are also some circumstances where regulation seems to enhance the free, diverse, and open flow of information, as happens with regulations which encourage educational and special interest broadcasting, balanced presentation on controversial positions, equal time for opposing candidates, and other policies aimed at ensuring content balance, equal and equitable access, and the protection of minority points of view. The points of view generated by precisely this kind of small but significant minority would have little if any public visibility were it not for some forms of protective regulation. Even the "rights" to freedom of speech and privacy which we take for granted are themselves the consequence of regulation. Rose (1983, 21), in fact, argues that regulation is a "citizen's sole means of exerting influence." While there are certainly a number of complex questions here, it would seem safe to conclude that the issue is not whether regulation is desirable or not, but what kind and by whom.

Often, media control and regulation are assumed to be primarily a consequence of government activity, in that governmental regulations, rules, and policies are overt. There is also, however, less obvious but not necessarily less impactful regulation of the flow of information inherent in so-called free, market-controlled communication systems. In such systems the regulation which may lead to prior restraint or censorship is not that of the government, but rather of the audience for a particular medium on the "demand" side of the equation, and the media owners and information providers on the "supply" side.

In considering the sources that limit the free flow of information in mass media, there are many elements to be examined. In commercial television, for instance, are the limitations and constraints on media fare more the product of broadcasting regulation or public preferences? Would a system with less government regulation assure more diversity of points of view, more educational programming, the inclusion of more minority voices, or more innovation? What are the functions and dysfunctions of regulation— implicit as well as explicit? What alternatives are available? What are the most appropriate agents and mechanisms of control and regulation? What values should guide our regulatory decision making?

Equally critical are issues relating to information ownership, privacy, and security (Bell 1983; Broad 1983; Hixson 1985; Rose 1983). The pages of our newspapers, magazines, and scholarly publications are filled with expressions of concern regarding the ownership and protection of proprietary rights related to the new communication technologies and accompanying software—computer programs, videotapes, and audio cassettes.

In an age where information is increasingly regarded as a commodity to be owned and possessed, one should not be surprised to find the emergence of legal arguments even about the propietary rights to information about an individual's own life, as is the case in recent legal discussions involving the rights to Elizabeth Taylor's life story (Hixson 1985). Instances such as this raise what Rose (1983) characterizes as a tension between property rights and human rights in the matter of information usage. The same concerns are central in widespread fears regarding the perceived threat to privacy and security posed by new technologies (e.g. Bell 1983; Pelton 1983; Rose 1983; Salvaggio 1983), providing a broad agenda of concerns which have already become central to the dialogue of the age.

The Relationship between Technology, Available Information, and Use

In discussion of the impact of the new technologies and the programming they provide, a fundamental issue as to whether any technology or class of technologies has an inherent capability to foster particular outcomes must be considered. Is it the automobile, the handweapon, or the television set that causes its specific use or abuse? Or is it the preferences and predispositions of the individual, group, or society which would have played themselves out with another technology if a particular one were not available? There is a certain comfort that can be taken in the position of mutual causality: That available technology—itself the consequence of human activity—sets certain limits within which the predispositions and preferences of potential users operate.

In considering communication and information technologies, issues involve not only hardware—which is directly analogous to other technology—but also software, which is not. Communication and information technologies are devices which create, transport, or use data or information (computer or television programs, news articles, photos, etc.). In addition to questions relative to the impact of the hardware, there are issues relative to the selection, interpretation, retention, and use of information. It is here that analogies between communication and information technologies and other technologies may break down.

Even if particular new media offer new and significantly diverse material, the relationship between the availability of data and its selection and use by individual members of an audience is by no means a simple or predictable one. In repressive situations where information flow is highly controlled, diversity, innovation, and freedom are nevertheless often spawned. Among the topics on which communication theorists, advertising practitioners, cognitive information-processing theorists, and political propagandists would agree is the proposition that there is no simple formula to explain whether and how information available in the environment affects the behavior of individuals who have been exposed to it. The relationship between available information and its use is central to all discussions of the impact of information and new technologies. For this reason the topic seems certain to play a major role in theory and research in the period ahead.

Variety and Constraint, Diversity and Control, Divergence and Convergence

As noted previously, the new technologies are increasingly interactive and user controlled. Are they therefore inherently more likely to foster variety and diversity as many suggest (e.g. Bell 1983; Cleveland 1982; Jennings 1982; Masuda 1980; Pool 1983)? More channels and more electronic devices provide more choices to be sure. But does quantity necessarily lead to diversity? Did the emergence of magazines and television greatly enrich the range of information available from the media? Even with cable television, where there are possibilities for greatly expanded media fare, one may question the impact on the breadth of what is available to viewers. Much the same may be said of the computer, at least in its at-home application. More and newer media, while adding flexibility in the forms of information and the places and times that information is available, do not seem to assure added substance to the information alternatives already available in the environment.

Even if one accepts the position that new user-centered technologies increase diversity, questions still remain concerning the extent to which these technologies serve desirable functions. Particular care is required in conceptualizing notions such as *variety, constraint, diversity,* and *control* as they relate to information and communication.

To the extent that one accepts the basic premises of the sociology of knowledge (e.g. Berger and Luckmann 1966), symbolic interaction (e.g. Blumer 1969), general semantics (e.g. Korzybski 1933), and communication theory, one can hardly overemphasize the role of communication and information in the discovery, social construction, internalization, and verification of reality, all of which are central to the establishment and maintenance of order, organization, and predictability within human social systems (Ruben 1979). For the social system, a group or a society, diversity or variety at the extreme is little more than a collectivity without order, pattern, coherence, or structure. It would, in fact, not be a social system at all but a collection of random actions and reactions, with no regularity or points of convergence other than the norm of nonnormativeness.

Despite the improbability of such an extreme, it is important to consider the issue in generic terms. This serves to remind us of the interdependency and delicate balance which exists between variety and constraint, diversity and conformity, randomness and order. From such a perspective, diversity may well be a positive attribute only in the context of ample, assumed nondiversity.

Technologies of information and communication play a central role in contemporary society. They contribute to the formation of the web of interconnections that link us together in relationships, groups, organizations, and society upon which we depend for our existence (Budd and Ruben, 1979). In this largely unintentional process, our verbal and nonverbal language conventions, social rules and roles, attitudes and values, and agenda of shared concerns emerge and are perpetuated in what might be termed the convergence function of information processes.

It would also seem to be these same mechanisms through which change, growth, and differentiation occur, perhaps more often by accident than through planning. From this perspective, it is in terms of this divergence function—which is undoubtedly far more difficult to serve in an institutionalized way—that diversity, pluralism, and freedom are critical. It may be that stripped of their emotional outerwear, *variety, freedom, diversity,* and *pluralism* are best thought of as wholly positive attributes only when in an optimum tradeoff with constraint and convergence. At any rate, the issues raised in achieving such a balance are worthy of substantial exploration in the dialogue of the information age.

Disciplinary Structure and the Information Age

The information age gives testimony to the ability of our species to create and use tools that enhance our genetic capacity to create, transmit, store, retrieve, and use information for a variety of human purposes, yet these achievements are not without complications. As seems so often to be the case with human advances, our accomplishments are mixed-blessings—solving some problems while creating new ones.

Such is clearly the case in the information age. The issues discussed in the preceding section are numerous and complex. Each cuts across traditional disciplinary boundaries and blurs traditional lines of scholarly inquiry. Whether one considers the prospects and problems created by broadening concepts of literacy, concerns about freedom and privacy, or the relationship between available information and use, the same pattern is apparent: It is not possible to meaningfully or comprehensively examine any of these issues from a single disciplinary vantage point. None is simply a communication issue, a computer science issue, a sociological issue, a media studies issue, a psychological issue, or an information science issue.

The past century has been notable as a period of progressive specialization in the study of behavior. Whereas early religious teachings and Greek thought approached the study of human activity holistically, subsequent developments saw the segmentation of the whole into what were thought to be the most crucial components. The study of the individual became the domain of psychology, while political science concerned itself with politics and political behavior, sociology centered on social organization and society, communication on messages, media, and their impact, and so on.

In many areas of inquiry, the constraint of disciplinary boundaries has led to the emergence of hybrid terminology for fields of study including social psychology, medical sociology, neurophysiology, psycholinguistics, and psychobiology. Other areas of investigation such as general systems theory, sociology of knowledge, symbolic interaction, sociology of science, and general semantics have sought to overcome disciplinary boundaries by advancing generic theories that span the domains of concern of several traditional disciplines.

Nowhere are the limitations of the traditional disciplinary configuration more dysfunctional and constraining than in the context of the information age. The dialogue suffers greatly from being broadly uninformed by theories, concepts, and analytic frameworks from fields which are relevant but untapped. Largely as a consequence of the present disciplinary structure, information professionals in decision-making and policy roles are unable to draw upon the broad range of potentially useful literature and intellectual resources of the academic community because those resources are not conceptually unified or terminologically organized in ways that would facilitate their use.

The realization that traditional disciplinary structures are limiting and often inadequate for addressing the issues of the information age has led a number of writers to argue the need for new terminology. *Compunications* is the label suggested by Anthony Oettinger (Harvard University Program on Information Technology 1976, 1977) for the convergence of telephone, computer, and television. *Telematique* is the term advanced by Nora and Minc (1978). Pelton (1982) offers the label *telecomputerenergetics* to suggest an even broader set of relationships.

While these new labels do serve to underscore the convergence of disciplinary perspectives in the information age, more is required. Needed are integrative, cross-disciplinary approaches which identify and give focus to relevant scholarship not only from information studies, communication, computer science, and media studies—but also from psychology, political science, medicine, economics, zoology, artificial intelligence, journalism, library studies, marketing, and management.

The Concept of Information

Of the various concepts that could be central to the development of a cross-disciplinary framework for the information age, *information* is particularly promising. The term has long been used in the literature of library and information science where it refers to data or documents (e.g. Lipetz 1970), decision making and problem solving (e.g. Yovitz 1975; Whittemore and Yovitz 1973), a commodity (e.g. Artandi 1973; Williams 1979), a physical form or property (e.g. Ursal 1968), a representation of internal knowledge (e.g. Farradane 1976), a collection of purposefully structured signs capable of changing the image-structure of a recipient (Belkin and Robertson 1976), and uncertainty, choice, or constraint on alternatives (e.g. Pierce 1961; Cherry 1966).

Information has also been used in the field of psychology, where it refers to stimuli (e.g. Bruner 1973; Jones 1969), learning (Bruner 1973; MacKay 1952a), and thinking, cognition, and memory (e.g. Loftus and Loftus 1976; Shiffrin, Castellan, Lindman and Pisoni 1975). Communication scholars have used the term to refer to messages (e.g. Fisher 1978; MacKay 1952a), meaning and personal knowledge (e.g. Boulding 1956; Goffman 1974), media (e.g. McLuhan 1964), and a linkage between living system and environment (e.g. Thayer 1968; Ruben 1984). Writers in the area of journalism and media use *information* to refer to news (e.g. Boorstin 1961; Williams 1982a). The term is also used in a variety of other ways (see Table 1.1).

Beyond functioning as a semantic bridge across disciplines, *information* can also help to focus common dimensions of concern and scholarship. For instance, one can reorganize the perspectives presented in Table 1.1, and thereby identify several major dimensions of the information concept that are central to a number of fields. More specifically, as indicated in Table 1.2, there are approaches to *information* as: (1) *data*—a code, commodity, pattern, etc.; (2) a *process* by which data is acquired, transmitted, transformed, stored or retrieved; (3) *a channel* or *technology* through which data is transmitted, transformed, stored, or retrieved; and (4) *uses, functions,* or *outcomes* of data processing, organization, transmission, storage, or retrieval.

The organization, management, storage, and retrieval of data are central topics of study in information and library studies, where the concern is primarily with purposeful, verbal, data transfer, through formal channels and institutions. In such a framework, the outcomes of the activity are analyzed and judged based largely on the quantity and quality of information transfer or storage.

In the field of communication, emphasis is on the process whereby data of all varieties are originated, transmitted, and transformed and used by individuals. As such, the concern is as much with nonverbal as with verbal data, informal as with formal channels, and unintentional as with purposeful information generation. Accordingly, outcomes such as the construction and labeling of social reality, relational development, the development of social networks, culture formation, and rule and ritual formulation are of interest. Similarly, the processes involved in the acquisition, transformation, storage, retrieval, uses, and outcomes of information are also significant in psychology, sociology, political science, and a number of other fields.

Each of the major issues discussed previously in this article involves considerations relative to data, process, channel/technology, and uses/functions/outcomes. It seems quite likely that scholarship on each of the

TABLE 1.1
Concepts of Information

Commodity (economic): Artandi 1973; Williams 1979
Consequence of action: Rogers and Kincaid 1981
Code or Pattern (commonly understood): Cannon and Luecke 1980
          (genetic): Maruyama 1968; Masuda 1981; Ruben 1984
Data or Documents (physically, mechanically, or electronically transmitted or
    stored): Lancaster and Gillespie 1970; Lipetz 1970
Decision making and Problem solving: Thayer 1968; Yovitz 1975; Whittemore and
    Yovitz 1973
Entropy: Wiener 1961
Knowledge (formal or recorded): Boorstin 1961; Davison, Boylan, and Yu 1976;
        Machlup 1962; McCroskey 1968
        (subjective or personal): Boulding 1956; Goffman 1974; Ruben 1984;
        Whittemore and Yovitz 1973
        (cultural, sociological, or psychological): Miller and Steinberg 1975
Learning: Bruner 1973; MacKay 1952a, 1952b; Pratt 1977, 1980; Thayer 1979
Linkage (between living organisms and their environment): Buckley 1967; Masuda
    1980; Miller 1965; Ruben 1972, 1984; Thayer 1968, 1979; Young 1960
Meaning (assigned to data based on conventions or rules): Horton 1979; Ruben
    1984
Medium: McLuhan 1964
Messages: Fisher 1978
    (message content): MacKay 1952a, 1952b
    (nonverbal): Rapaport 1982
    (message selection): MacKay 1952a, 1952b; Shannon and Weaver 1949
News (and other data of transitory interest): Boorstin 1961; Williams 1982a
Physical form or Property: Rogers and Kincaid 1981; Ursul 1968, 1971, 1973
Process of being formed: Pratt 1977; Thayer 1979
Processed sensory data: Boulding 1956; Budd and Ruben 1979; Loftus and Loftus
    1976; Ruben 1972, 1984; Thayer 1968
Product (to be used): Artandi 1973
Product of social interaction: Deetz and Mumby 1984; Ruben 1975
Regulation and Control: Ashby 1964; Laszlo 1969; Milsum 1968; Watzlawick,
    Beavin, and Jackson 1967; Young 1960
Representation of internal knowledge (surrogate): Farradane 1976
Resource: Horton 1979
Service: Artandi 1973
Signal (used in communication): Cherry 1966; Watzlawick, Beavin, and Jackson
    1967
Stimuli: Bruner 1973; Jones 1969
    (interpretation of stimuli): Otten 1975
Structure or Organization: Laszlo 1969; Watzlawick, Beavin, and Jackson 1967;
    Wiener 1961
    (capable of changing the image-structure of a recipient): Belkin and
        Robertson 1976
Symbols: Lin 1973
Technology (means of transmission): Dizard 1982; Salvaggio 1983; Schiller 1983
Text (collection of purposefully structured signs): Belkin and Robertson 1976
Thinking, Cognition, and Memory: Bruner 1973; Hunt 1982; Laszlo 1969; Loftus
    and Loftus 1976; Masuda 1980; Pratt 1977; Schroder, Driver, and Streufert 1967;
    Shiffrin, Castellan, Lindman, and Pisoni 1975
Uncertainty, Choice, or Constraint on alternatives: Artandi 1973; Cherry 1966;
    Jones 1969; Krippendorff 1977; Lin 1973; Pierce 1961; Rogers and Kincaid 1981;
    Rapoport 1966

TABLE 1.2
The Concept of Information: Major Dimensions

1. *Data* (property, code, pattern):
   commodity
   code/pattern
   documents
   knowledge
   messages
   news
   physical form or property
   processed sensory data
   product
   resource
   service
   signal
   stimuli
   structure or organization
   symbols
   text

2. *Process* (through which data are transmitted, transformed, or stored):
   learning
   linkage
   process of being formed
   thinking, cognition, memory

3. *Channel* or *Technology* (means through which data are captured, transmitted, transformed, stored, retrieved):
   medium
   technology

4. *Uses, Functions,* or *Outcomes* (of data transmission, transformation, organization, management, or storage):
   consequence of action
   culture formation
   decision making/problem solving
   entropy (decrease in)
   meaning
   management
   network development
   personality development
   product of social interaction
   reality construction, labeling, and validation
   regulation and control
   relational development
   rule and ritual formulation
   structure or organization
   therapy
   thinking, cognition, memory

four dimensions of information could do much to illuminate the dialogue of each topic. Thus, at a time when a number of complex issues loom large in our future, the concept of *information* can be an extremely powerful and useful vehicle. It can provide a means for linking disciplines with shared concerns, for focusing extant research and theory, and for delineating a cross-disciplinary research agenda for the future. More significant still, is the possibility that the dialogue of the age may serve as the catalyst for the development of a broad, new information-centered paradigm of human behavior.

References

Artandi, S. Infomation concepts and their utility. *Journal of the American Society for Information Science,* 1973, (July/August), 24(4), 242-245.

Artandi, S. Man, information, and society: New patterns of interaction. *Journal of the American Society for Information Science,* 1979, (January), 15-18.

Ashby, W.R. *An introduction to cybernetics.* London: Chapman & Hall, 1964.

Belkin, N.J. and S.E. Robertson. Information science and the phenomenon of information. *Journal of the American Society for Information Science,* 1976 (July-August), 27(4), 197-204.

Bell, D. *The coming of post-industrial society.* New York: Basic Books, 1976.

_____. Communication technology—for better or for worse? In J.L. Salvaggio (ed.), *Telecommunications: Issues and choices for society.* New York: Longman, 1983, pp. 34-50.

Berger, P.L. and T. Luckmann. *The social construction of reality.* Garden City, N.Y.: Doubleday, 1966.

Blumer, H. *Symbolic interactionism.* Englewood Cliffs, N.J.: Prentice-Hall, 1969.

Boorstin, D.J. *The image: A guide to pseudo-events in America.* New York: Harper & Row, 1961.

Boulding, K. *The image.* Ann Arbor: University of Michigan Press, 1956.

Broad, W.J. Rising use of computer networks raises issues of security and law. *New York Times,* August 26, 1983.

Bruner, J.S. *Beyond the information given: Studies in the psychology of knowing.* New York: Norton, 1973.

Buckley, W. *Sociology and modern systems theory.* Englewood Cliffs, N.J.: Prentice-Hall, 1967.

Budd, R.W. and B.D. Ruben. *Beyond media: New approaches to mass communication.* New York: Hayden, 1979.

Cannon, D.L. and G. Luecke. *Understanding communications systems.* Dallas: Texas Instruments Learning Center, 1980.

Cherry, C. *On human communication.* 2d ed. Cambridge, Mass.: MIT Press, 1966.

_____. *World Communication.* New York: Wiley, 1971.

Chesebro, J.W. Computer-mediated interpersonal communication. In B.D. Ruben (ed.), *Information and behavior, Vol. 1.* New Brunswick, N.J.: Transaction, 1985.

Ciano, J.M. and E.B. Carne. Telecommunications: The next generation. In H.F. Didsbury, Jr. (ed.), *Communications and the future: Prospects, promises, and problems.* Bethesda, Md.: World Future Society, 1982, pp. 143-156.

Cleveland, H. People lead their leaders in an information society. In H.F. Didsbury, Jr. (ed.), *Communications and the future: Prospects, promises and problems.* Bethesda, Md.: World Future Society, 1982, pp. 167-173.

Davison, W.P., J. Boylan and T.C. Yu. *Mass media: Systems and effects.* New York: Praeger, 1976.

Deetz, S. and D. Mumby. Metaphor, information, and power. In B.D. Ruben, (ed.), *Information and behavior,* Vol. 1. New Brunswick, N.J.: Transaction, 1985.

Dizard, W.P., Jr. *The coming information age: An overview of technology, economics, and politics.* New York: Longman, 1982.

Dordick, H., et al. *The emerging network marketplace.* Norwood, N.J.: Ablex, 1981.

Edelstein, A.S., J.E. Bowes, and S.M. Harsel. *Information societies: Comparing the Japanese and American experiences.* Seattle: University of Washington, 1978.

Farradane, J. Towards a true information science. *Information Scientist,* 1976, 10, 91-101.

Feather, F. and R. Mayur. Communications for global development: Closing the information gap. In H.F. Didsbury, Jr. (ed.), *Communications and the future: Prospects, promises, and problems.* Bethesda, Md.: World Future Society, 1982, pp. 198-210.

Fisher, B.A. *Perspectives on human communication.* New York: Macmillan, 1978.

Goffman, E. *Frame analysis: An essay on the organization of experience.* Cambridge, Mass.: Harvard University Press, 1974.

Hammer, D.P., ed. *The information age: Its development and impact.* Metuchen, N.J.: Scarecrow Press, 1976.

Harkness, R.C. and J.T. Staudel. Telecommunications alternatives to transportation. In H.F. Didsbury, Jr. (ed.), *Communications and the future: Prospects, promises, and problems.* Bethesda, Md.: World Future Society, 1982, pp. 229-233.

Harvard University Program on Information Technology. Cambridge, Mass., 1976, 1977.

Hixson, R.F. Whose life is it anyway: Information as Property. In B.D. Ruben (ed.), *Information and behavior:* Vol. 1. New Brunswick, N.J.: Transaction, 1985.

Horowitz, I.L. and M.E. Curtis. The impact of new information technology on scientific and scholarly publishing. *Journal of Information Science.* Volume 4, 1982.

Horton, F.W., Jr. *Information resources management: Concepts and cases.* Cleveland: Association for Systems Management, 1979.

Hunt, M. *The universe within: A new science explores the human mind.* New York: Simon & Schuster, 1982.

Jennings, L. Utopia: Can we get there from here—by computer. In H.F. Didsbury, Jr. (ed.), Communications and the future: Prospects, promises and problems. Bethesda, Md.: World Future Society, 1982, pp. 48-52.

Jones, A. Stimulus-seeking behavior. In J. Zubek (ed.), *Sensory deprivation: Fifteen years of research.* New York: Appleton-Century-Crofts, 1969, pp. 167-206.

Kerr, E.B. and S.R. Hiltz. *Computer-mediated communication systems.* New York: Academic Press, 1982.

Korzybski, A. *Science and sanity.* Lakeville, Conn.: International Non-Aristotelian Library, 1933.

Kraut, R.E. Telework: Cautious pessimism. Paper presented at the Conference on Technology and the Transformation of White Collar Work. Bell Communications Research, New Brunswick, N.J., June 1984.

Krippendorff, K. Information systems theory and research: An overview. In B.D. Ruben (ed.), *Communication yearbook 1.* New Brunswick, N.J.: Transaction International Communication Association, 1977, pp. 149-171.

Lancaster, F.W. and C.J. Gillespie. Design and evaluation of information systems. In C. Cuadra (ed.), *Annual review of information science and technology,* Vol. 5. Chicago: Encyclopaedia Britannica, 1970, pp. 33-70.

Laszlo, E. *System, structure, and experience: Toward a scientific theory of mind.* New York: Gordon and Breach, 1969.

Lin, N. *The study of human communication.* Indianapolis: Bobbs-Merrill, 1973.

Lipetz, B. Information needs and uses. In C. Cuadra (ed.), *Annual review of information science and technology, Vol. 5.* Chicago: Encyclopaedia Britannica, 1970, pp. 3-32.

Loftus, G.R. and E.F. Loftus. *Human memory: The processing of information* Hillsdale, N.J.: Erlbaum, 1976.

Machlup, F. *The production and distribution of knowledge in the United States.* Princeton, N.J.: Princeton University Press, 1962.

MacKay, D.M. In search of basic symbols. In H. von Foerster (ed.), *Cybernetics: Transactions of the Eighth Congress.* New York: Macy Foundation, 1952a.

_____. The nomenclature of information theory. In H. von Foerster (ed.), *Cybernetics: Transactions of the Eighth Congress.* New York: Macy Foundation, 1952b.

Maruyama, M. Mutual causality in general systems. In J. Milsum (ed.), *Positive feedback: A general systems approach to positive/negative feedback and mutual causality.* New York: Pergamon, 1968, pp. 80-100.

Masuda, Y. *The information society as post-industrial society.* Bethesda, Md.: World Future Society, 1981.

McCroskey, J.C. *An introduction to rhetorical communication.* Englewood Cliffs, N.J.: Prentice-Hall, 1968.

McLuhan, M. *Understanding media: The extensions of man.* New York: McGraw-Hill, 1964.

Miller, G.R. and M. Steinberg. *Between people: A new analysis of interpersonal communication.* Chicago: Science Research Associates, 1975.

Miller, J.G. Living systems. *Behavioral Science,* 1965, 10, 193-237.

Milsum, J.H., ed. *Positive feedback: A general systems approach to positive/negative feedback and mutual causality.* New York: Pergamon, 1968.

Naisbitt, J. *Megatrends.* New York: Warner Books, 1982.

Neustadt, R.M. Politics and the new media. In H.F. Didsbury, Jr. (ed.), *Communications and the future: Prospects, promises, and problems.* Bethesda, Md.: World Future Society, 1982, pp. 248-254.

Nora, S. and A. Minc, *L'Informatisation de la société.* La Documentation Française, January 1978.

Olson, M. Remote office work: Changing work patterns in space and time. Paper presented at the Conference on Technology and the Transformation of White Collar Work. Bell Communications Research, New Brunswick, N.J., June 1984.

Otten, K.W. Information and communication: A conceptual model as framework for the development of theories of information. In A. Debons and W. Cameron (eds.), *Perspectives in information science.* Leyden: Noordhof, 1975, pp. 127-148.

Oettinger, A.G. Information resources: Knowledge and power in the 21st century. *Science,* 1980, 209(4).

_____. P. Bergman, and W. Read. *High and low politics: Information resources for the 80's.* Cambridge, Mass.: Ballinger Publishing Co., 1977.

Pelton, J.N. Global talk and the world of telecomputerenergetics. In H.F. Didsbury, Jr. (ed.), *Communications and the future: Prospects, promises, and problems.* Bethesda, Md.: World Future Society, 1982, pp. 78-87.

_____. Life in the information society. In J.L. Salvaggio (ed.), *Telecommunications: Issues and choices for society.* New York: Longman, 1983, pp. 51-68.

Pierce, J.R. *Symbols, signals and noise.* New York: Harper & Row, 1961.

Pool, I. de S. *Technologies of freedom.* Cambridge, Mass.: Belknap, 1983.

Porat, M. *Information economy: Definition and measurement.* Washington: U.S. Department of Commerce/Office of Telecommunications, May 1977.

Pratt, A.D. The information of the image. *Libri,* 1977, 27, 204-220.

_____. Information and emmorphosis: An attempt at definition. In Ole Harbo and Leif Kajberg (eds.), *Theory and application of information research: Proceedings of the Second International Research Forum on Information Science.* London: Mansell, 1980, pp. 30-33.

Rapoport, A. What is information? In A. Smith (ed.), *Communication and culture.* New York: Holt, Rinehart & Winston, 1966, pp. 41-55.

_____. *The meaning of the built environment: A nonverbal approach.* Beverly Hills, Calif.: Sage, 1982.

Read, W.H. The First Amendment meets the information society. In J.L. Salvaggio (ed.) *Telecommunications: Issues and choices for society.* New York: Longman, 1983, pp. 85-104.

Rogers E.M. and D.L. Kincaid. *Communication networks: Toward a new paradigm for research.* New York: Free Press, 1981.

Rose, E.D. Moral and ethical dilemmas inherent in an information society. In J.L. Salvaggio (ed.), *Telecommunications: Issues and choices for society.* New York: Longman, 1983, pp. 9-23.

Ruben, B.D. General system theory: An approach to human communication. In R.W. Budd and B.D. Ruben (eds.), *Approaches to human communication.* New York: Spartan, 1972, pp. 120-144.

_____.Intrapersonal, interpersonal, and mass communication processes in individual and multi-person systems. In B.D. Ruben and J.Y. Kim (eds.), *General systems theory and human communication.* Rochelle Park, N.J.: Hayden, 1975, pp. 164-190.

_____. General system theory. In R.W. Budd and B.D. Ruben (eds.), *Interdisciplinary approaches to human communication.* Rochelle Park, N.J.: Hayden, 1979, pp. 95-118.

_____.*Communication and human behavior.* New York: Macmillan, 1984.

_____, J.R. Holtz, and J.K. Hanson. Communication systems, technology, and culture. In H.F. Didsbury, Jr. (ed.), *Communications and the future: Prospects, promises, and problems.* Bethesda, Md.: World Future Society, 1982, pp. 255-266.

_____ and B. Kwasnik. In M Voigt and B. Dervin (eds.), *Progress in communication sciences,* forthcoming.

Salvaggio, J.L. The telecommunications revolution: Are we up to the challenge? In J.L. Salvaggio (ed.), *Telecommunications: Issues and choices for society.* New York: Longman, 1983, pp. 148-153.

Schiller, H.I. Information for what kind of society? In J.L. Salvaggio (ed.), *Telecommunications: Issues and choices for society.* New York: Longman, 1983, pp. 24-33.

_____. Privatizing the public sector: The information connection. In B.D. Ruben (ed.), *Information and Behavior,* Vol. 1. New Brunswick, N.J.: Transaction Books, 1985.

Schroder, H.M., M.J. Driver, and S. Streufert. *Human information processing.* New York: Holt, Rinehart & Winston, 1967.

Shannon, C.E. and W. Weaver. *The mathematical theory of communication.* Urbana: University of Illinois Press, 1949.

Shiffrin, R.M., N.J. Castellan, H.R. Lindman, and D.B. Pisoni, (eds.), *Cognitive theory,* Vol. 1. Hillsdale, N.J.: Erlbaum, 1975.

Thayer, L. *Communication and communication systems.* Homewood, Ill.: Irwin, 1968.

_____. Communication: *Sine qua non* of the behavioral sciences. In R. Budd and B. Ruben (eds.), *Interdisciplinary approaches to human communication.* Rochelle Park, N.J.: Hayden, 1979, pp. 7-31.

Toffler, A. *Future shock.* New York: Random House. 1970.

_____. *The third wave.* New York: William Morrow, 1980.

Turock, B. The public library in the age of electronic information. *Public Library Quarterly,* 1983, (Summer), 4(2), 3-11.

Ursul, A.D. *Priroda informatsiia.* Moscow: Izd-vo Politicheskoi Literatury, 1968.

_____. *Informatsiia: Metologicheskie aspekty.* Moscow: Nauka, 1971.

_____. *Otrazhenie i informatsiia.* Moscow: Mysl, 1973.

Watzlawick, P., J.H. Beavin, and D.D. Jackson. *Pragmatics of human communication: A study of interactional patterns, pathologies, and paradoxes.* New York: Norton, 1967.

Whittemore, B.J. and M.C. Yovitz. A generalized conceptual development for the analysis and flow of information. *Journal of the American Society for Information Sciences,* 1973, 24, 221-231.

Wiener, N. *Cybernetics: Or control and communication in the animal and the machine.* 2d ed. Cambridge, Mass.: MIT Press, 1961.

Williams, F. Communication in the year 2000. In D. Nimmo (ed.), *Communication yearbook 3.* New Brunswick, N.J.: Transaction/International Communication Association, 1979.

_____. *The communications revolution.* Beverly Hills, Calif.: Sage, 1982a.

_____. Doing the traditional untraditionally. In H.F. Didsbury, Jr. (ed.), *Communications and the future: Prospects, promises, and problems.* Bethesda, Md.: World Future Society, 1982b, pp. 102-112.

Work, W. Communication education for the twenty-first century. In H.F. Didsbury, Jr. (ed.), *Communications and the future: Prospects, promises, and problems.* Bethesda, Md.: World Future Society, 1982, pp. 113-120.

Young, J.Z. *Doubt and certainty in science: A biologist's reflection on the brain.* New York: Oxford University Press, 1960.

Yovitz, M.C. A theoretical framework for the development of information science. In *Problems of information science.* Moscow: VINITI, 1975, pp. 90-114.

Zubik, J.P., ed. *Sensory deprivation: Fifteen years of research.* New York: Appleton-Century-Crofts, 1969.

2

Information and Communication: Information Is the Answer, but What Is the Question?

James D. Halloran

The article argues for the application of a critical, sociological approach to mass communication to current discussions of information and "the information revolution." Central to the case is emphasis on the importance of studying social implications of the introduction and development of communication technology and systems. Assessment and evaluation should be conducted in terms of basic social and communication needs. A research program is outlined which addresses central questions raised by this perspective. Included are questions such as: What sort of information do different groups or the population as a whole need? Why do they need it? Who decides what is needed by whom? Who selects and presents what is provided? What use is made of what is provided? What are the consequences of that use for individuals, groups, institutions, and societies?

While one might assume that mass communication researchers and information scientists have much in common, this assumption is not entirely warranted. Often, the differences between information scientists and mass communication researchers—both with regard to central interests and approaches to what some might regard as common problems—are at least as much in evidence as the similarities.[1]

What follows is an attempt to address those issues, problems, approaches, and research findings that seem most likely to improve our understanding of the relationship between the mass communication researcher and the information scientist, and between information and communication.

27

The Provision of Information vs. the Functions of Information

Who needs information? What sort of information do different groups or the population as a whole need? Why do they need it? Who decides what is needed by whom? Who selects and presents what is provided? What are the aims and intentions of the providers? What use is made of what is provided? What are the consequences of that use for individuals, groups, institutions and societies? Could what is functional for one group be dysfunctional for others, or for society as a whole? What criteria are used in determining what is functional and dysfunctional? These are just a few of the questions that spring to mind when examining the relationship between information and communication. They are the sort of questions that interest sociologists working in the field of mass communication, who study media and related institutions as social institutions and examine communication as a social process.

But these institutions and processes are not studied in isolation. They have to be studied in the wider social context, in terms of their relationship to other institutions and processes, to society as a whole, and to communication needs.

A major difference between mass communication researchers—as selectively defined above—and information scientists is that the former do not take as given facts apparently accepted by information scientists without question. The sociologically-oriented mass communication researchers ask questions about the factors that govern the production, dissemination, use, and consequences of information. They think in terms of the functions of information or communication for all involved in the process, and do not take for granted one of the assumptions which apparently underpins the thinking of several information scientists, namely that information necessarily reduces uncertainty.

In fact, it is reasonably clear from research that in particular circumstances the provision of information could well lead to increased uncertainty. Moreover, it must not be forgotten that ignorance can have its positive functions, that information could be a burden, and that the possession of knowledge may be used as a substitute for action.

Information—A Necessary but Not Sufficient Condition for Action

In many cases, information is provided in the hope that some appropriate action may follow, but it is clear from the results of research across a wide range of topics that, at best, information may be a necessary, but never a sufficient, precondition for the desired social action. These questions and problems—central to the terms of reference of the mass communication researcher—are rarely raised by the librarian or information scientist.

With them the emphasis is on information and its provision; but in so many cases this emphasis is to the exclusion of any consideration of communication. There is little or no reference to the context and complexity of the communication process. Of course, there is a notion of process implicit in the overall approach, derived—albeit unconsciously—from an acceptance of a one-way, purposive, top-downwards model of the communication process. At base, however, it is a version of a vertical process which is essentially sender-to-provider oriented, and it looks at the whole matter from the point of view of the message giver—asking questions about getting the message across, conversion, persuasion, and effectiveness rather than meeting basic needs or asking whether the message is worth getting across. This is communication in the service of the accepted, unquestioned system.

This outdated model—which still has its adherents, implicitly and explicitly, in certain areas of communication, education, and psychology—is a totally inadequate one. For example, much communication is not purposive, nor is it one-way, and effects are not the same thing as effectiveness. Some of the most interesting consequences of a given act of communication need never have been intended, and what is intended is not always a clear guide to what actually takes place.

Over the past twenty years or so, developments in the contributory streams of psychology and sociology have been fed into the study of the communication process so that today we have more sophisticated and refined models than were available in the past. But old habits die hard, and narrow, one-way, psychologically-oriented approaches are still with us.

Communication Content and Consequences. Whether we wish to study the effects or effectiveness of an act of communication, many factors apart from the content of the message and the intention of the communicator have to be taken into consideration. We cannot expect to accurately predict consequences from content—no matter how thoroughly the content is analyzed—unless we know something about the source of the information or message giver, the nature of the audiences or publics and their perception of the source, the medium used, the communication situation, and the nature of feedback (if there is any). All these variables—the factors that govern production, provision, and reception—are part of the communication process and must be taken into account in an attempt to assess the relationship between information and communication.

Information may be provided, and it may be readily available, but provision and availability tell us little about use and consequences. Additionally, one of the factors that governs use, influence, and consequences has to do with the nature of the source and how it is perceived. Perceived credibility in the wider political, historical contexts is often very important in this connection. Therefore, it is essential not only to have information on this score, but to study all the factors—historical, economic, legal, technological, organizational, structural, professional, personal—that may impinge on the production and provision processes. Put crudely, communication has functions for both sender and receiver, and these are not independent of each other, but are related in various ways within the overall social system.

Information and the Communication Process

Let us look at some possible implications of the prevailing naive beliefs in the power or value of information provision coupled with a relatively crude approach to the communication process. We may examine a simple act of speech. Following the observation of an event or an available piece of information, the event or information is selectively attended to, perceived, interpreted, classified, categorized, translated, and arranged by the person who wishes to speak. At each stage of the process, personal, social, and cultural factors intervene. The event observed is filtered and processed through a mesh of past experiences, present relationships and associations, and future hopes and expectations. In a way, we are always talking about ourselves. We cannot escape our experience. Our words represent a joint product of "reality" and self. What information we provide in such circumstances is bound to be removed from what we observed, or what was provided in the first place. Moreover, what the receiver makes of what is provided, given selective perceptions and interpretations, is a further source of change or distortion.

It is not surprising, therefore, that exchanges, discourse, and dialogue frequently lead to confusion and misunderstanding. Many people have a naive faith in the power and value of information provision, and others, perhaps exhibiting a little more sophistication, have more confidence in the constructive possibilities of discussing what information is made available. As it happens, both provision and discussion, in themselves, do not necessarily lead to understanding. Progress in this direction depends on a recognition of the complexities of the communication process.

Technology. It is frequently suggested, these days, that many of the communication problems that are mentioned here might be overcome by technological developments in information provision, developments which have no shortage of well-financed advocates. Technological innovations may indeed solve problems—the potential is undoubtedly there—but what really matters is how and in whose interests the innovations are introduced and applied. The solutions depend on economic and political decisions and, of course, such economic and political considerations also help to determine what is regarded as problematic in the first place.

If an overly optimistic approach to the perceived benefits of technology were coupled with the naiveté and the lack of knowledge about the communication process mentioned earlier in this paper, several difficulties might arise. Communication operations will always entail the selectivity of classification, interpretation, arrangement, and presentation, and these processes do not operate in a political or economic vacuum. One of our tasks is to seek to ascertain the wider implications of the political and economic factors that govern the organization and operation of communication/media systems.

It is important to note in this connection that the provision of information in symbolic form can only by meaningful to the recipient if that person has experiences corresponding to the symbol. Put simply, the fields of experience, the world of discourse of sender and receiver, must overlap before communication can take place. It is important to take this into account and to attempt to ascertain the degree of coincidence in fields of experience in societies that are stratified, fragmented, and differentiated (all are in some way or other). We need to recognize that most communication systems are establishment-oriented, reflecting the power structure of the societies in which they operate, generally serving to legitimize and maintain the status quo.

The flow of information and the direction of most communication, both nationally and internationally, is from the top downwards. Invariably the vertical prevails over the horizontal. It is important to consider the implications of communication being so often a matter of the few talking to the many, directly or indirectly, about the perceived needs or problems of the many, from the unquestioned standpoint of the few. We must ask if it need always be like this. Rather than go along with Marshall McLuhan's superficial catch phrase that "the medium is the message," we might more usefully attempt to discover to what degree the medium is the system.

It is worth emphasizing that studying information and communication systems or processes in isolation, apart from other institutions and processes in society, rather than in the wider context is inadequate. Any changes or improvements we may have in mind are not likely to be brought about simply by confining our attention and activities to the media. The changes must be on a much broader front, involving nonmedia institutions and processes.

Effects and Influence. The limitations of the models of the communication process that, occasionally explicitly but usually implicitly, guide much of our thinking in communication research have already been mentioned. These shortcomings are particularly marked in approaches to the study of influence or effects. Although this is not the place for a detailed examination of this important question, there are some points that must be made.

Often, in attempts to study the impact or effectiveness of a campaign or program to put across a certain message, the evaluation is based on such criteria as verbal responses to questionnaires or interview schedules administered some time after transmission or exposure. These responses are regarded as indications that the message has been registered, remembered, understood, and is capable of being recalled. (Claims about understanding or comprehension are often based on data which more accurately reflect retention and ability to articulate.)

In some evaluations members of the public may also be asked if they have changed their attitudes and/or behavior since receiving the message, or as a result of the campaign. It is important to recognize that the answers to these questions are essentially self-reports. Rarely is substantiation required, and even more infrequently is actual social behavior examined. Even if we accept the validity of the concept of attitude and are prepared to believe that attitude change may be accurately assessed by interview or questioning in this way, we still need to remember that attitude change is not the same as behavioral change, nor necessarily even an indication of it. This qualification applies, all the more, to the mere possession of information or knowledge, as well as to the ability to simply remember and articulate what has been received.

As a result of these forms of evaluation, success has often been claimed for campaigns and educational programs in the absence of any valid evidence to support such claims. On any given subject, a person may possess the necessary information, but many factors may intervene between the information and what might be regarded as the related action. These intervening factors might include physical, financial, and geographical barriers; inappropriate attitudes (perhaps relating to past experience); conflicting attitudes (even when the appropriate attitude is present there may be others pushing in the opposite direction—attitudes do not exist in isolation); and lack of skill, competence, or opportunity to translate the information into the appropriate social action.

There are examples from recent research in health education where people certainly had the necessary information, but where such factors as definition of sickness and health, expectations with regard to the medical profession, hospitals or clinics, and past experience in related areas—all of which vary according to social class—came between the possession of knowledge and the desired action. Other research on information and the welfare services indicates that some people who had the necessary basic information about benefits and allowances just did not know what to do about it, or how to use it. Additionally, we know of television programs that have stimulated members of the audience to the threshold of pro social action, only for them to be left waiting on the threshold because other institutions in the support system had not been alerted.

On the other side of the coin, there are many examples from research of people being so conditioned by past experiences that they only receive information—if, in fact, they receive it at all—about specific objects, people, nations, and so on in a heavily conditioned manner. There can be no good news about country X because we have all been geared, over time, in so many ways, to accept it as a bad country. Clearly, an admirable formula for international understanding.

Let it be emphasized once more that information may be the necessary starting point, but rarely, if ever, a sufficient condition for action. As much attention should be given to the conversion and utilization of information as to its provision. Yet, currently, simple provision seems to be the sole concern—even the obsession—of so many. What determines what is provided and what comes between knowledge and action is not given the attention it requires.

Media Influence. Several times references have been made to the poor quality of our thinking about the communication process, to our conceptual inadequacies with regard to the process of influence, including the misplaced emphasis on attitude change. There is much more to media influence than can possibly be covered by approaches that rely on attitude change or imitation, or by other conventional psychological approaches. Undue emphasis on these approaches has contributed to our inadequate understanding of the complexities of the communication process.

We know that the media may exert influence in many ways. For example, they may set the social-political agenda by defining problems, amplifying them, and suggesting solutions. They may confer status and associate certain forms of behavior with specific groups or individuals, and label both behavior and groups as good or bad. They may legitimize certain practices, values, structures, and systems and, in so doing, promote consensus and solidarity and reinforce the establishment or status quo.

On the other hand, their influence may be disruptive, in some cases exacerbating conflict. But this, as with other forms of disruption, dysfunctional from one point of view, need not be so from the standpoint of those who seek change. In any case, as indicated earlier, the overall implications will depend, not only on the media or supply of information, but on a wide range of nonmedia factors.

Communication and Information Research and Policymaking

A major task of social scientists is to examine, in different circumstances and with different issues, the interplay between media and nonmedia experiences, and to attempt to assess the significance of this interaction for social consciousness. But to do this along the lines briefly indicated above is no easy matter. It is much easier to stick to information-testing and attempts to measure attitudes and opinions; easier—but misleading—because these approaches rely more on available techniques, perceived susceptibility to conventional measurement, and false interpretations of so-called scientific focus on methods, quantification, and reliability rather than on validity. Such approaches, which have been criticized as concerned with doing rather than with thinking, are theoretically impoverished, conceptually crude, and substantively sterile. The stand adopted here is that they need to make way for more sociologically-oriented approaches which attempt to study both sender and receiver in terms of their relationship to each other, to their own and other institutions, and within the framework of the appropriate social systems.

Research also shows that so much of the information that is provided by the media is just not understood (and this is stated here knowing full well that our understanding of *understanding* leaves much to be desired). Research seems to suggest that only a small proportion of media news presentations—print and electronic—is understood or capable of being reproduced intelligently.

The decoding of messages by the audience or public does not always coincide with the encoding by the media practitioner, and frequently, as indicated earlier, this is because the worlds of experience of encoder and decoder do not overlap to the required degree. The importance of social class and associated cultural and linguistic differences may be obvious. But it may not be so obvious that there are also other important differences which impinge on communication and understanding, and that these are an integral part of media professionalism.

In a manner of speaking, broadcasters and other media practitioners and providers of information speak to and for themselves according to internal and internalized professional criteria, creating and perpetuating a form of professional tyranny. All too often programs and information campaigns are considered successful if they satisfy internally-defined professional criteria, yet they may have no known relationship to effective communication. There is a self-fulfilling prophecy about much of this—the reinforcing message is repeated and the system is maintained and strengthened. Independent criteria are not invoked. "It is news because it's news because it's news; and if you were a journalist you would recognize this."

Decisions are being made every day about information and communication policies and practices. In many Western industrialized countries we prefer not to think in terms of communication policies, for it smacks too much of central planning. Nevertheless, policy decisions are made by many agencies, and there is a communication policy, albeit fragmented and latent rather than overt and consciously articulated.

When we examine policymaking with regard to information and communication we may find that the vast majority of decisions may be classified under three headings: 1) private profit; 2) political expediency; and 3) the need to maintain existing structures. At times all three coincide or overlap in some way or other.

Ideally, a fourth criterion should prevail and should be related to the basic information or communication needs of individuals and/or societies. Technology should not be allowed to determine needs. Information and communication needs should be identified and evaluated from a specific value position, and then technological developments, communication policy, and political and economic decisions should be formulated to meet those needs. A major task of research is to identify such needs and to provide the information essential for intelligent policy formulation. This is no easy matter, but it represents an approach to information and communication problems preferable to an approach that stems from an unholy mixture of technological and market determinism. It is certainly one that must be fully explored, for it should be clear to all of us that basic communication needs will never be met by the unrestricted operation of market forces. The concept of public serrvice is as essential in communication as it is elsewhere.

An Agenda for Research and Theory Development

The title of this article is "Information Is the Answer, but What Is the Question?" Let us conclude, then, by being more specific about the questions we should ask and which might be addressed in research.

Some years ago, seeking to anticipate what some people have referred to as the communication revolution, I attempted to prepare a research program to deal with these changes. The program was based on the assumption that there was a high probability that technological and organizational developments in the media, and the communication industry generally, would be quite widespread. These developments might cover electronic publishing, video games, home computers, teletext, viewdata, video cassette recorders, video disc players, cable systems, and satellites. It was further assumed that some countries, such as Great Britain, would experience some organizational and institutional changes; for example, obligations to serve new regions, and the development of new TV channels and breakfast television. Additionally, there would be an increase in the possession of remote control sets and more than one set in any given household.

It was projected that the application of the technology would be governed by a range of factors, including political and economic considerations, which might or might not be related to perceptions of needs. The rates of development, domestic and business, individual and societal, would differ from country to country, from innovation to innovation, and from institution to institution.

The social implications of the developments would also differ in similar fashion and would depend, among other things, on stages of development, institutional structures and arrangements, existing cultural patterns, prevailing social trends, financial resources, and political decisions. An appreciation of the importance of this wider, social context is vital. A basic principle of the research approach, reflecting the argument presented earlier in this article was that the media, new communication technologies, innovations, and the communication process generally, including social implications, can only be studied adequately within the appropriate wider contexts.

The research program, as originally formulated, had as its general aims:

- to examine the introduction, application, and development of the new electronic and communication technologies, the factors (e.g., finance, control, and organization) that influence these processes, and related institutional changes;
- to study communication behavior generally, and to identify and evaluate the wider social—including economic, political, cultural, and religious—implications of the technological and institutional developments at several levels (e.g., attitudinal and behavioral) and in relation to other social and leisure pursuits, and other institutions such as the family and education.

The questions and points listed below are primarily concerned with this second part, with social implications. They may be seen as clear indications of research possibilities motivated by social concern. They are not presented here in any order of priority:

- Will the nature and scale of the new communication operation promote tendencies towards centralization and metropolitanization of information sources?
- Will people generally be more informed?
- What changes, if any, will there be in information-seeking behavior by different groups in the population?
- Will new audiences emerge for both information and entertainment?
- Will new educational needs emerge and, if so, how will they be met?

- Will increased privatization, greater concentration of home- or family-based activities stem from the changes?
- Could the changes lead to reduced interpersonal relationships and more human/machine interaction?
- What will be the influence of the innovations on the relationship of the media to other institutions (e.g., church and school), and on the role of support systems and their complementary functions with regard to education, social services, and social action?
- How will the new technologies be used in relation to agricultural developments, medical and health services, and family planning programs?
- Will there be an increase or decrease in the gap between the information-rich and the information-poor, and the gap between the leisure-rich and the leisure-poor? Will there be more for those who already have all the facilities, the knowledge of where to go, what to do? Will we see the further development of elites?
- What about the question of dependency? Will the technologies be introduced internationally in such a way that Third World countries become increasingly dependent?
- Will the innovations, and the media operations generally, impede or facilitate the development of a national identity and national culture and language?
- What about the important relationship between multiplicity and diversity in both provision and use? Will more mean more of the same thing, or will there be more variety, more real choice?
- Will the changes lead to more access, more participation, more involvement, more democratization, etc. (including democratization of management and editorial functions)?
- Will the innovations lead to changes in community, and to local or regional identification at both attitudinal and behavioral levels?
- How will the new technology interact with traditional forms of communication, and what will the implications be for opinion leaders and gatekeepers?
- Will the special needs of minority and temporary membership groups be more effectively met as a result of the increase in channels?
- Will the development of consumerism be facilitated?
- Might society become information leaky as well as information-rich or over-rich, and what will be the implications of this for more open government and more participation in political decision making?
- How will the status and role of women be influenced?
- Will there be any overall change in values, belief systems, or life styles?

Conclusion

This list of questions and issues is not exhaustive, and additional ones could be added in light of the circumstances and conditions in any given society.

At the risk of oversimplification, we might say that the focus of the concern, as itemized above, is the quality of life and, more specifically, how this quality of life is being influenced, or could possibly be influenced, by the new communication technologies. How is the communication system being used—by whom, for whom, and for what purposes? Communication may be seen as: 1) persuasion; 2) transfer (passive) of information; 3) a means of personal expression and social relationships; and 4) an instrument of social change. These functions are not mutually exclusive, but where is the emphasis? What are the priorities? Ideally, we need to ask these questions in all countries to obtain a comparative picture.

A concern to promote the development of opportunities to facilitate the exercise of real choice and self-expression and increase creativity, autonomy, and participation underlies these questions. The possibility of narrowing the gaps between the information/communication rich and poor is also a central concern.

On the other hand, there is the fear that, if applied in certain ways—and the signs are not encouraging—the new technologies could serve to maintain the existing divisions and discrepancies. They might even add to our problems by reinforcing the reliance on commerce and cultural production and creating needs that can only be met by increased production.

Present indications, such as they are, taken together with predictions based on past related experience, indicate that in many societies commercial criteria will predominate, and that the new media technology is likely to be introduced in such a way as to reinforce the old rather than create the new.

We must also remember the vitally important point that, at present, the opportunities and potential for beneficial and creative use are unevenly distributed among different groups and classes within societies, as well as among different societies. Intervention is required at more than one level if we are to do more than reinforce the status quo. The new international information and communication order is not unrelated to the new international economic order.

It is essential that the ambitious claims of those who advocate the rapid adoption of new communication technologies should be seen in light of these considerations and put firmly to the test. History may not always repeat itself, but it certainly has lessons for us. There is no room for undue optimism in this connection.

Much of what has been written in this article has been based on the assumption that innovations will take place within existing socio-political-economic frameworks. At a different level, and in terms of the real potential of the new technology, it could be argued that, appropriately introduced, applied, and operated (regardless of how "appropriate" is defined), the micro-electronic revolution might represent the only means at our disposal to make it possible for us to transcend present conditions of dependency (nationally and internationally) and to combat what is often referred to as the alienation brought about by the consumer society.

Presumably, technological innovations and developments (although not on their own) could facilitate this progress by making available the information for people to act collectively, with full knowledge of the implications of their actions. They could also make possible more access, more participation, more involvement, and more democratization. Unfortunately, there are grounds for disquiet, for each of the above developments will depend on economic and political decisions which govern not only the innovation and application of new technologies but also the development of the appropriate skills and competencies within the general population. These will influence how and with what implications the new technologies are used.

If we are concerned, then, with the potential of technological innovations as they impinge on the quality of life in our societies, we need to recognize that the way this potential becomes activated and is eventually realized will depend on other considerations. For example, any attempts to develop the active, positive, creative, expressive, liberating use of the media will inevitably come up against past experience and the inertia of present behavior patterns and attitudes, both of which are closely related to past and present media provision. How do we transcend our experiences? How can we escape the past and make it clear that, despite commercial claims to the contrary, we are not being given what we really want (still less what we need) but what we have been conditioned to expect?

To secure the optimum use of the media and the system we must take into account other, nonmedia factors. Desired changes in the former will not be achieved without changes in the latter.

Note

1. In making these generalizations it is recognized that they may beg as many questions as they answer. Mass communication researchers are by no means a homogeneous group. The standpoint adopted here is that of the critical, sociologically oriented communication researcher, and the implications of this should become clear during the course of the article.

3

Conflicts of Theory and Issues of
New Information Media Policy

Denis McQuail

This article explores the relation between media theorizing and media policymaking, and tries to answer the question of whether theoretical choices correspond with policy options. Divisions of theory are held to turn on four global issues: (1) Whether society or media should be considered as a prime mover in social and cultural change; (2) whether the media are essentially hegemonic or pluralistic; (3) whether media have mainly unifying or mainly fragmenting effects in society; and (4) whether the proper approach to questions of communication should be "culturalistic" or "scientific." The options for new media policy are numerous, but reducible to a small number of basic choices. An overview of the situation in Europe would suggest the following issues: Should the potential of new media be actively encouraged or not? What should be the balance between private and public investment? Whether to promote diversity or allow forms of monopoly? Whether to regulate or deregulate? If regulation, whether to attend to content or only to technical and distribution matters? The whole will be influenced by whether there reigns a cultural or an industrial policy climate. It is suggested that each theoretical position does imply some policy choice, and that there is some correspondence between theory discourse and policy discourse.

As the media have developed within the last century or so into a major institution of society, they have brought with them two other kinds of institutionalized activity which seem invariably to have accompanied similar developments in other fields: theorizing by academics and policymaking by government and other public agencies. Thus, in addition to those who actually carry out the tasks of the institution, whether it be teaching, prescribing drugs, fighting wars, or making films, there are to be found others, either speculating on the task—theorists—or setting directions, guidelines or limitations—regulators. Those involved in these secondary activities are fewer, less visible and varying in their significance for the institution as a whole, from place to place and case to case. The media have been somewhat reluctant to acknowledge that they have their share of theorists and regulators, two breeds toward which the media have some antipathy. They tend to treat the former as irrelevant and the latter either as an enemy to be vanquished or a temporary ally out of expedience.

If we consider the triangle of relationships between media, theory, and policy, the general character of the relation between media and theory and also media and policy is clear enough: Theory is a reflection of and on the activities of the media, and policy is a reaction to, or anticipation of, what the media do or might do. While there is no necessary direct connection, it would be surprising if the third link—that between policy and theory—were not also very much in evidence. There are many possible questions regarding the divergence and convergence of policymakers and theorists, but the question to be considered here is the extent to which theoretical choices correspond with policy options and, underlying that, whether the study of media theory has much to contribute to policymaking (one can hardly envisage a relationship in the opposite direction).

The task of tackling this question is made methodologically more feasible by the circumstance that both theory and policymaking have undergone and are undergoing a period of change and ferment. To judge from academic history, a necessary stage in the maturation of a scientific discipline and a mark of its arrival is a period of division and conflict among its practitioners. By that criterion, something like a science of mass media would seem to have arrived. We are, in any case, faced with several real choices in interpreting the direction and significance of mass media, and scholars clash heartily and not always predictably over a wide range of subjects.

On the policy side, the growth and change of media in almost all their aspects—organizationally, technically, culturally, and in scale—has triggered either demands for appropriate policy or spontaneous interventions, which can also be represented as a series of options. At its simplest, therefore, the matter at issue concerns the degree and kind of correspondence to be found between alternative theoretical positions on the one hand and available policy options on the other. While any such comparison involves gross oversimplification, one should be able to reach a general answer to the question posed, if only in the form of a negative. Thus it is quite possible that there is little or no correspondence between the practical options considered for policy and the divisions of theory. The remainder of this article is concerned with testing this null hypothesis.

It is one thing to assert that theory is in a state of flux and to adduce telling instances of clashes between scholars which are to be found in the literature. It is quite another thing to give a profile of the differences which is likely to be recognizable and acceptable to the various protagonists. There are too many alternatives of terminology, emphasis, and personal definition to expect a tidy result even in this task. As to the clashes of a fundamentally theoretical kind which have surfaced in recent years, one can cite: the clash between Gerbner's (1976) school of cultivation analysis and its critics (e.g. Hirsch 1980; Hughes 1980); the debate over the viability and usefulness of the tradition known as "uses and gratifications" research (e.g. Elliott 1974; Carey and Kreiling 1974); the reassessments of the significance of personal influence and opinion-leader theory (Katz and Lazarsfeld 1955; Gitlin 1978); the long-running argument about the power of the media (Lang and Lang 1981); the question of whether or not international media flow is free and balanced and whether it ought to be so (Boyd-Barrett 1982); the broad struggle between proponents of Marxist or critical theory and those who have come to be labelled as "pluralists" (Bennett 1982); the debate about whether or not news reflects social reality and whether it sets or responds to the agenda of public or politics (Glasgow Media Group 1980; Becker 1982).

As soon as one begins on such a list, the possibilities will seem endless, ranging from the level of irreconcilable differences of world view to the minute, but intense, disagreement about some micro-process of communication or the appropriate method for its investigation. Nevertheless, it is reasonable to suppose that there is a structure to disagreements and choices over theory and even to disagreements over the interpretation of evidence which often conceal underlying differences of theory. It is to be expected that such differences will be related to broader matters of philosophy, ideology, and world view, given the sensitive position usually occupied by mass media in relation to politics and public morality.

The version of an underlying structure in this article is one attempt to provide an economical account of alternatives and to relate these to each other, but it has to be admitted that they could be ordered in different ways and multiplied according to other dimensions of variation. To give a brief explanation in advance, four central issues are discussed according to the main alternative possibilities and these are related in turn to each other. The question of policy options is reserved for a later point in the discussion.

Four conflicts of theory

Media or Society as Mover?

The first dispute concerns the choice between media and society as first mover in the relationship between the two. Some theory is media-centered, some is society-centered, the former stressing the media as a force for change, either through technology or through the typical content carried, the latter emphasizing the dependence of the media on other forces in society—especially the power structure and the pattern of social and class stratification (Murdock and Golding 1978).

The difference is quite fundamental, since upon it depend most of the answers concerning the relationship between mass communication and social change (McQuail 1983; Rosengren 1980). From a society-centered perspective, the media are to be seen as the outcome of historical change—a reflection and consequence of political liberation, industrialization, secularization and as a means of servicing the requirements of other social institutions. From a media-centered view, however, the dominant forms of communication technology—and the typical message systems which are carried—are likely to serve, as they have in the past, as an independent causal factor in social change. The most fully worked-out theories of medium- or communication technology-determination (and they are not very worked out) are those of the "Toronto School" of Innis (1950, 1951) and McLuhan (1964). They have been supported, not entirely independently, by others such as Gerbner (1967) and Gouldner (1976) who have argued that forms of public communication do matter a good deal for what happens in history. Thus the rise of mass media since the age of printing has been credited by several theorists with some element of causation of the rise of liberalism, political freedom, individualism, nationalism, rationality, and, more recently, irrationality and the decline of ideology.

The society-centered view of media and social change has, not surprisingly, been the orthodoxy of sociologists and political scientists who are, by definition, committed to a belief in fundamental social and political forces of a structural kind as prime movers. The forms and techniques of mass media are viewed as the typical elements of the superstructure—essentially aspects of culture which owe their origin to more fundamental changes. The more materialist the view, the more firmly is this held to, as with classical Marxism. The difference, however, is more than one between idealism and materialism. Media determinism, in the tradition of the Toronto School, is itself quite materialist, since it is the technology and not the message which is seen as the cause of change.

The contrast between society and media as causal factor is much too global to account for several of the varieties of theory which have already been referred to, but it can be improved as a classification by a simple internal subdivision. On the side of society, we can broadly separate potential influences which are structural from those which flow from individual social and psychological differences. On the media side, it is also helpful to separate those theories which emphasize technology and the nature of the medium from those which have more to do with typical contents carried by a medium or media, whatever the cause.

Thus, by *social-structural* theory, we are likely to refer to such matters as the power and class structure of society, to economic influences on the structure of the media, and on the behavior of audiences. The emphasis is more on the objective determinants of the context in which media messages are produced, disseminated, and consumed. From the *individual differences* perspective, attention is paid more to the chosen motives and expectations of an audience, to matters such as prior knowledge and interest which affect behavior, to processes of selective attention, perception, and interpretation and to chances for individuals (and eventually audiences) to resist or avoid influence of whatever kind. Roughly speaking, the former kind of theory is more materialist and determinist and the latter more idealist or mentalistic, and places more stress on personal and variable free choice.

The essential difference between technology-determination and message-determination is also one between a more materialist and a more idealist concept of change and influence and also between a more determinist or a more voluntarist conception of the process at work. The *medium/technology* version stresses the constraining and directing effect of a dominant technology on the kind of messages to be carried and on the capacity and purpose of a given communication system. It also stresses the pressure on the receiver to adapt processes of thought and behavior to the lines indicated by the meaning forms and the communication system. McLuhanism is the extreme form of expression of this view. The *medium as content* theory of influence involves no rejection of the view that technology influences content, but it is not the only or necessarily dominant influence, and the meaning of the message will be seen as having more to do with the cultural and institutional context of those who produce and those who receive. Gerbner's (1976) cultivation theory would be an example.

There are a number of consequences which flow from the broad choice of theory as outlined in this way. In general, media-centered theories emphasize powerful media, and society-centered theories lead to a much lower estimate of media power and stress either the fact that the power of media comes from society (the structural position) and is mistakenly attributed to media on their own, or that individuals and audiences have considerable defenses and that effects are a function of the predispositions, needs, and choices made by receivers (the individual differences perspective). Thus, effects from media as such are unlikely to be large, random, uncontrollable, or unwanted. Media-centered theory, in either version, does not necessarily involve an assumption of large, direct effects on individuals, but as it has usually been formulated, either in its medium-technological or its message-centered forms, it does presuppose important long-term institutional effects, either causing or preventing change.

Each of the four variants named also point in rather different directions for research, emphasizing either the structure of society and its institutions, or the audience, or the medium, or the message system, and each has its own school of theorists. These are inclined to disagree, yet it would have to be said that the available evidence is both too incomplete and inconsistent to sustain any one version of the direction of cause-and-effect in respect to media and society (Rosengren 1980) and the pragmatist might conclude that the reality is too complex to admit or oblige an exclusive choice of theory.

Dominance versus Pluralism

This second broad division of theory has a more openly political and normative character—separating, as it does, the media pluralists from those who view the media as a significant instrument of dominance in a class divided society (Bennett 1982). The second view has manifested itself in a number of familiar forms: in classical Marxism; in more or less conspiratorial versions of mass society theory; in the work of the Frankfurt School; in current political economic media theory (e.g. Murdock and Golding 1977), in much "new-left" or "critical" media theory.

Despite variations of emphasis, the shared point of view is that mass media, directly or indirectly, are always in the hands of, or serve the interests of, a dominant class or elite who use their influence to shape the consciousness of the majority, and to subordinate any potential class opposition by ideological means. Thus, they are able to disseminate a view and definition of the world more or less in their own interests and omit alternative interpretations which might cast doubt on their legitimacy. The theory draws on and emphasizes several features of mass communication—its tendency to standardization of content and homogenization of audience taste, its unidirectionality, its great attraction and prestige. The media are seen, in this view, as more or less unified and in ruling class or power elite ownership or control, offering relatively undifferentiated and ideologically colored content to dependent mass audiences, with predictable consequences for control of a manipulative kind. In contrast, the pluralistic view of the media emphasizes the diversity and fragmentation of media structures in liberal societies, the range of choice of message and ideology offered, the multiplicity of audiences with freedom to select, reject, answer back, and apply what is received to their own needs rather than those of any elite or ruling class.

These two views are separated not only by an interpretation of evidence but also a set of political values, on the one hand favoring liberalism for its own sake, on the other hand doubting whether liberal forms lead to democratic-pluralistic outcomes and favoring elements of collectivism, popular control, and limitations on private ownership. They also involve alternative interpretations of likely effects from mass media, since pluralistic theory emphasizes the unpredictability and diversity of effects in the consistent direction of favoring established forms of society and the interests of the class or elite which happens to own or have control.

Again, while ultimately opposed in their tendency, both global versions of what are fundamentally different social and political theories indicate hypotheses and elements which might, under some circumstances, prove reconcilable. Many media systems are likely to exist in mixed forms, with elements of monopoly and hegemonic tendency, but also showing some variety and meeting resistance from afar from helpless mass audiences. Many adherents of pluralism are also critical and fearful of tendencies toward concentration and dominant control. For these reasons, it is appropriate to treat this division as a continuum rather than an absolute division.

Given the very broad nature of this theoretical division, it is not surprising that it encompasses, partly or wholly, a number of other oppositions and distinctions which are encountered in mass communication theory. It largely subsumes the opposition between critical and administrative research, first formulated by Lazarsfeld (1941), since the critical perspective is likely to go with a dominant media perspective, while administrative research is more justifiable under pluralistic assumptions. Beyond this, we can suggest a tendency at least for dominance theory to go with assumptions that mass communication is purposive rather than purposeless, consistent rather than random, sender-directed rather than receiver-chosen, closed rather than open, consensual rather than divisive. The breadth and generality of the division sketched in this way reduces its value as a tool of analysis, but testifies also to its potency and is a reminder that it is a good deal more than a distinction between "left" and "right" or "consensus" and "conflict" approaches to the study of society.

Centrifugal versus Centripetal Effect

A third dimension of media variation, which also has deep roots in other traditions of social thought, distinguishes the view that mass communication contributes to change, fragmentation, diversity, and mobility from the alternative view that it is a unifying, stabilizing, integrating, and homogenizing force (Carey 1969). This, in turn, reflects the larger sociological dilemma of accounting for the occurrence of social change without loss of continuity and order.

Different theorists have associated the mass media very closely with both tendencies. On the one hand, mass media have, historically, brought messages of what is new, fashionable, and advanced in terms of goods, ideas, techniques, and values from city to country and from the social top to the base, and they have seemed to challenge established ways and value systems. They have also probably motivated people to move in search of better material conditions and stimulated demands for consumption. Beyond that, they have some potential, at least, for weakening the hold of traditional values enforced by group sanctions and have helped to free individuals from the ways of thinking of their own limited social milieu, consequently "privatizing" certain areas of social life.

On the other hand, the media have been credited with replacing diverse and long-established value systems with new and homogeneous sets of values which are not very complex or constraining, but which, nevertheless stress conformity and order and, to the extent that they are so widely held, contribute to the binding together of a large-scale, differentiated society more effectively than would have been possible through older mechanisms of religious, family, or group control. Such, at least, might be a functionalist explanation of media tendencies.

The positions seem far apart, the one stressing centrifugal, the other centripetal tendencies, although in fact, as several authors have pointed out, it is not inconsistent to suppose that both forces are at work, the one compensating to some extent for the other (McCormack 1961). Nevertheless, the attempt to hold both views at the same time, or to reconcile them theoretically, can be confusing and it helps to think of both versions of media theory—centrifugal and centripetal—as each having their own dimension of evaluation, so that there are, in effect, four different theoretical positions. In the case of the centripetal proposition about media, there is a positive version which stresses the media as integrative and unifying (essentially the functional view) and a negative version which represents this effect as one of homogenization and manipulative control (critical theory or mass society view). For the centrifugal proposition, the positive version stresses modernization, freedom, and mobility as the effects to be expected from media (individualism in general), while the negative version points to isolation, alienation, loss of values, and vulnerability (a "dysfunctional" view of change as social disorder).

Each of these four positions can be identified in the literature of mass media theory, although sometimes without clear distinctions being made. For example, in some versions of mass society theory (e.g. Mills 1956), the greater central control (a centripetal effect) is seen partly as a result of the isolation and alienation of the individual and lack of solidarity (a centrifugal effect). Thus, a pessimistic view of society can reconcile seemingly opposed processes of unification and dispersion. An optimistic view can do the same for the "positive" policies of the two tendencies, treating media as providing the still-necessary but much looser kinds of consensus and integration for a more mobile and changing society. Despite this complication of a simple picture, it is useful to identify the four positions, bearing in mind the possibility of recombining them either according to the type of change involved or the direction of valuation (optimistic or pessimistic view of society).

Cultural versus Scientific Theory

Finally, we can make a distinction which has more to do with conception and method than with the substance of what the media are or what they do in society, yet which has far-reaching consequences, both for research and for assessing the role of media. As usual there are alternative ways of naming the division, but it broadly corresponds to a difference between culturalistic or humanistic and scientistic approaches, between subjective and objective methods, between attention to content and attention to systems.

In respect of communication, it is well rendered by Carey's (1975) contrast between a "ritual" and a "transmission" model of communication. The former directs attention to the part played by communication (of all kinds) in the expression of the culture of a society and with maintaining it over time. Thus, communication connotes other concepts, such as commonality, community, and sharing. According to Carey, " a ritual view of communication is not directed towards the extension of messages in space, but the maintenance of society in time; not the act of imparting information, but the representation of shared beliefs." By contrast, the transmission view, which is especially associated with industrial cultures, has as its center the "transmission of signals or messages over distances for purposes of control." It derives from a metaphor of geography or transportation and is defined as imparting, sending, or transmitting information to others, thus emphasizing speed and effect in the sending of messages. There is more to be said of this, but Carey illustrates the distinction by explaining its implications for the study of media news. The transmission view of news emphasizes objectivity, facticity, speed, and effectiveness. From a ritual view, "news is not information, but drama; it does not describe the world, but portrays an arena of dramatic forces and action." As with the other three theoretical choices already named, the distinction is rich in associations with other dimensions. The ritual or culturalist approach goes with an attention to meaning rather than effect, to interpretation rather than measurement, to expressive forms and functions of communication rather than the factual and informative, to interactive rather than one-way communication relations, to open rather than closed systems, to a view of communication as consummatory (an end in itself) rather than instrumental.

An Integrating Framework

To this point, four varieties of theoretical distinction or conflict have been characterized. Each has a potential for further elaboration and each with several strands of normative judgement as well as some basis in objective interpretation of evidence and potential for use in forming testable hypotheses of a rather global kind. It has also to be acknowledged that the dimensions named tend to overlap and that rather similar predictions about the media tend to flow from different theories or theoretical positions.

Before looking at their implications for media policy, it is worth trying to fix them in relation to each other somewhat more securely, at the risk of some arbitrariness. To do so, a suggestion is borrowed from K.E. Rosengren, which is to adapt to the present purpose a scheme developed by two sociologists of organizations who mapped out various schools of sociology relevant to their work. Burrell and Morgan (1979) mapped out schools of sociology according to two master dimensions, of which the vertical axis extends from "the sociology of radical change" to "the sociology of regulation," while the horizontal axis polarizes "subjective" and "objective" theory and research. Rosengren (1983) used the same framework for a somewhat similar purpose, comparing and assessing traditions of research in mass communication.

While the parallel is not exact, approximately the same two dimensions have appeared in the foregoing discussion and the other positions named may be located in the space so mapped out. Thus, in the following figure, the vertical dimension represents the choice between a dominance view (left, critical, committed to radical change) and a pluralistic view (optimistic, somewhat functionalist, liberal-reformist) of mass media systems. The horizontal dimension corresponds even more closely to what was named as a cultural-science distinction (ritual versus transmission view). Within the four cells so identified, we can locate the other positions which have been named: two alternative versions of society-determined and of media-determined perspectives and two versions of both the centrifugal and the centripetal views, giving each a positive and a negative form of expression.

FIGURE 3.1
Varieties of Media and Social Theory

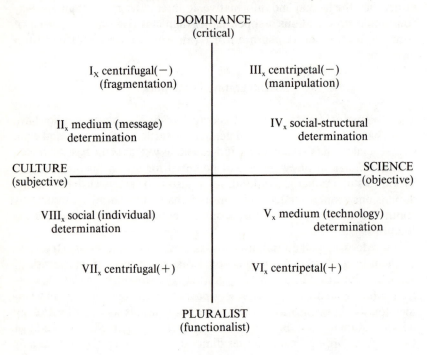

DOMINANCE
(critical)

I_x centrifugal($-$)
(fragmentation)

III_x centripetal($-$)
(manipulation)

II_x medium (message)
determination

IV_x social-structural
determination

CULTURE
(subjective)

SCIENCE
(objective)

VIII_x social (individual)
determination

V_x medium (technology)
determination

VII_x centrifugal($+$)

VI_x centripetal($+$)

PLURALIST
(functionalist)

Issues of Policy in Relation to New Media

What follows is written from a European rather than a North American—United States—perspective. It refers to widely varying national situations, but, nevertheless, to a broadly similar context characterized by a long tradition of media policymaking and currently considerable scope for making and applying policy decisions.

The technological possibilities opened up by the new media are much the same as in the United States and the social consequences can at least be discussed in the same terms, although the options available to governments and electorates are effectively wider and the pressures from various sources are more diverse. The main difference lies, of course, in the greater extent of public and state intervention in decisions about media systems and structures and sometimes in the actual operation of media services. Direct public intervention is largely confined to the broadcast media. But it is precisely here that technological innovation has most effect and where the question of how to respond is faced most immediately by the policymaker. In the case of media systems which are fundamentally commercial, public policy is usually only confronted, if at all, with the need to deal at a late stage with the consequences of innovation, aside from certain matters of licensing and standards.

Distribution

What, briefly, are the major policy issues? The starting point is a set of inventions and developments which mainly affect distribution possibilities for electronic messages of all kinds. These are by now well enough known and can be referred to under the headings of cable, satellite, videotext and video. *Cable* refers to the enormously enlarged capacity of electronic message-carrying on land; *satellite* to equally increased range and capacity for distributing television and other messages over the air; *videotext* to the systems which link individuals to each other and to computers and databases, allowing search and interchange and involving graphics and sound; *video* to possibilities for individuals to make, record, hire, or buy their own "programs" for showing on their own television receivers.

The implications for existing media and for expansion into new areas of service are considerable and still very hard to calculate as to time scale, degree, and type of change, since the pattern of demand is still unclear, the social forms within which the new media will operate have yet to be settled, and the new kinds of use have yet to be institutionalized. Even so, it is clear that, between them, the new developments offer potential abundance of audiovisual media in place of scarcity, greater individual choice and thus a chance of more differentiation and certain kinds of freedom, interactive possibilities which are not currently present in most mass media, and eventually innovation in forms and applications of communication. If any significant degree of this potential is realized, it has obvious relevance for the theory discussed since it seems to promise a future in which society could be more free, changeable, and interactive, with less chance of monolithic control, but also more risk of individual isolation and social fragmentation or privatization.

It has been noted that, in most countries, governments are being faced with the need to take decisions and make choices. Even inaction is a form of policy, since it usually means preserving the status quo and holding back change. There are a large number of considerations involved in what is, first of all, a broad choice between encouraging and advancing media development or holding back from decisions and effectively limiting development. The relevant considerations can be scarcely more than named here, but they include: the interests of existing media, public as well as private; the possibilities for some of financial gain from new hardware, software and distribution services; the interests of telecommunication production industries; the established cultural policies relating to media, the arts, and education; competing political views; technical standards; diplomatic and international commitments and consequences. These various considerations can be broadly grouped together according to whether they fall within the scope of cultural policy or industrial policy. The former covers those matters which have to do with social purpose, national culture, and political principles of freedom, equality, and diversity, the latter with economic consequences of media change.

While there are a good many varieties of cultural policy, in most European states they are characterized by a number of the following components: the wish to protect the national language and cultural traditions and achievements; the intention that media should serve broad public goals of general information, education, and entertainment; the principle of equal rights for different regions, minorities, and social groups; a concern with moral standards; a (more variable) concern to protect freedom of expression. Under industrial policy can be grouped: the wish to stimulate national electronic and related industry; a general concern with job-creation and widening investment opportunities; a greater attention to financial and economic aspects of new media developments and also to matters of technical standards.

In nearly all countries, there are elements of both cultural and industrial policy, but often one predominates over the other and sets its mark on the scope and nature of media development. In general, the more cultural the policy, the more tendency there is to hold back from rapid change. The more it is industrial, the more enthusiasm exists to encourage change. In any case this is the first main choice to which social theory concerning the media will be relevant.

Support

The second major policy issue concerns the manner of encouragement and the form or forms which are likely to be involved. Given the background of television and radio in Europe, the main choice is between public initiative and private (commercial) enterprise. Here the balance of policy climate will also play a part, with cultural policy favoring public and industrial policy usually leaning toward private and commercial development.

The choice is not usually that simple. For instance, it is possible to give new powers to existing public service media in order to expand new services, with or without commercial aspects, or it is possible to create new kinds of authority or license new private bodies. Alternatively, the market can be allowed to cater to whatever new demands arise for new kinds of service.

Central to the choice between public or private initiative is the question of finance, and again there are a number of options which are available to both public and private media operators. There are also issues to be faced over the cultural consequences of advertising, the political feasibility of raising extra public funds, and the implications for existing media of any chosen financial path.

Diversity

The third broad issue, whether the choice has been made for public or private expansion or a mixture of both, has to do with monopoly and concentration, or, to put it more positively, the protection of diversity, which is a prominent value in most democracies. This issue indeed arises at an earlier point in the process, since the decision to encourage new media or not may well be a choice for or against an existing monopolistic situation.

There is nothing very new about the basic considerations, although new media have different possibilities for enlarging diversity. The very nature of new media seems to offer some guarantee against monopoly and concentration of the old kind—that based on centralized large-scale dissemination, whether of television, radio or print. Certainly, the new electronic media seem to promise in Europe a decrease in the degree of public or commercial monopoly and of centralization. Thus, the monopoly issue finds more expression in the demand to decentralize the operation, control the new media, and widen the range of access for social groups as senders or information suppliers in a way which matches the widening of consumer choice. As far as private commercial exploitation of new media is concerned, the same issues arise together with the familiar matters of cross-media ownership and concentration of ownership. In a sense, the prevention of monopoly may be viewed as a "coalition goal" from widely divergent political standpoints, and what is more at issue are the means by which diversity can best be realized.

The "private enterprise" way favors the market as the best and most efficient mechanism for achieving a suitable diversity. The alternative is one of several kinds of artificial or "constructed" forms of pluralism, based on control or subsidy and designed to achieve equal or proportional representation in media according to region, political or social divisions, or language and ethnic origin (McQuail and van Cuilenburg 1983).

Control

There is a set of matters to do with the degree and kind of control to be exercised in the interests of any of the principles which have been opted for

FIGURE 3.2
The Nature and Sequence of New Media Policy Choices

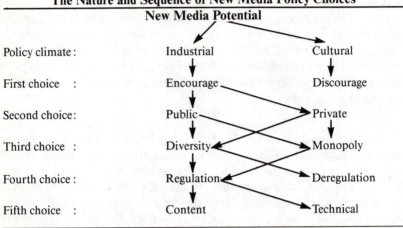

at an earlier stage in the policymaking process. Regulation can relate to the technical standards and type of equipment, to the setting of positive goals for media (e.g. minority representation or balance) and the monitoring of performance, or to setting minimum cultural and moral standards applicable to content which is transmitted. At issue here will be chiefly the balance of proscription over prescription and the question of who sets and polices whatever regulations are chosen. On the latter point the choice is between media industry self-regulation and special legal instruments externally applied. These choices are summarily presented in Figure 3.2.

Bringing Theory and Policy Issues Together

Even when policy issues and options are described in these rather global terms, it is difficult to see a close relationship between the policy options as named (encouragement or discouragement of new media, industrial policy or cultural policy, private or public development, choice of form of diversity, and high regulation or low regulation) and the several media theories mapped out in Figure 3.1. However, although a one-to-one relationship between theory and policy option is unlikely, a closer look at what is involved on either side—theory and policy—does show a web of connections.

No theory on its own delivers a coherent and complete policy, yet, each one contains a number of pointers to policy and, in turn, each broad policy option often implies a theoretical perspective. The connections can be explored either from the theory or the policy vantage point. The former may give a better idea of the extent to which a theory is likely to lead to a consistent media policy, even though the theories discussed are unlikely to be held as exclusive and self-sufficient guides to the reality of the connections between media and society.

We can begin with the two main dimensions which were used to locate other theoretical visions: dominance vs. pluralism and culture vs. science. In general, dominance expresses a view of the media as a powerful influence in a few hands, while pluralism suggests a view of media as a complex and diverse set of activities, under fragmented control, involving change and exchange and adaptation to circumstances and to the needs of the time.

Leaving aside the normative elements which are picked up at the next level of theory specification, dominance theory seems to: favor social, cultural, and political considerations; be cautious about rapid expansion; prefer public rather than private control and even public monopoly over private (given the dangers of irresponsible use of private power); and favor detailed regulation, with particular respect to matters of social power and access. Pluralist theory, on the other hand, is likely to lead to a more favorable view of media expansion, with less fear of unwanted or unpredictable consequences. Pluralism goes with some concern: for a cultural policy which would favor diversity of access; for a structure which would inhibit monopoly; for a minimum of external regulation; but without a clear choice between public and private systems.

The culture-science dimension has less direct implication for policy, but we might expect theories nearer the cultural end to invoke considerations of cultural policy rather than industrial policy and to be associated with regulation directed at content rather than technical matters, and to favor public rather than private solutions. The reverse tendencies would, consequently, tend to go with theory at the science end, since the market might seem the appropriate test of, and stimulus for, effective development of new media for a variety of communication ends.

These general observations have a bearing on the implications of the more specific theories mapped out by the two main dimensions. If we use the labels, as entered in Figure 3.1, and move in a clockwise manner from top left quadrant, following the sequence of Roman numerals, we can make a series of sketches for policy frameworks based on the eight entries.

The Pessimistic Version of Centrifugal Theory (I)

This theory, it is recalled, involves the view that mass media contribute to the fragmentation of society, the isolation of individuals, and loosening of social bonds. The theory, thus, would urge caution in any expansion of new media on cultural and social grounds and would lend support to public rather than private exploitation, with little intrinsic objection to monopoly (which could help to guarantee the desired form of diversity). It would be likely, however, to favor a decentralized policy which would give regions and local communities some say, and preferably their own media.

In principle, the atomizing tendencies of modern societies could be slowed or reversed as a result of interactive and locally-based communication. There would also need to be limits set to the import and access of foreign, especially commercial, cultural content across national borders, since preservation of a national culture is also a protection against alienation and fragmentation.

The Medium-as-Message Deterministic Theory (II)

This theory framework is also likely to involve a major reference to cultural considerations and a certain caution about encouraging new media and media abundance in general, in advance of clear plans for their use or evidence of demand for specific services. Much media development has taken place in the past without any real definition of purpose and on the basis of technological development for its own sake, or for its industrial and commercial benefits. In order to avoid this, public frameworks of exploitation would be preferred in which content-relevant criteria of performance could be effectively applied.

The objection to monopoly which is common to most of the theories (although for varying reasons) would here focus on dangers of uniformity of content rather than unity of structures. It has more reference, thus, to software than to hardware, to the range of sources and kinds of content, to the diversity or homogeneity of the professional groups of media producers and to opportunities for access. As with theory position I, there would be justification for regulation of content on such matters as the balance of functions to be fulfilled, the filtering of imported or direct broadcast satellite content, the control of film or video material sold or rented. Thus, this perspective is gradualist, protectivist, and somewhat paternalist.

The Pessimistic Version of Centripetal Theory (III)

Perspective III represents the most critical version of media theory in the present catalogue. It stresses potential manipulative control by monopolistic and centralized media operating in the interests of a ruling class or elite. Such an approach offers more than one perspective on the question of encouraging new media. On the one hand, they might be seen as having some potential for weakening the concentration of media-productive power and for allowing the development of more diverse, even deviant, media beyond the control of ruling interests. The characteristics of smallness of scale, interaction, and flexibility all offer some potential for change and freedom. They seem able to offer the conditions for what Enzensberger (1970) hailed as "emancipatory" uses, as against "repressive" uses of media.

On the other hand, this theory would also lead some to a more cynical view that all such developments would sooner or later be colonized and controlled by or for the same interests that control the centralized broadcast media and that commercial exploitation of hardware will inevitably confine innovative and oppositional uses to the margins of society, where they will make little difference. Moreover, the conditions of publicly guaranteed safeguards against commercial monopoly—and for minority access—might themselves open the way to state or capitalist control. Thus, the regulation necessary to secure social and political benefits from new media developments would in the end militate against the potential for liberation contained in the new media. The outcome of this ambiguity is to leave critical media theory somewhat on the sidelines, watching the development of new media policy with little capacity to suggest positive directions.

The Society-Centered Media Dependence-Determination Theory (IV)

This version, which emphasizes media dependence and determination by the social structure, is also somewhat inadequate as a guide to taking directions in policy-making. Extensive regulative and interventionist measures can be justified on the grounds that society always comes first. And, since media development must reflect the balance of social and political forces in society, it is better that this be open and explicit.

This theory gives little guidance on the question of whether to encourage or discourage the proliferation of new media and new communication technologies. It does not support the view that such proliferation will, in itself, be a major innovative stimulus to society economically or culturally, but it offers no compelling basis for resisting this view. If development is to take place, it should, however, be within a framework of public control and with adequate regulation to achieve ends set by society. In the matter of monopoly, attention is directed to structures rather than to content and to economic or political aspects of monopoly associated with market forces. Hence this theory generally leads to a strategy for constructed, or planned diversity.

The Technology-as-Medium Determination Theory (V)

This view is unambiguous on the question of abundance, lending support to unfettered development of new technology as a contribution to a more innovative, efficient, and productive society. It is relatively empty of conscious ideology and certainly would not prove a source of objection to state-sponsored innovation—on grounds of industrial policy—of the kind observed in France. On the other hand, if private exploitation were to be seen—as in Britain—to offer the best chance of new media growth, then it would be equally acceptable. The theory contains its own grounds of objection to monopoly, which probably slows down change and inhibits diversity. It does not provide support, however, for regulation of content or of structures.

The Positive Version of Centripetal Theory (VI)

This theory, it will be recalled, referred to the effect of mass media in integrating a modern society and encouraging a consensus of norms, values, and objectives which help to hold a pluralistic society together at the level of local community as well as the whole society. This theory values pluralism, but also continuity, order, and efficiency. From this perspective, the potential of new media is likely to be greeted with some caution. While it is not to be supposed that there are malignant forces at work in society, the value placed on an achieved stability and consensus argues for a gradual and conditional extension of communication facilities, with industrial/commercial interests balanced by attention to social-cultural implications, especially the dangers of privatization and social fragmentation. Thus, public service frameworks are likely to be favored as well as measures to secure the kind of diversity which will reflect the existing variations of society in terms of regional cultures, differences of belief, ethnicity. In other words, development should take place with respect for cultural and social continuity and integration, and this makes regulation within a framework of public provision more desirable than unfettered commercial exploitation, if that should be the alternative. Those aspects of new media which contribute to the efficient working and coordination of society are also likely to be favored and that is likely to mean videotex and computer-based networks and services.

The Positive Version of Centrifugal Theory (VII)

This perspective has a good deal in common with technological (medium) determinism (V), since it offers a welcome to any widening of individual freedom of choice as a contribution to change and diversity of society. Pluralism and diversity are valued for their own sake, and there is no suspicion of conspiratorial forces lurking in society to pervert the course of progress through the working of the market. Consequently, there is reason to encourage the development of new media on cultural even more than on industrial grounds, especially on the basis of market mechanisms, which are most sensitive to individual wants and most likely to maximize satisfaction. It is clear that such an approach entails little fear of monopoly and would only favor decentralization if there were clear evidence of regional or local demand and of viable markets. Regulation should be kept to a minimum, especially in relation to content.

The Individual-as-Determinant Theory (VIII)

This last perspective was separated out so as to clarify certain elements of the broader theoretical position, according to which the media are dependent on society—an effect and phenomenon of society, rather than a prime mover. Briefly, it identifies the individual as "in charge," able to choose sensibly and resist unwanted influence or pressure. This is an element in the more pluralistic theories which have already been described and needs no detailed explication of its own.

Like these other theories, it is welcoming to innovation in, and extension of, media provision, according to clear expression of demand. Hence it also favors guidance by the market and opposition to paternalistic regulation, whether structurally or in the details of content. The media offer materials and services to be used, and the abundance of media should in itself present no major problems, given the built-in defenses of the individual.

Those new media which open up possibilities for interaction and consultation are especially to be valued as an extension of the potential for individual development and satisfaction. The adoption of what is, in effect, an individual-differences theory does not preclude a wish to achieve broad cultural objectives, but this is likely to be seen as better achieved through cultural policies which directly aid individuals and groups (by education or subsidy, which can increase communication potential) than by policies of controlling media content or structures of provision. Thus, the given cultural and communication resources are a result of two factors—supply of content and the capacity of receivers to choose or respond wisely. The theoretical option under discussion points to action at the receiving end rather than the sending end, if there is any ground for intervention. The social structural determinist option (IV) would, in contrast, lead one in the direction of acting on the side of supply of content.

Old and New Media; Old and New Theory

It would be fruitless to pursue the question of possible correspondence between versions of media theory and policy options much further, since too much depends on variations of definition and formulation and the location of lines of division. Insofar as there was a hypothesis to be explored, one may cautiously reject the view that there is no connection between theory and policy.

If nothing else, the discussion should have been a reminder of the strongly normative elements which are contained in most versions of media theory, however objectively one tries to formulate them. It should also be a reminder of the equally normative character of most policy choices in relation to new media. The exercise also helps to draw attention to the degree to which much current media theory is rooted in somewhat dated concepts of the nature of mass communication, especially in a view of it as a powerful and monolithic flow delivered to mass audiences and in turn ensuring the delivery of these audiences to the mass communicators. Abundance and diversity of media choice are already features of the societies under discussion, even if there are limits to the diversity of what usually reaches most people. Audiences are already fragmented and profiled in many different ways and the journey to those conditions opened up by the new media has already begun for many.

This is not to conclude that the concerns at the heart of the "old" media theory are now irrelevant, especially when they relate to freedom, equality, and the exercise of power in society. But there is a need to bring the formulation of theory and the expression of underlying normative viewpoints somewhat more into line with the complex reality of modern media systems, partly in anticipation of the future heralded by the innovations under discussion. The theoretical struggle is still engaged with old enemies, and theoreticians have not yet really focussed clearly on either the positive or the negative features of the unfolding media landscape. It is natural and even inevitable for theory to result from reflection on a given reality, and thus, to be always somewhat late in arriving, but it might be useful to remap the theoretical space in anticipation of a range of media futures.

References

Becker, L. The mass media and citizen assessment of issue importance. In D.C. Whitney et al. (eds.), *Mass communication review yearbook*, Vol. 3. Beverly Hills and London: Sage, 1982.

Bennett, T. Theories of the media: Theories of society. In M. Gurevitch et al. (eds.), *Culture, society and the media*. London: Methuen, 1982.

Boyd-Barrett, J.O. Cultural dependency and the mass media. In M. Gurevitch et al. (eds.), *Culture, society and the media*. London: Methuen, 1982.

Burrell, G. and G. Morgan, *Sociological paradigms and organizational research*. London: Heinemann, 1979.

Carey, J. The communications revolution and the professional communicator. In P. Halmos (ed.), *The sociology of mass media communicators*. Sociological Review Monographs, 13, 1969.

———. A cultural approach to communication. *Communication*, 1975, 2(1), 1-22.

———. and A.L. Kreiling. Popular culture and uses and gratifications. In J.G. Blumler and E. Katz (eds.), *The uses of mass communications*. Beverly Hills, Calif.: Sage, 1974.

Elliott, P. Uses and gratifications research: A critique and a sociological alternative. In J.G. Blumler and E. Katz (eds.), *The uses of mass communications*. Beverly Hills and London: Sage, 1974.

Enzensberger, H.M. Constituents of a theory of the media. *New Left Review*, 1970, 64, 13-36.

Gerbner, G. Mass media and human communication theory. In F.E.X. Dance (ed.), *Human communication theory*. New York: Holt, Rinehart, 1967.

———— and L.P. Gross. Living with television: The violence profile. *Journal of Communication*, 1976, 2, 173-199.

Gitlin, T. Media sociology: The dominant paradigm. *Theory and Society*, 1978, 6 (2), 205-253.

Glasgow Media Group. *More bad news*. London: Routledge & Kegan Paul, 1980.

Gouldner, A. *The dialectic of ideology and technology*. London: Macmillan, 1976.

Hirsch, P.M. The "scary world" of the non-viewer and other anomalies—a re-analysis of Gerber et al.'s findings in cultivation analysis. *Communication Research*, 1980, 7(4), 403-456.

Hughes, M. The fruits of cultivation analysis: A re-examination of some effects of TV viewing. *Public Opinion Quarterly*, 1980, 44(3), 287-302.

Innis, H. *Empire and communication*. Oxford: Clarendon, 1950.

————. *The bias of communication*. Toronto: University of Toronto Press, 1951.

Katz, E. and P.F. Lazarsfeld. *Personal influence*. Glencoe, Ill.: Free Press, 1955.

Lang,G.E. and K.Lang. Mass communication and public opinion: Strategies for research. In M.Rosenberg and R.H. Turner (eds.), *Socal psychology*. New York: Basic Books, 1981.

Lazarsfeld, P.F. Remarks on critical and administrative communication research. *Studies in Philosophy and Social Science*, 1941, 9, 2-16.

McCormack, T. Social theory and the mass media. *Canadian Journal of Economics and Political Science*, 1961, 4, 479-489.

McLuhan, M. *Understanding media*. London: Routledge, Kegan Paul, 1964.

McQuail, D. *Mass communication theory*. Beverly Hills and London: Sage, 1983.

———— and J.J. van Cuilenberg. Diversity as a media policy goal: A strategy for evaluative research and a Netherlands case study. *Gazette*, 1983, 31(3), 145-162.

Mills, C.W. *The power elite*. New York: Oxford University Press, 1956.

Murdock, G. and P. Golding. Capitalism, communication and class relations. In J. Curran et al. (eds.), *Mass communication and society*. London: Edward Arnold, 1977.

————. Theories of communication and theories of society. *Communication Research*, 1978, 5, 339-356.

Rosengren, K.E. Mass media and social change: Some current approaches. In E. Katz and T. Szecsko (eds.), *Mass media and social change*. Beverly Hills, Calif.: Sage, 1980.

————. Communication research: One paradigm or four? *Journal of Communication*, 1983, 33, 185-207.

4

The Functions of Human Communication

Frank E.X. Dance

This article provides a consideration of the recent history and current state of theory concerning the functions of human communication. It calls for a strict division between treatments of functions *and treatments of* purposes *in human communication theory building. A survey of various contributions to functional theory over the past fifty years is presented. The author suggests distinct but interrelated functions for information, communication, vocalization, speech, and spoken language. The claim is made that of the three most defensible functions of spoken language (1. linking, 2. mentation, and 3. regulation) the primary function is mentation.*

This article (1) sets forth some of the varying usages of the term *function*; (2) suggests a narrowing of the term's usage for the purposes of the study of human communication; (3) points out the woes attendant upon collapsing the terms and concepts *function* and *purpose*; and (4) briefly reviews the treatment of the *function* concept since 1927, after discussing the problems created by the lack of consistency concerning exactly what it is of which we are positing the function(s). Are we talking about the function(s) of information, or of communication, or of speech. One possible solution is set forth and suggestions are offered as to the functions of spoken language.

Some Uses of the Term *Function*

1. *Function* as a ceremonial event. Example: "Are you going to attend that university function this evening?" In this context *function* is being used to mean, "Are you going to attend that university ceremony this evening?"

2.1 *Function* as activity appropriate to something, someone, or some institution. Example: "This darn typewriter isn't functioning correctly!" Here, *functioning* means: "This darn typewriter isn't working correctly."

2.2 "He fulfills the function of host very well." In this example, *function* means: "He does a good job of fulfilling those duties expected of a host."

3. *Function* as the behavior or activity proper to something. Example: "How is this instrument supposed to function?" Here, the meaning is: "How is this instrument supposed to work?"

4. *Function* as being useful, as utility or practicality. Example: "This isn't a very functional recipe." In this example, *functional* is used as follows: "This recipe doesn't work very well, it's too difficult for the results achieved."

5. *Function* as "a relationship wherein one quality is so related to another quality that it is dependent on and varies with it" (Gould and Kalb 1964, 277-79). Example: "X is a function of Y and varies directly with it." In this context, *function* means: "When X is present, Y also will be present, and however X acts, Y will also be affected."

When discussing communication (or members of its conceptual family) we can find examples of most if not all of the above usages given to *function*. If we are not more selective in our use of the term *function*, and of the concepts to which we apply it, then the term is useless for the purposes of rigorous examination or explication of the phenomenon of communication.

Narrowing Our Usage

Certainly we are free to make a choice among the usages given to the term *function*, to select the usage that promises to be most productive in our search for a better understanding of human communication. In keeping with such an informed and free choice there seems to be a relatively natural and familiar division of happenings which may point the way to a more appropriate and fruitful use.

Some things happen whether or not we wish them to happen. The earth is besieged by natural calamities such as drought, floods, tornadoes, hurricanes, and earthquakes. In fact, even if we anticipate such an undesirable event there is little we can do to stop its occurrence although we can mitigate its impact by a variety of self-protective measures. In terms of the original event there is no evidence of human intentionality.

Other things happen because we plan for them to take place. We plan to attend a play. We plan to complete the course requirements for a degree. We plan to let someone know that we appreciate their efforts on our behalf.

We plan to host a birthday party. Many of the things we plan for would not take place without our planning and our subsequent action to fulfill the plan we made. A nuclear holocaust demands human participation in its inception and enactment. In these cases the presence of human intentionality is apparent. (Searle 1983)

So there are incidents which may fairly be characterized by the presence or absence of intent. There are also happenings which seem always to involve other occurrences. When, in nature, there is the dissipation of energy, there is also always the presence of heat. When flames are present, light accompanies them. These are "if this, then that" types of relationships—if the dissipation of energy, then the creation of heat; if open flames, then light.

In addition to being "if this, then that" types of relationships, they are also independent of human intent. The heat accompanying the dissipation of energy may be used to warm food, or oneself. The light accompanying the open flame may be used to read by, or to illuminate a canvas for painting. There is nothing about the presence of heat that necessitates the presence of either food or oneself. There is nothing about the presence of an open flame that necessitates its use as a reading aid or as an illumination device. To be put to such uses requires human intent. Such uses are purposeful acts, acts revealing a particular intentional purpose on the part of the human agent.

The usage of *function* given in example 5 (function as a relationship wherein one quality is so related to another quality that it is dependent on and varies with it) best serves our search for a deeper understanding of human communication. If we restrict our use of *function* to those instances in which there is an "if this, then that" relationship, then we may use the term *purpose* to refer to our intentional use of communication.

This distinction between *function* and *purpose* may sharpen both our examination and our explication of the phenomenon of human communication. The distinction between the terms enhances our scholarly precision. On the other hand by collapsing, or conflating, the terms and the concepts to which they refer we muddy the conceptual waters, thus increasing the difficulty of discerning the correct outlines and dimensions of the behavior in question.

As an example of the muddiness induced by using *function* and *purpose* interchangeably consider the following excerpt:

> An intention to communicate is not, however, the only intention implicit in communicative acts. At another level, there is also the intention to accomplish a *specific type* of communication. For example, one may wish to *plead with* someone, or *to greet* someone, or *warn* someone, or *inform* someone,

and so on. Thus consideration must also be given to the purpose for which a communicative act is performed. (Chalkey 1982, 80)

By collapsing the two terms we have lost the opportunity both for precision and for a more productive analysis of the behavior in question. This becomes even more important later in the same essay when the author categorizes pre-verbal and early verbal functions but lumps together both functions and purposes, thus clouding the structure of the behavior and depriving herself of the opportunity for more refined analysis.

For the purposes of this article the terms function and purpose shall be distinguished on the bases of: 1) necessity of relationship (if this, then that); and 2) presence or absence of intent.

Functions underlie, and undergird, purposes. Purposes are built upon functions. A hammer functions as a pounder. It may be used for the purpose of pounding nails or of pounding earth. If we use something else (like a shoe) for pounding, common usage says, "I am using my shoe as a hammer." Obviously we need to study both the functions and the purposes of human communication. By conflating the terms and the underlying concepts we rob ourselves of a critical differentiation. Clarity concerning the function(s) of human communication may contribute much to clarity concerning the rhetorical purpose(s) of human communication.

A Brief History of the Treatment of *Function*

In 1927, Grace Andrus De Laguna ([1927], 1963) anticipated much of the subsequent work on functions with her analysis:

- The social function of speech. . . . The same fundamental function of coordinating the activities of the members of the group (p.ix)
- The function of speech in . . . its relation to the higher forms of intellectual life (p.x)
- The function of speech to control the behavior of others (p. 327)

K. Buhler (1934) distinguished three functions of speech communication—representation, expression, and appeal. Lev S. Vygotsky ([1934], 1962, 42) identified emotional release and socialization as two functions of speech. There is in Buhler some evidence of the confusion of function and purpose since Buhler's "appeal" certainly suggests the kind of intentionality not found in functions.

Joshua Whatmough (1956) illustrated function/purpose confusion when he listed as functions "formative," "dynamic," "emotive," and "aesthetic." Eisenson, Auer, and Irwin (1963), who treated the functions under the dual headings of "noncommunicative purposes" and "communicative purposes," did not even attempt to make the function/purpose distinction.

Joyce Hertzler (1965, 38-57) mixed functions and purposes when she listed twelve general functions of communication such as the identification function, the categorization function, language as the means of thinking, language as record and as a human memory, and so on.

Harold Lasswell, with roots in political science, and Wilbur Schramm, with roots in English and in journalism, brought their disciplines to bear on that extension of human communication generally referred to as mass communication. Both scholars arrived at sets of functions that seem quite similiar in essence though differing in nomenclature. Lasswell (1948) stated:

> The communication process in society performs three functions: (a) *surveillance* of the environment, disclosing threats and opportunities affecting the value position of the community and of the component parts within it; (b) *correlation* of the components of society in making a response to the environment; (c) *transmission* of the social inheritance. In general, biological equivalents can be found in human and animal associations, and within the economy of a single organism. (p. 51)

Wilbur Schramm (1964) discussed information roles (functions) within society.

> The three information roles in society [are] the Watchman role (to scan the horizon and report back); the Policy role (to decide policy, to lead, to legislate); and the Teacher role (to socialize new members, by which we mean to bring them into society with the skills and beliefs valued by the society). (pp.38-39)

> [They are] the watchman, decision maker, and teacher functions. (p.126)

Lasswell's suggested functions[1] of surveillance, environmental response, and transmission of the social inheritance seem to match comfortably Schramm's suggested roles of Watchman, Policy, and Teacher.

Note, however, that Lasswell is using the term *communication* while Schramm uses the term *information*. Are we to presume that they are discussing the same concept? Are we to presume that the functions and roles of which they speak are the functions and roles of the same phenomenon? Is there no distinction to be made between communication and information? We know that such distinctions exist and that the distinctions are important, and we know that scholars of the stature of Lasswell and Schramm are also aware of the existence of such distinctions, distinctions which are not trivial.

Perhaps this muddling of terms and concepts simply mirrors the confusion attendant upon the beginnings of a new area of study. The shape of the study of communication seems similar to the shape of a bow-tie, since the field is composed of some scholars who came from diverse fields (as exemplified by Lasswell and Schramm) and who then focused their diverse backgrounds onto the study of communication, whereas other scholars—the one's from speech—coming from a single area, widened their interest to the broader concern of communication. The effects of this dual influx is still observable in a seeming distinction between the field of communication and the field of speech communication.

Scholars like Lasswell and Schramm coming to the study of communication from other disciplines brought with their interest a usage of terms most likely originating in their primary disciplines. If this is so, perhaps we should expect to find a more rigorous usage on the part of those scholars with their roots in the field of speech since the academic discipline of speech, a direct descendant of the ancient academic discipline of rhetoric, had a more unified tradition and set of terms. But alas, such has not been the case.

Given that those interested in human communication must be sensitive to the important relationship of verbal precision to conceptual precision, it is remarkable how loosely we have used and continue to use some of the terms and concepts central to our subject matter. Given that we are so flabby in our distinctions among the terms and concepts relating to the study of human communication (such as communication, information, human communication, language, speech, spoken language, speech communication) perhaps we should not be surprised at the history of confusion surrounding *function* and *purpose*.

Susanne K. Langer (1967, 1972, 1982) suggests that in the evolutionary development of human spoken language the first function of spoken language to evolve was the representation and potentiation of conceptual thought and that only much later did the function of exchange of ideas (what is commonly called the communicative function) arise. As will be argued later in this article, Langer's thesis is quite compelling.

Ernst Cassirer (1944), defining man as *animal symbolicum*, discusses the roles played by spoken language in human beings. Without labeling these roles *functions*, he shows how spoken language binds society together and how spoken language interacts with and structures higher mental processes.

Frank E.X. Dance (1967a, 1967b, 1972, 1976, 1978, 1982) argues for the presence of three functions of spoken language: 1) the linking of the individual with the milieu—linking; 2) the development of higher mental processes—mentation; and 3) the regulation of human behavior—regulation.

The data in support of the three functions is drawn from many fields and is presented most fully in Dance and Larson (1976). Although Dance states that current evidence supports the three functions he sets forth, he also admits that there may be other functions as well.

While the linking and regulatory functions have precursors in animal communication, the mentation function seems to be unique to humans. With the presence of spoken language the human being is able to link with other human beings in a particularly human fashion. Obviously humans link with other animals, as do other animals with each other, using sign communication. Spoken language also enables the human being to regulate his or her own behavior and the behavior of other human beings through the medium of spoken symbols. Communication in general, that is communication by signs rather than by symbols, may be used by humans to regulate their own behavior as well as the behavior of other animals and by animals in general to regulate the behavior of other animals.

Dance uses *mentation*, a relatively old term according to the *Compact Edition of the Oxford English Dictionary*, to distinguish between ratiocination, or closely-knit reasoning, and cognition or thinking in general. Cognition has a broader domain than does thinking, and thinking a broader domain than does mentation.

Animal communication seems to lack a functional precursor to mentation. By means of spoken language the human being is able to reason closely, to deal with conceptual rather than only perceptual matters, and to elevate animal interaction, animal linking, animal regulation to uniquely human levels through the power of mentation and the use of higher mental processes. Mentation, resulting from spoken language, is uniquely human, according to Dance.

The term *function*, although appearing in recent and current publications, is still seldom sufficiently distinguished from *purpose*. Cummings, Long, and Lewis (1983, 24) in their organizational communication text state that "regardless of their goals or purposes all organizations make use of the four basic functions of communication—information management, problem and solution identification, conflict management, and behavior regulation—to assist in problem solving and problem management." While at least acknowledging the existence of purpose aside from function, this statement fails to make the distinction clear; all management is a form of behavior regulation—a function—and the specification of the setting for that regulation (conflict, information management) is purposeful. By separating function and purpose in organization settings the researcher, as well as the practitioner, may better focus on the underlying problems and their possible treatment.

Williams (1984, 88ff), focusing on what he refers to as communications, chooses to use the term *function* in a more general manner than advised here. He refers to the social functions of communication—surveillance, correlation, socialization, and entertainment—similar to the functions set forth by Lasswell and by Schramm and as discussed above. However, he centers on the use of *functional* and *dysfunctional* in relation to the manner in which we usually use something or the purposes for which we usually use something and to the outcomes of that utilization. Although a legitimate use of the term *function*, it seems to encourage conceptual slippage and at the very least adds little rigor to the analysis of the matter at hand. *Outcomes* may have been just as useful a term for Williams's general purposes and would have left *function* free for use in tighter analyses.

Masterson, Beebe, and Watson (1983) center on speech communication rather than communication in general and adhere to the "if this, then that" usage and list the functions as presented by Dance (1967a) and by Dance and Larson (1976).

Ruben (1984, 20) discusses the basic functions of communication by rooting them in the basic biological functions they serve: survival of the species and adaptation. Although this is a rigorous use of the term *function* and serves us well by bringing us back to baseline biological considerations, it raises questions concerning evolutionary continuity/discontinuity and the erroneous suggestion, common in years past, that human communication is an overlaid behavior (Routhier 1979).[2]

The continuity/discontinuity question pervades much of the discussion of recent efforts to teach chimpanzees and gorillas a form of human language. Are the differences between human and infrahuman communication of degree or of kind (Adler 1967), and how do whatever differences exist relate to functions? The argument about continuity/discontinuity is most important to those who are committed to a strict Darwinian evolutionary interpretation of the development of the human species. There exist evolutionary hypotheses which do not find it implausible to posit a number of discontinuities in evolution. The development of spoken language may well be one of those discontinuities.

In later chapters of his new basic text Ruben treats what he considers to be "human" about human communication and by implication suggests behaviors which he may consider to be functions. Unfortunately, since Ruben neglects to directly call these behaviors functions, we are left without his guidance as to whether or not he sees them as functions in the narrow sense we recommend in this article.

Functions of What?

As noted above, there is an astonishing lack of consistency as to the object of our functional analyses. Are we interested in the functions of information, communication, human communication, spoken language, symbolization, speech, inner speech, or what? This thicket can appear impenetrable. However, if scholars could agree on what they mean by the various terms they use, what concepts those terms point to, perhaps they would have more success in assessing the functions of those various concepts.

In compiling this brief survey of the treatment of functions in the literature the problem, functions of what, came up again and again. The term *function* was used in such different ways and seemed to refer to such different targets that it was easiest to deal with the multiple usages by looking not at the term but at the concept to which it was applied. When this was done, using the taxonomy presented in Dance (1982, 125-27), transformations were made that seemed to allow for appropriate rigor and consistency in interpretation. Taking into consideration these transformations and then submitting the entire subject to re-analysis yields the following observations concerning the functions of various of the terms and concepts making up that family of concepts we call communication.

It seems likely that in the process of evolution and development of the various functions we may well expect to find differentiations and transformations taking place; for example, the differentiation from general interaction to more specific linking and regulation. One may also speculate that as the function(s) of each of the more general behaviors are passed on to the more specific behaviors (as from information to communication to vocalization to speech to spoken language) there may well be a movement towards parsimony resulting in the absorption of one behavior's function into a function of a more complex behavior (for example, the possible absorption of vocalization's functions of "expression of affect" into spoken language's function of "linking").

It may also be suggested that in the development of the human individual the functions probably take on differing relationships to one another at different stages of individual development (for example, in the earliest years linking may well be the dominant function, then with further development regulation may come into greater prominence, and finally for the long run we would hope that mentation would prevail as the dominant function).

The Functions of Humans Communication

In this discussion of functions it is implicit that the functions of human communication are distinct, either in degree or in kind or in both, from the function(s) of other forms of communication or its consituent elements, such as information. Human communication is unique. Of course this position is neither new nor idiosyncratic. Among many others who have

TABLE 4.1
The Functions of Various
Concepts in Communications

THE WHAT	THE FUNCTION(S)
INFORMATION	The reduction of uncertainty
COMMUNICATION	Interaction: The primal interaction being that between a stimulus and an organism. This interaction then generalizing within, among, and between organisms. This is the level of communication in general. With further differentiation and refinement incipient linking and regulation may be seen in this general interaction.
VOCALIZATION	Expression of affect: The utterance of sound carries with it an unintentional revelation of the affective state of the sounding organism. Of course this revelation, in higher organisms, may be consciously controlled for the purposes of the organism.
SPEECH	Incipient mentation: The argument for this function of speech derives from the role played by speech in the human infant's development of a sense of contrast upon which is built a complex cognition, ratiocination, or mentation. (Dance 1978, 1982)
SPOKEN LANGUAGE	Linking, mentation, and regulation. These suggested functions of spoken language have been mentioned earlier in this article and will be once again at the conclusion. Spoken language may be examined as it exists within the human being as well as its use between and among human beings. When discussed in its intrapersonal usage it is often referred to as inner speech (Vygotsky 1962). More recently, Kuczaj and Bean (1982) have suggested a further differentiation between inner speech (silent, unconscious spoken language for oneself), internal speech (silent, conscious spoken language for oneself), and private speech (vocalized spoken language for oneself). Kuczaj and Bean consider such aspects of spoken language for oneself as essentially noncommunicative. Unfortunately the authors conflate function and purpose in their listing of the functions of what they refer to as noncommunicative spoken language.

treated the subject, Mortimer Adler (1967) spends an entire, closely reasoned book, spelling out what he considers to be the differences in kind, rather than in degree, between animal and human communication. Susanne K. Langer (1967, 1972, 1982) also opts for the uniqueness of human language, which she states accounts for the great shift between other forms of animal life and human life.

Not only is human communication unique, but that which is unique about it derives from the primary fact that humans have speech (Dance 1975, 1982). Speech, because of its unique capacity as the earliest sense developmentally to blend subjectivity and objectivity in the human infant (Dance 1979), leads to the development of a sense of contrast, to the birth of the symbol (Dance 1983), to the development of natural languages, and finally joins with language to constitute uniquely human spoken language with the functions and other developments flowing therefrom (Dance 1982).

Since there is evidence that the symbol derives from speech, symbolic language always derives from spoken language—although after spoken language has been developed, symbols may well take other forms such as print, graphics, or gestures. It is this ability to render symbols in forms other than vocal that some refer to as nonverbal communication. Actually, such stimuli are nonvocal rather than nonverbal since they are always interpreted in terms of their symbolic referents. Truly nonverbal communication is best characterized as animal communication. Since as humans we have in our communicative repertoire the evolutionary heritage of animal communication as well as the uniquely human communication represented by spoken language, we may find in the communicative repertoire of humans examples of both verbal and nonverbal communication. Uniquely human communication is, however, always verbal—meaning always symbolic—and in that reality is always rooted in spoken language.

There is compelling evidence to recommend three functions of human communication—communication by humans using spoken language. Two of these—linking and regulation—may also be found in animal communication and are thus different only in degree. The third function—mentation—seems to be unique to spoken language, is a function radically different in kind, and, after its developmental emergence, elevates the animal functions of linking and regulation to a peculiarly human level.

Given the uniqueness of the function of mentation and the parsimony of nature, it is quite possible that the raison d'être of spoken language is not interaction, as many seem to think, but mentation. Although interaction seems to be the manifest and primary function of and reason for communication in general, a new and complicated behavior such as spoken language would seem to be evolutionarily extravagant and superfluous if it only served the same functions satisfactorily served by an extant behavior. In fact, it is not difficult to recall instances in which spoken language sometimes gets in the way of, rather than furthers, linking and regulating.

Spoken language, used in human communication, gives us something to think with and about, mentation, and then the functions of linking and regulation enable us to share these thoughts with others.

Human beings project these functions upon their organizations. Thus, we find the three functions of spoken language present in the organization structures of the family, school, business setting, government, and so on. The functions are presented within roles and can only be assessed as such. Certain roles call heavily on certain functions—a line manager needs to be able to regulate, a research and development manager to mentate, a personnel manager to link—although all functions are interrelated and work together (Johnson 1979).

We must clearly specify the terms and concepts we use to discern these functions, perform the appropriate analyses, including possibly experimentation, and then proceed to an examination of the interrelationship of functions and purposes in human communication. A more careful analysis of these functions as well as of the functions of related concepts (such as information and spoken language) will increase our understanding of the phenomenon of human communication.

Notes

1. Lasswell identifies his functions on the mass communication level and then suggests that these functions, in their biological equivalents, can be found in the individual organism. This tendency to work from the artifact inward is similiar to the current suggestions that the brain is like a computer rather than that the human brain projects itself into its creations.
2. Routhier effectively counters the old belief that spoken language or speech (depending on the source) is overlaid on structures and organs that serve more fundamental or basic human needs (such as respiration or digestion).

References

Adler, M. *The difference in man and the difference it makes.* New York: Holt, Rinehart & Winston, 1967.

Buhler, K. *Sprachtheorie.* Jena: Fischer, 1934. Quoted in D. McCarthy, Language development in children. In L. Carmichael (ed.), *Manual of child psychology,* 2d ed. New York: Wiley, 1966.

Cassirer, E. *An essay on man.* New Haven: Yale University Press, 1944.

Chalkley, M. The emergence of language as a social skill. In S.A. Kuczaj II (ed.), *Language development.* Vol. 2 Hillsdale, N.J.: Lawrence Erlbaum, 1982.

Cummings, H.W., L.W. Long, and M.L. Lewis. *Managing communication in organizations.* Dubuque, Iowa: Gorsuch Scarisbruck, 1983.

Dance, F.E.X. The function of speech communication as an integrative concept in the field of communication. In *Proceedings of the fifteenth international convention of communication*. Genoa, Italy: International Institute of Communication, 1967a.

_____. Toward a theory of human communication. In F.E.X. Dance (ed.), *Human communication theory: Original essays*. New York: Holt, Rinehart & Winston, 1967b, pp. 288-309.

_____. Speech communication: The sign of mankind. In *The great ideas today: 1975*. Chicago: Encyclopedia Britannica, 1975, pp. 40-57, 98-100.

_____. Human communication theory: A highly selective review and two commentaries. In B.D. Ruben (ed.), *Communication yearbook 2*. New Brunswick, N.J.: Transaction-International Communication Association, 1978, pp. 7-22.

_____. An acoustic trigger to conceptualization. In *Health Communications and Informatics*. 1979,5, 203-213.

_____. A speech theory of human communication. In F.E.X. Dance (ed.), *Human communication theory: Comparative essays*. New York: Harper & Row, 1982, pp. 120-146.

_____. The role of speech in symbol formation. In J. Sisco (ed.), *The Jensen lectures: Contemporary communication studies*. Tampa, Fla.: University of South Florida Press, 1983, pp. 44-49.

_____ and C. Larson. *Speech Communication: concepts and behavior*. New York: Holt, Rinehart & Winston, 1972.

_____ and C. Larson. *The functions of human communication: A theoretical approach*. New York: Holt, Rinehart & Winston, 1976.

DeLaguna, G.A. *Speech: Its function and development*. Bloomington: Indiana University Press, 1963.

Eisenson, J., J.J. Auer, and J.V. Irwin. *The psychology of communication*. New York: Appleton, 1963.

Gould, J. and W.L. Kalb, eds. Function. In *Dictionary of the social sciences*. New York: Free Press, 1964, pp. 277-279.

Hertzler, J.O. *A sociology of language*. New York: Random House, 1965.

Johnson, S.H. The development of an instrument to measure human communication functional dominance. Ph.D. dissertation, University of Denver, 1979.

Kuczaj II, S.A. and A. Bean. The development of noncommunicative speech systems. In S.A. Kuzcak II (ed.), *Language development*, Vol. 2. Hillsdale, N.J.: Lawrence Erlbaum, 1982, pp. 279-300.

Langer, S.K. *Mind: An essay on human feeling*. Baltimore: Johns Hopkins University Press, Vol. 1, 1967; Vol. 2, 1972; Vol. 3, 1982.

Lasswell, H.D. The structure and function of communication in society. In L. Bryson (ed.), *The communication of ideas*. New York: Harper & Row, 1948, pp. 37-51.

Masterson, J.T., S.A. Beebe, and N.H. Watson. *Speech communication: Theory and practice*. New York: Holt, Rinehart & Winston, 1983.

Routhier, M.E. A critical analysis and examination of the issue of speech as an overlaid function. Ph.D. dissertation, University of Denver, 1979.

Ruben, B.D. *Communication and human behavior*. New York: Macmillan, 1984.

Schramm, W. *Mass media and national development: The role of information in developing countries*. Stanford, Calif.: Stanford University Press, 1964.

Searle, J.R. *Intentionality: An essay in the philosophy of mind*. Cambridge, England: Cambridge University Press, 1983.

Vygotsky, L.S. *Thought and language.* New York: MIT Press and Wiley, 1962.
Whatmough, J. *Language: A modern synthesis.* New York: New American Library, 1956.
Williams, F. *The new communications.* Belmont, Calif.: Wadsworth, 1984.

5

Whose Life Is It, Anyway?
Information as Property

Richard F. Hixson

Celebrities, such as Elizabeth Taylor and Frank Sinatra, seek to protect their privacy through the use of the new legal concept known as right of publicity—*the exclusive right of individuals to make money out of their prominence. The article traces the development of the right of publicity through various cases to the present, and the author concludes that the Copyright Revision Act of 1976 provides the most realistic protection for both plaintiff and the public.* Zacchini *is discussed at length as the most significant Supreme Court decision to date.* Sony *may also have application, since the dissenting justices distinguished between* productive *and* ordinary *fair use. Perhaps the same distinction can be applied to right of publicity when an individual's persona is in conflict with society's right, or need, to benefit.*

Elizabeth Taylor and Frank Sinatra are among the recent additions to a long list of famous and not-so-famous persons who have sought to protect their privacy through the use of a relatively new legal concept known as the right of publicity—the exclusive right of individuals to make money out of their prominence. Strict privacy law does not provide the protection they seek, for the courts have consistently ruled that the First Amendment protects, instead, what is said about public figures in the interest of free speech, unless, of course, it is a knowing or reckless falsehood. As reported in the *New York Times*, "What Elizabeth Taylor wants to know—and what the law doesn't clearly tell her—is whose life is it, anyway? More specifically, can ABC go ahead and produce a docu-drama based on Taylor's life even over the movie star's vociferous objections?" (Lewin 1982, 1).

Sinatra's suit against an "unauthorized" biographer, Kitty Kelley, is tempered somewhat by the singer's friendship with presidents and other politicians who are normally beyond the protection of privacy law. His suit seeks punitive damages for Kelley's "misappropriation of name and likeness for commercial purposes." His suit, which is similar to Taylor's, is based on the right of publicity premise that nobody else can write the Sinatra Story without his permission.

In both cases, the basic premise is actually twofold—that Taylor and Sinatra own their personalities and that the docu-drama and biography would present them in a false light. They contend that their life stories are commercial properties and, further, that they alone have the right of exploitation. Anyone else who uses the material without permission is misappropriating personal property. "Someday I will write my autobiography, and perhaps film it, but that will be my choice," Taylor is quoted as saying. "By doing this, ABC is taking away from my income." Common law right of publicity, for many years a part of privacy law, basically asserts that an individual can control the commercial use of his of her name or image. It would be illegal, for example, to use Elizabeth Taylor's name or picture without her permission in an advertisement for a brand of cosmetics. With regard to false light, the second part of the premise, courts have held that a wrong impression of a person, even though it may not be unfavorable, is an invasion of privacy.

Right of Publicity: History

The right of publicity as a common law tort separate from the right of privacy has its roots, ironically enough, in the famous Warren and Brandeis article of 1890, which first projected the concept of right of privacy as independent from the rights of property, contract, and trust. Unlike these rights, long recognized in common law, the right of privacy, as envisioned by Warren and Brandeis, sought to protect the plaintiff's right "to be let alone." The essential injury was to one's feelings of consequent mental anguish. They said, in part: "Our law recognizes no principle upon which compensation can be granted for mere injury to the feelings" (Warren and Brandeis 1890, 197). They argued that, while some aspects of privacy may involve ownership or possession, privacy to them meant "inviolate personality," and what they sought was a way to protect the personality itself. They did not reject the property aspects of privacy, but rather pleaded for a distinct right "to be let alone." Theirs was a remedy for the mental anguish caused an individual where there had been no injury to property or contract rights. In short, Warren and Brandeis sought to extend the scope of jurisprudence to protect not the reputation—already covered by the law of defamation—but the feelings of individuals subjected to some form of intrusion.

But when the new concept was first applied, it failed. Right of privacy was also the issue in *Roberson v. Rochester Folding Box Co.* (1902), which involved the use of a girl's portrait on 25,000 posters advertising flour. The picture had been used without her consent, but the four to three court denied her an injunction on the ground that no right yet known to the common law had been infringed. The majority said that none of the cases cited by Warren and Brandeis applied to a plaintiff's relief solely on the basis of hurt feelings.

What makes the case of young Abigail Roberson significant to the right of publicity tort is Judge Gray's dissent, the earliest statement linking personality to property within the right of privacy concept as advanced by Warren and Brandeis.

> Property is not, necessarily, the thing itself which is owned; it is the right of the owner in relation to it. The right to be protected in one's possession of a thing or in one's privileges, belonging to him as an individual . . .*is property*, and as such entitled to the protection of the law. *I think that this plaintiff has the same property in the right to be protected against the use of her face for defendant's commercial purposes as she would have if they were publishing her literary compositions (Canessa v. J.I. Kislak, Inc. 1967, 68)* (Emphasis Judge Gray's).

Thus, within the concept of right of privacy is at least an implied right of property. Subsequent cases which recognized Warren's and Brandeis's concept also recognized the basic common law right of property in one's name and likeness. Most decisions held that a plaintiff's name and likeness could not be appropriated in a purely commercial venture for the profit and gain of the appropriator (Gordon 1960, 558). But, since most of the cases alleged humiliation and mental anguish as the principle injury, the courts usually applied the independent right of privacy theory. Under this theory, property ownership is not a factor.

Meanwhile, criticism of the *Roberson* ruling was directly credited with passage of the New York "Civil Rights" statute in 1903. Utah (1909) and Virginia (1919) enacted similar early privacy laws. The New York statute, which does not mention *privacy* by name, says, in part: "A person, firm, or corporation that uses for advertising purposes, or for the purpose of trade, the name, portrait or picture of any living person without having first obtained the written consent of such person or, if a minor, of his or her parent or guardian, is guilty of a misdemeanor" (Francois 1982, 197). Following passage of the statute, actions occurred in New York State designed to carry the interpretation of the right of privacy into nonadvertising parts of a newspaper. In *Moser v. Press Publishing Co.* (1908), the court said that the statute did not prevent publication of a person's photograph without his consent in connection with nonlibelous news about him. Two years later, another New York court upheld this position. A former prize fighter, Jim Jeffries, sought to enjoin a newspaper from using his picture in connection with the publication of his autobiography. He said the picture increased the paper's circulation and he asked $25,000 in damages. The court said no, ruling that advertising and trade purposes referred to commercial use, not to the dissemination of information. In *Sweenek v. Pathe News* (1936), a film distributed by Pathe News showing the plaintiff as one of a group of women, all weighing more than 200 pounds, going through reducing exercises in a gymnasium, the film accompanied by amusing comments, was held no grounds for action. The court said: "The publication of matters of public interest in newspapers or newsreels is not a trade purpose within the meaning and purview of this statute" (*Sweenek v. Pathe News* 1936, 747).

The view of the majority in *Roberson* was to be completely wiped out by what at the time had become the leading case, *Pavesich v. New England Life Insurance Co.* (1905), which also marked the first recognition of the right of privacy by a state court where no statutory law existed. The insurance company had published an advertisement in the *Atlanta Constitution* containing photographs of two men, one of whom was Paolo Pavesich, an Atlanta artist, depicted as one who had bought sufficient life insurance. The other photo, of an ill-dressed and sickly-looking man, illustrated one who had no insurance. Underneath Pavesich's photo appeared a testimonial ascribed to the artist in praise of the company. Pavesich sued for $25,000, and the Georgia Supreme Court ruled that "the form and features of the plaintiff are his own" (Pember 1981, 226). The case is doubly significant because of Pavesich's public position as a noted artist. "It is not necessary to hold . . . that the mere fact that a man has become what is called a public character, either by aspiring to a public office, or by exercising a profession which places him before the public . . . gives to every one the right to print and circulate his picture" (Pavesich 1905, 217-18). Even noncelebrities enjoy a property right in their identity. A New Jersey court ruled that a real estate firm could not use a family's name and picture in its advertisements without permission. The court said: "However little or much plaintiff's likeness and name may be worth, defendant, who has appropriated them for his commercial benefit, should be made to pay for what he has taken, whatever it may be worth" (Canessa 1967, 75).

While the pioneer New York privacy statute recognized the right to damages for personal harm suffered by misappropriation, it denied recovery when the public's interest in the otherwise private information was deemed more important than privacy or property. Thus, public interest appears crucial in right of publicity cases. Subsequent common law rulings are rather explicit in their attempt to inhibit serious intrusion into the integrity of personal identity. But they are also equally supportive of the public's First Amendment right to information. "A writer is left with the options of either paying closer attention to the facts or conceiving a more artfully designed fictional world," writes Erik D. Lazar (1980, 513).

Another early privacy case which had implications for the right of publicity is *Binns v. Vitagraph Co. of America* (1911). John R. Binns, a wireless telegraph operator aboard the passenger steamship Republic, obtained wide notoriety because of his heroism when the ship collided with another steamer in January 1909. The rescue of all 1,700 lives was attributed to messages sent by Binns, who subsequently sued Vitagraph for using his name and picture in connection with a "moving picture" purporting to show the wreck and exhibiting Binns in a "ridiculous posture." He had been "greatly disturbed in mind" and he said his feelings were injured. A New York appellate court noted that, although the plaintiff could not prevent a reasonable factual account from being aired, Binns had the right to enjoin a fictionalized treatment clearly designed for selfish, commercial purposes. In other words, the determination in *Binns* gave more impetus to the belief that a plaintiff's private interests should prevail where the commercial aspects clearly overshadow a depiction's public worth as news or commentary.

On the other hand, a simple fictional story with only the name of an institution and a few symbols associated with it may not constitute undue exploitation. When the University of Notre Dame and its president, Father Theodore M. Hesburgh, sought to enjoin the release and distribution of the film, *John Goldfarb, Please Come Home*, the court viewed the film and the novel on which it was based as not unreasonable intrusion. Besides, the film, produced by Twentieth Century-Fox, was clearly a spoof, much like parodies that are allowed under common law copyright.

Right of publicity finally emerged as a legal concept of its own with *Haelan Laboratories v. Topps Chewing Gum* (1953), when the distinguished American jurist Jerome Frank argued that plaintiffs may be as offended by commercial exploitation as by humiliation or embarrassment in appropriation suits. Judge Frank, whom President Roosevelt had appointed to the Second Circuit Court of Appeals, noted, in identifying the emerging right, that persons have a right to control the commercial exploitation of their name or likeness.

For it is common knowledge that many prominent persons (especially actors and ball-players), far from having their feelings bruised through public ex-

posure of their likenesses, would feel sorely deprived if they no longer received money for authorizing their countenances displayed in newspapers,
magazines, buses, trains and subways. This right of publicity would usually
yield them no money unless it could be made the subject of an exclusive
grant which barred any other advertisers from using their pictures (Haelen
Laboratories 1953, 868).

Judge Frank's opinion for the majority in *Haelan Laboratories* stands as
the real beginning of the right of publicity. Its influence has been
substantial.

In the late 1970s and early 1980s, the right of publicity, or property,
continued to move away from the traditional right of privacy. What seems
to distinguish the two legal concepts is this: privacy right focuses on how
intentional distortion (false light and fictionalization) affects the individual, particularly his or her feelings; publicity right focuses on what the
individual may own to the exclusion of others, an economic right of property, in other words. As Lazar (1980, 535-36) notes: "In publicity, performance, and copyright, it is the personality that creates the asset, and
consequently the personality that is being exploited." *Publicity figure* has
come to mean, or describe, a person who, by virtue of public exposure,
generates an interest in his or her life and likeness, and includes celebrities,
public figures, entertainers, and athletes.

Major Models and Issues

Although the heirs of actor Bela Lugosi were unsuccessful in their effort
to stop Universal Studios from exploiting the Count Dracula characterization created by the actor when he was employed by the studio, the case
generated four different views, or models, as to the ownership of the commercial rights to, in this instance, Lugosi's famed portrayal: 1) property
right; 2) privacy right; 3) employer-owned product of employment; and 4)
copyright. Kevin Marks's (1982, 786-815) assessment of the four models is
an important contribution to the evolving publicity right.

The publicity figure's interest is more than the right to profit from his
image; it additionally includes the right to manage his image as he sees fit.
As Marks notes: "Unless the public figure has marketing control by license
or assignment over the commercial use of his name or likeness, the asset's
value is reduced because it is subject to waste by self-interested entrepreneurs" (Marks 1982, 789). The value of the property model is that it
is assignable and descendible; that is, the figure is able to assign his name or
likeness to others for maximum exploitation, and his heirs may acquire
ownership of the name or likeness. The weakness of the model, on the
other hand, is its failure to recognize the countervailing social interests in

free enterprise and free expression. In *Memphis Development Foundation v. Factors, Etc., Inc.* (1980), the court reasoned, in a suit brought by Elvis Presley's heirs against a company selling replicas of the singer's statue: "The memory, name, and pictures of famous individuals should be regarded as a common asset to be shared, an economic opportunity available in the free market system" (Marks 1982, 789). In other words, any recognition of a right of publicity should take into account the First Amendment's guarantees of free speech and the exchange of ideas.

The privacy model, according to Marks, is inapplicable to cases involving celebrities and public figures who voluntarily expose themselves and their exploits to public view. "To such persons, public exposure is desirable and thus does not invade any dignitary interest protected by the right of privacy" (Marks 1982, 791). Confusion over the various court decisions may rest with whether privacy or publicity law is applied. If the former, there may be the conflict with the First Amendment. If publicity, or property, law is applied, no such conflict may occur.

Regarding the work-product model, one of the appellate judges indicated that Lugosi had no proprietary rights because it was the image of Dracula, not Lugosi, that was marketed and that an employee's creation in the course of his employment belongs to the employer "pursuant to California labor law." Therefore, as Marks explains, the creation of a marketable character leads to an inheritable property interest, but performance as a marketable character does not, much like the distinction between authorship and performance rights under copyright law, which does not recognize rights in performance. Arthur Miller, the playwright, owns all the rights to Willie Loman and not the score of actors who have performed the role in *Death of a Salesman*, not even Lee J. Cobb, the first to "create" the stage character. However, the court has ruled that, where an entertainer develops an original character, as Stan Laurel and Oliver Hardy did, for example, the right of publicity survives the death of the creator (Prince v. Hal Roach Studios, Inc., 1975). Lugosi's fictional character had been created, and was thus owned, by someone else.

Copyright law, as the basis for the model bearing its name, seeks to secure a fair reward for creative efforts in order to achieve two related goals: 1) the encouragement of artistic endeavor; 2) the assurance that the return to the creator is a fair one. California Chief Justice Bird, in dissenting in *Lugosi*, acknowledged both the artistic incentive and entitlement to the fruits of one's labors as compelling policy justification for recognizing a right of publicity. "Most courts, including a majority of the justices of the California Supreme Court deciding *Lugosi*, believe that a right of publicity exists for individuals seeking to market their own image," concludes Marks (1982, 813). In *Zacchini v. Scripps-Howard Broadcasting Co.* (1976), the

U.S. Supreme Court applied the copyright model as a way of protecting rights in performance and thereby encouraging participation in the performing arts.

The Zacchini Case

Despite some of its bewildering aspects for the news media, *Zacchini* is probably the most important right of publicity case to date. First, because of the judicial reversals it engendered and, second, because it involved a variety of judicial statements on privacy, free press, publicity, and copyright—a case, in short, that has all the appropriate ingredients for the scholar.

Hugo Zacchini, a human cannonball, was performing at the Geauga County Fair in Burton, Ohio, when, over his objections, a reporter from WEWS-TV of nearby Cleveland filmed his act. General admission was charged to the fairgrounds, but there was no additional fee to watch Zacchini's performance. News reporters were admitted to the grounds without charge. The reporter visited the fair on two occasions. On the first, Zacchini requested that his act not be filmed, but on his return visit the reporter filmed the performance. A fifteen-second film clip of Zacchini's stunt was broadcast by WEWS during its 11 p.m. news program on September 1, 1972. The news clip was accompanied by the following favorable commentary: "This . . . now . . . is the story of a *true spectator* sport . . . the sport of human cannonballing. . . . In fact, the great *Zacchini* is about the only human cannonball around, these days . . . just happens that, *where* he is, is the Great Geauga County Fair, in Burton. . . . And believe me, although it's not a *long* act, it's a thriller. . . . And you really need to see it *in person*. . . to appreciate it" (Zacchini 1977b, 564). (Emphasis in the original.) Zacchini sued the Scripps-Howard Broadcasting Co., owner of WEWS, for invasion of privacy.

In his complaint, Zacchini said that the broadcast was an unlawful appropriation of his professional privacy and sought $25,000 in damages. The *Second Restatement of Torts* (1967) states: "One who appropriates to his own use or benefit the name or likeness of another is subject to liability to the other for invasion of his privacy" (American Law Institute 1967, 109). This is the aspect of privacy that has come to be called the right of publicity. The trial court granted Scripps-Howard's motion for summary judgment, but the Ohio appellate division reversed the decision, holding that Zacchini had a cause of action for conversion of property and invasion of common law copyright. Two judges relied on the theories of conversion and copyright and the third on Zacchini's right of publicity. All agreed that the First Amendment did not protect the television broadcast.

The trial court ruling was reinstated by the Ohio Supreme Court, which held that, even though Zacchini enjoyed a performer's right to the publicity value of his performance under Ohio law, his act was nevertheless a matter of legitimate public interest which the station was privileged to broadcast. The court rejected conversion and copyright. Citing *Time, Inc., v. Hill* (1967), the court said "that freedom of the press inevitably imposes certain limits upon an individual's right of privacy" (Zacchini 1976, 460). The court also said, citing *New York Times Co. v. Sullivan* (1964), that the privilege could be lost only if the intent of the press "was not to report the performance, but rather to appropriate the performance for some other private use, or if the actual intent was to injure the performer" (Zacchini 1976, 461).

Justice Stern, writing for the majority in the six to one decision, said: "Certainly it has never been held that one's countenance or image is 'converted' [wrongful exercise of dominion over property in exclusion of the right of the owner] by being photographed" (Francois 1982, 237). The court also concluded that plaintiff's performance was "safely outside" the bounds of copyright designed to foster and protect literary and artistic expression. The case was decided on the basis of the common law's sanctions against appropriation of one's name or likeness for commercial purposes without that person's consent. In reviewing *Time, Inc.* and *New York Times*, Justice Stern noted that "The press has a privilege to report matters of legitimate public interest even though such reports might intrude on matters otherwise private. The same privilege exists in cases where appropriation of a right of publicity is claimed, and the privilege may properly be said to be lost where the actual intent of the publication is not to give publicity to matters of legitimate public concern" (Zacchini 1977c, 1203-4).

Justice Celebrezze, in dissent, questioned the court's reliance on the public interest standard of *Time* and *New York Times* in light of the subsequent public figure standard of *Gertz v. Robert Welch, Inc.* (1974). With the latter case, the U. S. Supreme Court held that the extent of the constitutional privilege given defamation defendants depends upon the status of the plaintiff, whether he or she is a public figure, rather than upon the public's interest in the subject matter. Celebrezze said he believed the *Gertz* standard may apply to false light privacy, despite the fact that Zacchini had charged appropriation of his economic livelihood. Celebrezze's greatest concern, however, was the "novel and delicate issues of fact and law" in a case that was devoid of facts necessary for a well-reasoned opinion *(Zacchini* 1976, 466). He agreed that Zacchini was protected by a right of publicity, but raised the question whether the performer was a public figure as defined in *Gertz*.

On appeal the U.S. Supreme Court reversed. The majority said, in the opinion written by Justice White, that certiorari was granted, first, because the state supreme court had based its decision on the First and Fourteenth Amendments and not on adequate and independent state grounds, and, second, because the amendments do not immunize the broadcasting company from liability for televising the performer's entire act. Further, the court held that *Time, Inc.* "involved an entirely different tort from the right of publicity, that is, a 'false light' privacy case." "In 'false light' cases," White said, "the only way to protect the interests involved is to attempt to minimize publication of the damaging matter, while in 'right of publicity' cases the only question is who gets to do the publishing. An entertainer such as petitioner usually has no objection to the widespread publication of his act as long as he gets the commercial benefit of such publication" (Zacchini 1977a, 975).

Likening the right of publicity to common law copyright protection, White said: "The Constitution no more prevents a state from requiring respondent to compensate petitioner for broadcasting his act on television than it would privilege respondent to film and broadcast a copyrighted dramatic work without liability to the copyright owner." Broadcasting the entire act "poses a substantial threat to the economic value of that performance" (*Zacchini* 1977a, 975-76). White argued that, if the public can see the act free on television, it will be less willing to pay to see it at the fair. In a footnote to the decision, White acknowledged, however, that television exposure could have increased the value of the performance by stimulating public interest in seeing the act live. If that had been the decision, Zacchini would have to prove damages. One therefore suspects that had the television station not broadcast the entire act, the high court would have required a show of damages, as *Gertz* stipulates in private as well as public person defamation suits.

> The broadcast of petitioner's entire performance, unlike the unauthorized use of another's name for purposes of trade or the incidental use of a name or picture by the press, goes to the heart of petitioner's ability to earn a living as an entertainer. Thus, Ohio has recognized what may be the strongest case for a "right of publicity"—involving, not the appropriation of an entertainer's reputation to enhance the attractiveness if the very activity by which the entertainer acquired his reputation in the first place (Zacchini 1977a, 975-76).

While Justice White's majority opinion, joined by Chief Justice Berger and Justices Stewart, Blackmun, and Rehnquist, discussed the importance of protecting a performer from an unauthorized broadcast of his or her entire act, the opinion fails to take into account the possible impact of the "entire performance" rule on freedom of expression. This was taken up by Justice Powell, in a dissent joined by Justices Brennan and Marshall. Alluding to the rule's chilling effect, Powell said:

> Hereafter whenever a television news editor is unsure whether certain footage received from a camera crew might be held to portray an "entire act" he may decline coverage—even of clearly newsworthy events—or confine the broadcast to watered-down verbal reporting, perhaps with an occasional still picture. The public is then the loser. This is hardly the kind of news reportage that the First Amendment is meant to foster. Rather than begin with a quantitative analysis of the performer's behavior—is this or is this not his entire act?—we should direct initial attention to the actions of the news media: What use did the station make of the film footage? (Zacchini 1977b, 579)

Also dissenting, Justice Stevens would have interpreted the Ohio Supreme Court's ruling as a definition of "the substantive reach of a common-law tort rather than . . . as a limit on a federal constitutional right." He thought the basis of the state court's decision to be "sufficiently doubtful" as to require a remand "for clarification of its holding before deciding the federal constitutional issue" (Zacchini 1977b, 583).

Implications of Zacchini

Zacchini is important for a number of reasons. For example, the U.S. Supreme Court upheld both the value of protecting personal privacy and the specific right of publicity. One may conclude, therefore, that the court reaffirmed its recognition of both torts. Further, the case helped to establish the difference between false light privacy and appropriation. The court also determined that the public interest standard, as applied in *Time, Inc. v. Hill,* is not applicable to areas other than false light. Finally, the First Amendment is not an absolute protection for factual news reports that infringe on a performer's right of publicity. But the decision left several questions unanswered.

Justice Powell's notion that intention to exploit someone else's performance solely for commercial gain would be enough to support a finding of liability is remarkably similar, as Bennet D. Zurofsky (1978, 294) observes, to the majority's point that the right of publicity protects a performer from the theft of his or her good will for the unjust enrichment of another. Yet the majority of the court failed to deal with an "unjust enrichment" in *Zacchini,* nor did the justices rebut their colleague's conclusion that there had been none. The mere appearance on a late-evening newscast of the performance, albeit the entire brief act, hardly seems to indicate enrichment beyond that of public interest in an unusual, if not unique, daredevil stunt.

In defamation cases, the court has ruled that public figures who voluntarily inject themselves into an event or issue where public interest is clearly an element must show that the defendant had acted with actual malice. In *Curtis Publishing Co. v. Butts* (1967), the Supreme Court said that, in order for the famous football coach to collect damages in his libel suit, he had to show that the *Saturday Evening Post* had been severely reckless. The libel directly affected Wally Butts' ability to make a living. As Zurofsky (1978, 295) correctly asserts: "Were Butts to litigate his case today, his attorney might suggest an action based on the invasion of Butts' right of publicity." Justice Powell's view on intention to exploit is similiar to the actual malice standard, which may, in the future, become a critical factor in determining fault.

And speaking of fault, what is to become the standard when less than an entire act is appropriated? The *Zacchini* decision strongly suggests that the appropriation of an entire act does not require a finding of fault. Something less, perhaps, as an entire act is clearly exploitative and parts of a performance require more investigation and adjudication. One assumes that the television station's knowledge of Zacchini's objections created the requisite fault for a finding of liability consistent with *Butts* and *Gertz*.

Truthful speech, normally a media defense in defamation cases, is not an absolute protection in privacy invasion. Courts have not allowed tort liability to arise from a truthful publication, but have permitted it when actual malice, or at least negligence, was also proved. Likewise, the publication or broadcast of a falsehood is not an element in appropriation.

Perhaps what is most troublesome about *Zacchini* is Justice White's view that right of publicity is analogous to copyright protection, as it is also for patent, unjust enrichment, and unfair competition laws, where inevitably the concept of fair use comes into play. White said that the considerations underlying the right of publicity are identical to those underlying copyright! This would also mean, as we have seen, that the right is descendible. Zurofsky's (1978, 306-7) final analysis is probably accurate: "One possible explanation of the court's opinion is that it simply wanted to place its imprimatur on the right of publicity either to encourage acceptance of what it viewed as a worthwile development of the common law or to indicate that the First Amendment could not be relied upon by the media as a defense to torts against intellectual property. The facts of *Zacchini* were unusual, but its application is not likely to be limited to human cannonballs."

Assessment of the Concept

An analysis of the right of publicity tort may begin by attempting first to identify some of the confusion and ambivalence surrounding this recent legal concept. Warren and Brandeis, whose pioneer treatise reads today more like an historic anomaly than a viable guide to protecting privacy, set out to protect an individual's right to what might be called selective anonymity—the principle that each of us should be able to control, with few exceptions, the circles within which details of our lives and characters are disseminated (Zimmerman 1983, 338). However, neither the courts nor the mass media of communication have been able since then to apply the concept of selective anonymity with much success. It is simply too broad to set realistic legal limits on the dissemination of private information. Further, the sheer size and pervasiveness of the mass media have shattered the line between fact and fancy, form and substance, and, to borrow from the late Marshall McLuhan, the difference between media and message. Barbara Goldsmith (1983, 75) recently observed: "Today we are faced with a vast confusing jumble of celebrities: the talented and untalented, heroes and villains, people of accomplishment and those who have accomplished nothing at all, the criteria for their celebrity being that their images encapsulate some form of the American Dream, that they give enough of an appearance of leadership, heroism, wealth, success, danger, glamour and excitement to feed our fantasies."

Alexander Meiklejohn, who once identified protected speech as that which helps citizens in a democracy to govern themselves, considered the ownership of property a constitutional right limited in terms of public, or governmental, restrictions: "The liberty of owning and using property is, as contrasted with that of religious belief, a limited one. It may be invaded by the government. And the Constitution authorizes such invasion. It requires only that the procedure shall be properly and impartially carried out and that it shall be justified by public need" (Meiklejohn 1960, 89). That is, the government may take whatever part of one's income it deems necessary for the promoting of the general welfare.

Meiklejohn said we confuse freedom of belief and freedom of property as having the same "freedom." We are in constant danger, therefore, of giving to one's possessions the same dignity, the same status, as we give to the person himself. This careful distinction has some bearing on publicity as it is viewed as a property right—that society's right to information is as strong as is the individual's right to anonymity. Zechariah Chaffee (1947, 138) noted that a major source of ambivalence about privacy versus exposure is the pleasure that many people derive from receiving publicity in the media. "Times have changed since Brandeis wrote in 1890. Seeing how society dames and damsels sell their faces for cash in connection with cosmetics, cameras, and cars, one suspects that the right to publicity is more highly valued than any right to privacy."

In theory, it may be possible to separate form from substance, or, in the case of Elizabeth Taylor and Frank Sinatra, fact from fiction; in other words, the private person from the public one. It is silly to assume that such celebrities, whose prominence is directly the result of vast media exposure, have the exclusive rights to their life stories. But it is also silly to say that they have absolutely no claim to personal privacy. Their status as public personalites, however, provides little real means for determining the legal boundaries of exploitation. The words *celebrity* and *personality* have become interchangeable in our language, Goldsmith argues, and, if true, Taylor and Sinatra claims of ownership of their personalities are tenuous indeed. Roscoe Pound (1915, 354) wrote many years ago, in adapting William James's proposition to law, ". . . that all demands which the individual may make are to be met so far as they are not outweighed by other demands of (a) other individuals, (b) the organized public, (c) society." And, since it is impossible to satisfy all demands equally, there is need for systematic criteria for determining which demands—those of the individual or of society—are to be met.

Early right of publicity, or appropriation, cases still provide some guidelines. *Roberson, Jeffries,* and *Pavesich* identified "commercial purposes" as a necessary element for damages. The latter decision said that public persons may retain some protection against what the court called "blatant exploitation." All sought to balance the concern of the personality trying to protect the publicity right against the public need and right to know—privacy vs. public interest. In *Haelan Laboratories,* the court said that the right of publicity had little to do with privacy, because the bruised feelings of prominent persons is simply the price paid for public exposure over time. Rather than tying television docu-dramas and unauthorized biographies to commercial exploitation, the offended personality and the public's right to know would be better served if the principle of knowing or reckless falsehood were applied. In this case, defamation law, rather than exploitation or false light, would be the determining element. And, since most plaintiffs in such situations are public figures, the *Times* malice standard would come into play, as urged by the Ohio Supreme Court in *Zacchini.*

Publicity and Copyright Law

What is most problematic about the right of publicity tort, as with privacy invasion in general, is the element of prior restraint, which places a heavy burden on freedom of expression. With libel, an offense is dealt with only after the fact. The same is true with copyright. The offending party is protected by the First Amendment, but later may be held accountable in court if defamation of character has occured or copyright has been in-fringed. Elizabeth Taylor and Frank Sinatra are attempting to head off a possible offense before any has taken place. Howard Hughes lost that argu-ment in 1966, when he tried unsuccessfully to stop publication of a biogra-phy about him by Random House. The court said that public interest should be the primary concern in any fair use determination under copyright law. "Whether an author or publisher reaps economic benefits from the sale of biographical work, or whether its publication is motivated in part by a desire for commercial gain, or whether it is designed for the popular market . . . has no bearing on whether a public benefit may be derived from such a work" (Francois 1982, 685).

Copyright law and the right of publicity share a common goal—intellec-tual, literary, and artistic creativity. As one observer put it, they are "one pea in two pods." Both doctrines protect the unauthorized use of original works and insist on financial rewards for their creators. However, while they have the same end in mind, there are substantial differences between them. Right of publicity is designed to protect the creator's fame, or per-sona; copyright protects more tangible creative expression. "Under copyright law, a work is controlled by its creator. Under the right of pub-licity . . . control is often placed in the hands of the subject depicted in the creative work (Apfelbaum 1983, 1567).

Copyright protection, including financial compensation to the creator, is but a means to the stimulation of "artistic creativity for the general public good." The courts, in fact, have ruled that the author's reward is secondary to the goal of encouraging literary and artistic works. Copyright law seeks to maximize societal enrichment by protecting only the author's expression of ideas, not the ideas themselves. "Through the idea/expression di-chotomy, copyright law provides monopolies in expression while simul-taneously encouraging competition in ideas" (Apfelbaum 1983, 1570). When it passed the Copyright Revision Act of 1976, Congress took into account both the interests of creators and the broader interests of society.

Under the revised copyright law, the principle of fair use was statutorily enacted for the first time in U.S. history, emphasizing the government's commitment to preserving society's interest in the creative endeavors of its members. Fair use provides significant limitations on the exclusive rights of copyright holders and attempts realistically to balance the interests of both parties. Fair use is determined when a work is used "for purposes such as criticism, comment, news reporting, teaching . . . scholarship or research" (U.S. Code 1976). The common theme in previous case law is that each of these uses is a productive use, resulting in some benefit to the public. As Associate Justice Blackmun wrote in *Sony v. Universal City Studios*, fair use permits works to be used for "socially laudable purposes." Blackmun, in dissent, said that the making of a videotape recording for home viewing, however, is an ordinary use of the copyrighted work and, thus, is an infringement (New York Times 1984). Perhaps the same distinction can be applied to right of publicity when an individual's persona is in conflict with society's right, or need, to benefit.

As we have seen, even celebrities have legal protection under libel law, if the material reported is deliberately inaccurate or untrue, and under privacy law if they are depicted in a false light. They also enjoy the benefit of copyright law, especially if their performances or works are fixed in a tangible form. The rub comes because right of publicity extends to personal attributes, such as name or likeness. But that's too broad a legal concept for the public good to be served as well. *Zacchini*, which may have been a poor decision in terms of First Amendment law, serves to explain why copyright law is the more appropriate protection for both parties. The case involved Zacchini's own "work of authorship," which, though not in fixed form, is nonetheless authorized by the 1976 Copyright Act. His persona was of little consequence, the price he and others like him pay for fame.

References

American Law Institute. *Restatement (second) of torts*, Tentative Draft No. 13, April 27, 1967.

Apfelbaum, M.J. Copyright and the right of publicity: One pea in two pods. *Georgetown Law Journal*, 1983, 71, 1567.

Berkman, H.I. The right of publicity—Protection for public figures and celebrities. *Brooklyn Law Review*, 1976, 527.

Binns v. Vitagraph Co. of America, 132 N.Y.S. 237 (1911).

Burgoon, J.K. Privacy and communication. In M. Burgoon (ed.), *Communication yearbook* 6. Beverly Hills, Calif.: Sage, 1982.

Canessa v. J. I. Kislak, Inc., 235 A. 2d 62 (1967).

Chaffee, Z. *Government and mass communications: A report from the commission on freedom of the press.* Chicago: University of Chicago Press, 1947.

Curtis Publishing Co. v. Butts, AP v. Walker, 388 U.S. 130 (1967).

Douglas, W.O. *Go east, young man.* New York: Vintage, 1983.

Factors, etc., Inc. v. Pro Arts, Inc., 579 F. 2d 215 (1978).

François, W.E. *Mass media law and regulation*, 3rd ed. Columbus, Ohio: Grid Publishing, 1982.

Gertz v. Robert Welch, Inc., 94 S.Ct. 2997 (1974).

Goldsmith, B. The meaning of celebrity. *New York Times Magazine*, December 4, 1983, p. 75.

Gordon, L. The right of property in name, likeness, personality, and history. *Nw. Law Review* 1960, 55, 553-590.

Haelan Laboratories, Inc. v. Topps Chewing Gum, 202 F.2d 866 (1953).

Jeffries v. New York Evening Journal, 124 N.Y.S. 780 (1910).

Lazar, E.D. Towards a right of biography: Controlling commercial exploitation of personal history. *Comm/Ent Law Journal*, 1980, 2, 489.

Lewin, T. Whose life is it anyway? Legally it's hard to tell. *New York Times*, Nov. 12, 1982, Sec. 2, p. 1.

Lugosi v. Universal Pictures Co., 160 Cal.App.3d 323 (1979).

Marks, K. An assessment of the copyright model in right of publicity cases. *Calif. Law Review*, 1982, 70, 786-815.

Meiklejohn, A. *Political freedom.* New York: Harper & Row, 1960.

Memphis Development Foundation v. Factors, etc., Inc., 5 M.L. Rept. 2526 (1980).

Moser v. Press Publishing Co., 109 N.Y.S. 963 (1908).

New York Times. Justice Blackman for the dissent. Jan. 18, 1984, Sec. D., p. 20.

New York Times Co. v. Sullivan, 376 U.S. 254 (1964).

Pavesich v. New England Mutual Life Insurance Co., 122 Ga. 190 (1905).

Pember, D.R. *Mass media law* (2d ed.). Dubuque, Iowa: Wm. C. Brown, 1981.

Pound, R. Interests of personality. Part 1. *Harvard Law Review*, 1915, 28, 343-363.

Prince v. Hal Roach Studios, Inc., 400 F. Supp. 836 (1975).

Prosser, W. Privacy. *Calif. Law Review*, 1960, 48, 383.

Roberson v. Rochester Folding Box Co., 171 N.Y. 538 (1902)

Rosemont Enterprises, Inc. v. Random House, 256 F. Supp. 55 (1966).

Shipley, D. E. Publicity never dies; It just fades away: The right of publicity and federal preemption. *Cornell Law Review*, 1981, 66, 673.

Swan, P.N. Publicity invasion of privacy: Constitutional and doctrinal difficulties with a developing tort. *Oregon Law Review*, 1980, 58, 483.

Sweenek v. Pathé News, Inc., 16 F. Supp. 746 (1936).

Time, Inc. v. Hill, 385 U.S. 374 (1967).

University of Notre Dame Du Lac v. 20th Century-Fox Film Corp., 15 N.Y. 2d 940 (1965).

U.S. Code, Congressional and administrative news. Vol. 2, 94th Congress, 2d session (P.L. 94-553, 107, 90 Stat. 2546). Minneapolis: West, 1976.

Warren, S., L.D. Brandeis. The right to privacy. *Harvard Law Review*, 1890, 4, 193.

Weisman, A.M. Publicity as an aspect of privacy and personal autonomy. *Southern California Law Review*, 1982, 55, 727.

Zacchini v. Scripps-Howard Broadcasting Co., 53 L. Ed.2d 965 (1977a); 351 N. E.2d 454, 462 (1976); 433 U.S. 562 (1977b); 2 Media Law Rptr. 1199 (1977c).

Zimmerman, D.L. Requiem for a heavyweight: A farewell to Warren and Brandeis's privacy tort. *Cornell Law Review*, 1983, 68, 291.

Zurofsky, B.D. Constitutional law—privacy torts—First amendment does not privilege violation of right of publicity—*Zacchini v. Scripps-Howard Broadcasting Co. Rutgers Law Review*, 1978, 31, 269.

PART II

COMMUNICATION AND INFORMATION-PROCESSING TECHNOLOGY: ISSUES AND IMPLICATIONS

6

Computers and Communication

Everett M. Rogers and Sheizaf Rafaeli

Computers have recently evolved into communication tools, especially as a means of linking individuals and organizations in networks. Such highly interactive communication systems mark a departure from the one-way mass media of broadcasting and film, and represent a hallmark of the emerging information society. Computer technology grew out of such important inventions as the transistor, the integrated circuit, and the microprocessor (which made possible the microcomputer). As semiconductor technology progressed toward further miniaturization and lower cost, computers became more widely available, which in turn facilitated their use as communication tools. Perhaps the basic change in communication technology will lead to a paradigm shift in communication research.

This article describes the historical development of the use of computers as a means of human communication, though the original function of this technology was quite different. The transformation of computers into a tool of communication consisted of: 1) a line of technological developments through which computers became much smaller in size and cheaper in cost, and hence widely accessible; and 2) an accompanying realization of the potential of computers for communication networking. Computers are interactive and thus are forcing a basic change in the previously one-way nature of mass media communication. One hallmark of the emerging information society will be its basis on highly interactive communication systems. To study and understand communication, scholars must shift from linear to convergence models of human communication and move away from their rather singular focus upon communication effects which has characterized most past research (Rogers and Kincaid 1981).

The Beginnings of Computer Technology[1]

Perhaps one cannot adequately understand the present-day impacts of interactive communication technologies without knowing something of the history of computer development. In a sense, the electronic era began in 1912 in Palo Alto, California, with the invention of the amplifying vacuum tube. Lee de Forest and two colleagues from the Federal Telegraph Company, an early radio engineering firm, leaned across a table and watched a housefly walk across a sheet of paper. They heard the fly's footsteps amplified 120 times, so that the steps sounded like marching boots. This event, marking the birth of the electronics era (at least one of the several births that are variously claimed), led to inventions of radio broadcasting, television, and computers.

Vacuum tubes used in electronic communication systems burned out easily, generated a lot of heat, and used considerable electrical power. These shortcomings were illustrated by the first mainframe computer, ENIAC, at the University of Pennsylvania. It filled an entire room and used so much electricity that the lights of Philadelphia dimmed when the computer was turned on. For computers to become cheaper, smaller, and more widely utilized, an alternative to vacuum tubes would have to be found.

The Beginnings of the Semiconductor Industry

Bell Labs is a valuable national resource, with the largest basic research program in electronics in the world. Located in Murray Hill, New Jersey, Bell Labs was the research and development arm of the American Telephone and Telegraph Company. For over fifty years, Bell Labs spawned a flood of technological innovations; currently it holds 10,000 patents and produces about one per day. Bell Labs's most significant discovery was the transistor. Some call it the major invention of the century. Arno A. Penzias, vice-president for research at Bell Labs, says "Without Bell Labs there would be no Silicon Valley." He's completely right. Nor would there be a Silicon Valley without Dr. William Shockley, co-inventor of the transistor with John Bardeen and Walter Brattain at Bell Labs in 1947. And if there were a microelectronics complex some place, such a "Silicon Valley" would not be in Santa Clara County, California, except for the fact that Palo Alto was Shockley's hometown.

The transistor (short for *transfer resistance*) was important because it allowed the magnification of electronic messages, as did vacuum tubes, but transistors required only a little current, did not generate much heat, and were much smaller in size. At the time of its invention in 1947, it was obvious that many useful applications of the transistor would be made. But it proved difficult to manufacture reliable transistors, and the first commercial use did not happen until 1952, five years later, when transistors were used in hearing aids. Gradually, transistor technology advanced, and by the time that Shockley received the Nobel Prize in 1956 (soon after he returned to Palo Alto), twenty companies were manufacturing transistors. One of these was Shockley Semiconductor Laboratory.

Shockley started the first semiconductor company in Silicon Valley to commercially exploit the technology that he had invented at Bell Labs. His stated goal was to make a million dollars through starting-up a new company to capitalize on technological innovation. His firm was short-lived and unsuccessful, but Shockley astutely recruited eight bright young men who became the cadre for the semiconductor industry that was to sprout in Silicon Valley. He was the founder of the founders. Shockley taught them the entrepreneurial spirit, which has since characterized the microelectronics industry in Santa Clara County, California. The eight were attracted to Shockley's research and development firm by his scientific reputation. But within a year they defected to start semiconductor manufacturing companies on their own.

The beginning of the semiconductor industry in Silicon Valley is illustrated by a famous photograph of a dozen men toasting Shockley in early 1956 on the occasion of his winning the Nobel Prize. Noteworthy about the photo is the youth of the men surrounding Shockley. Robert Noyce looks like a boyish college sophomore. Actually he was 31 at the time of the photo in 1956; in a year or two he was to become the chief of Fairchild Semiconductor, the company that he began with others of the Shockley Eight. Today, Noyce is the most admired entrepreneur- engineer of Silicon Valley, and a multimillionaire. The photo includes Gordon Moore, who left Shockley Semiconductor Laboratory with Noyce to found Fairchild in 1957, and then accompanied him in founding Intel in 1968. Next to Moore in the photo is Shelton Roberts, who left Fairchild in 1961 with Eugene Kleiner, Jay Last, and Jean Hoerni of the Shockley Eight to found another semiconductor company. Hoerni later participated in founding Union Carbide Electronics in 1964, and Intersil in 1967. Thus, Shockley's admirers shown in the toasting photo were to splinter off as founders of many of the eighty or so semiconductor firms and many other high-technology companies in Silicon Valley.[2] None of the firms, nor their founders, would have been in California had it not been for Shockley. He deserves credit for starting the entrepreneurial chain reaction, as well as for co-inventing the transistor.

While at Fairchild, Noyce developed the integrated circuit; the same concept had been invented by Jack Kilby at Texas Instruments in Dallas a few months previously. In July 1959, Noyce filed a patent for his con-

EXHIBIT 6.1

Toasting William Shockley on the Occasion of his Winning the Nobel Prize in 1955.

William Shockley is at the end of the table. At bottom left of the photo are Gordon Moore and Sheldon Roberts; at top, fourth from left, is Robert Noyce, who later (in 1968) left Fairchild with Moore to spin off Intel.

ception of the integrated circuit. Jack Kilby's employer, Texas Instruments, filed a lawsuit for patent interference against Noyce and Fairchild, and the case dragged on for some years. Today, Noyce and Kilby are usually regarded as co-inventors of the integrated circuit, although Kilby was inducted into the Inventors' Hall of Fame as the inventor. In any event, Noyce is credited with improving the integrated circuit for its many applications in the field of microelectronics.

Bob Noyce is noteworthy as a Silicon Valley entrepreneur for having founded two of the most successful semiconductor firms in the world. Launching the second was relatively easy. All it took was a couple of million dollars, some very talented people, and a certain degree of technical genius. The millions came mainly via Arthur Rock, a legendary venture capitalist of Silicon Valley. Rock had been impressed by Noyce since the 1950s, when Rock helped arrange the financing for Fairchild Semiconductor. So, says Noyce, "It was a very natural thing to go to Art and say, 'Incidentally Art, do you have an extra $2.5 million you would like to put on the crap table?'" Noyce and Moore invested about $250,000 each of their own money, amassed from their original investments of $500 in Fairchild. So Rock got on the phone, and in thirty minutes he lined up the $2.5 million to start Intel. Noyce did not even write a business plan for the new firm; such was his reputation with the venture financiers.

From the beginning, Intel concentrated on computer memory chips, and it was almost accidental that it got a big boost from microprocessors, invented at Intel in 1971. Just as invention of the integrated circuit made Fairchild a commercial success, so did the microprocessor boost Intel into the big time.

The Invention of the Microprocessor

Other than the invention of the transistor at Bell Labs and the co-invention of the integrated circuit by Noyce and Kilby, the most significant innovation in the microelectronics industry is the microprocessor, invented in 1971 by Marcian E. (Ted) Hoff, Jr. Although Intel's main emphasis was originally upon semiconductor memory chips, it welcomed customers like Busicom, a now-defunct Japanese manufacturer, who wanted Intel to design special chips for its proposed family of desk-top calculators. On June 20, 1969, a team of Busicom engineers arrived unexpectedly from Tokyo to meet with Hoff. That night he left for a long-planned vacation in Tahiti. When he returned, the Japanese engineers were still waiting. They presented their design for a set of six highly complex chips to drive their new calculator. Hoff told them their design was too complex for Intel to handle.

Near Hoff's desk was a PDP-8 minicomputer, which he used in his research. At one time he had thought about the possibility of designing something like a microcomputer, and the idea of a computer-on-a-chip was still on his mind: "I looked at the PDP-8, I looked at the Busicom plans, and I wondered why the calculator should be so much more complex." Then he worked out the design for a four-bit microprocessor. A microprocessor is a semiconductor chip that serves as the central processing unit (CPU) controlling a computer. In other words, a microprocessor is the computer's brains. In defining the world's first microprocessor, Hoff had the inspiration to pack all the CPU functions on a single chip. He attached two memory chips to his microprocessor—one to hold the data and another to contain the program to drive the CPU. "Hoff now had in hand a rudimentary general-purpose computer that not only could run a complex calculator (like Busicom's) but also could control an elevator or a set of traffic lights and perform many other tasks depending on its program" (Rogers and Larsen 1984).

One might expect that the Busicom engineering team would have been immediately impressed by Hoff's microprocessor design, but they weren't. Instead, the Japanese set to work redesigning their chip-set. However, Hoff continued to work on his design for a microprocessor, even if Intel's Japanese customers were not very interested.

Bob Noyce, chief at Intel, was a firm supporter of Hoff's project. In fact, he had anticipated the possibilities of the microprocessor several years before. At a conference in the late 1960s, when Noyce predicted the coming of a computer-on-a-chip, one of his critics in the audience remarked, "Gee, I certainly wouldn't want to lose my whole computer through a crack in the floor." Noyce responded, "You have it all wrong, because you'll have a hundred more sitting on your desk, so it won't matter if you lose one" (Rogers and Larsen 1984).

At this time most computer designers were not interested in working on small computers; they felt the real action was in bigger mainframes. But Hoff was able to convince Stan Mazor, another Intel employee, to join him in the microprocessor work. In October 1969, the manager of Busicom and his engineers presented their revised design for the calculator chip-set at Intel's headquarters. Then Hoff made his argument for the microprocessor, claiming that it would be more general in its potential applications. Busicom's top manager was convinced, essentially telling Hoff, "I hope it's as good as you say."

Hoff and Mazor now set to work in earnest to design the microprocessor. In March 1970, Federico Faggin joined Intel and did the chip layout and circuit drawings. Faggin came up with a "nice clear design," in Hoff's opinion. Meanwhile, Busicom was going ahead with their calculator in which the Intel microprocessor would be a key component. Less than a year later, in January 1971, Hoff and his colleagues had a working micro-processor, the Intel 4004.

Gordon E. Moore, president of Intel, described the elegance and power of Hoff's microprocessor: "Now we can make a single microprocessor chip and sell it for several thousand different applications." That was the beauty of the microprocessor—it could serve as a component in any electronic product where one wanted miniaturized computing power. This flexibility of application, of course, had tremendous commercial implications. In short, Ted Hoff had invented a means for Intel to get rich quick.

Marketing the Microprocessor

The Intel marketing division wasn't convinced that microprocessors were a viable product, worth the cost and effort to manufacture. Even though Intel had been selling the 4004 to Busicom since January 1971, this microprocessor had still not been announced publicly by Intel, so there was no demand for it. As 1971 went by, Hoff and Mazor and their little team of microprocessor enthusiasts would march down to the marketing people and urge them to announce the 4004. Hoff says, "We pleaded with them each month, but each time they decided not to announce it yet."[3] Finally, Edward L. Gelback came on board as Intel's marketing chief; he'd been at Texas Instruments and had a more positive view of the microprocessor as a product line.

There is a natural tension between technical people and the marketing department in every high-technology company. It is a battle between orig-inal creativity on one side and merchandisers who know customers' needs on the other. The conflict can arise in designing a new product, when the research and development people proudly present their concept of the innovation and the marketers tell them it won't sell. Or the conflict may occur over the allocation of company resources, such as whether funds should go to research and development or to marketing.

In November 1971, Intel decided to announce the 4004 by taking out an ad in *Electronic News*, the industry house organ. Intel's ad announced not just a new product, but "a new era of integrated electronics. . . , a micro-programmable computer on a chip." Semiconductor companies had made exorbitant promises in the past, and the industry reaction to the Intel 4004 was ho-hum. Typical was the reaction of an irate customer at the fall 1971 computer show in Las Vegas, who walked into the Intel display and ex-claimed, "The nerve of you people, . . . you can't have a computer on a chip." Stan Mazor gave him a data-sheet on the 4004; the customer grudgingly admitted, "Sure enough, it is a computer."

Shortly thereafter, increasing numbers of people in the electronic industry began to wake up to the fact that the invention of the microprocessor meant a whole new ball game in the field of microelectronics. But there were still plenty of doubters. Ted Hoff and Stan Mazor went on the road for three weeks in mid-1972, giving seminars and meeting with design engineers in various companies that might use their microprocessor. The Intel enthusiasts for the microprocessor encountered a range of doubt; a frequent question was, "Well how do you repair it?" People could not understand that a computer could be a throwaway item like a light bulb. Other customers scoffed, "How would you keep them busy?" Hoff remembers that at this early stage in the life of the microprocessor, what many customers really wanted was the power of the minicomputer at the price of a microprocessor chip. "We still could not do what a minicomputer could do in processing power. Not then."

In April 1972, Intel announced its eight-bit microprocessor and continued to stay one jump ahead of its competition. As Hoff explains, "The cost of a chip is a function, a strong function, of its size. The smaller a chip the more of them you can get on a wafer, so the price is correspondingly cheaper." This trend to further miniaturization is characteristic of memory chips, as well as semiconductors. The direction of progress in the semiconductor industry for the past decade or so has been to put more and more memory capacity on the same-sized semiconductor chips. This tendency has greatly reduced the cost per bit of computer memory. The most widely sold memory chip (RAM, or random access memory) in the semiconductor industry has moved from the 1K (one thousand bits of information) to the 4K to the 64K. The cost per bit of computer memory has decreased correspondingly about 28 percent per year. This lowered cost translates directly into more computers in society. Most of the millions of computers sold in the past decade are microcomputers, built around the microprocessor that Ted Hoff invented at Intel in 1971.

The Rise of Computerized Communication

Mainframe Time-Sharing

The word *computer*, with overtones of counting and calculation, correctly describes the machine's past uses, but does not convey its present and potential uses as a tool of communication (Feigenbaum and McCorduck 1983, 40). The practice of communicating with computers has its roots in the fact that mainframe computer power was, until recently, relatively expensive. Until the mid-1970s, when the microcomputer burst on the American scene, computers were owned and operated by the establishment: government, big corporations, universities, and other large institutions. These mainframe computers were mainly used for data-crunching tasks: accounting, record-keeping, research and data-analysis, and airline ticketing, for example. Single users could rarely afford to own a mainframe computer of their own.

Various solutions evolved in response to the economic challenge of the costly transistor-age, pre-microprocessor large computer. These solutions entailed a communication structure, whereby shared usage of the expensive capacity of the mainframe was made possible. Several users took part in a time-sharing system in which a single mainframe computer was wired so it would perform more than one task simultaneously, thus sharing its time across several users. The technology of time-sharing allowed users to communicate with large computers using either "dedicated" or public telephone lines. The user enjoyed the convenience of spatial independence from the central computer, as well as a large reduction in the costs of using a mainframe computer.

Although the cost of the central processing unit has since declined continuously, time-sharing has remained a very popular technique for computing. Time-sharing gave birth to an elaborate set of protocols, standards, software and hardware arrangements, all designed to allow many users to plug in at the same time to the omnipotent (but costly) single computer. These elaborate arrangements to access the computing power of mainframes are now widely used, so physical proximity to the computer is unnecessary. Telephone lines now replace proximity. The technology of transmitting data between the users and their central computer over telephone lines was improved by packet-switching digital communication technology. The communication process itself was now governed by a computer.

User Networks

More complex computing tasks began to require communication between users, as well as between their machines. It was soon realized that these same channels could now be utilized for user-to-user communication. The potential of computers as a special means of communication became evident.

A series of networks emerged whereby users and computers were linked. These networks provided an environment of pooled computer power. But they also became a kind of public commons, an arena in which users could now not only share mainframe power but could communicate among themselves and share a common collection of information like a data base. Computer-communication technology began to provide the users an opportunity to talk to remote machines, to converse with each other over temporal and spatial distance, and to share in the creation and utilization of data bases.

Microcomputers and Networks

Enter the factors of miniaturization, decreased cost, and the popularization of microcomputers and programming that were discussed in our earlier section. We began to see a tremendous explosion in computer networking. Computers had now become a special medium of communication.

Networks that have previously been the exclusive domain of computer users in industry and academia are now accessible to a rapidly-growing user-public. Microcomputers are particularly well-suited for replacing the big mainframes in networking communication functions. Microcomputers do it cheaper. The availability of computer networks brings about both the appetite for, and the recipes for, very diverse applications. In these applications the computer serves as both a participant and a medium of communication.

In late 1984, an estimated 12 percent of U.S. households owned a home computer, with a very rapid rate of adoption in recent years (rivaling the rate of adoption for black-and-white television sets in the 1950s). While this rapid adoption is impressive (the number of households with a home computer doubled during 1983), only a relatively small percentage of the 80 million households in the United States currently own microcomputers, compared to the 10 million color television sets, 20 million calculators, and 40 million radios sold each year. So it is important to remember that home computers are still at a very early stage of diffusion. Sales in 1982 were $2 billion, about half the size of pet food sales. The doubled sales of home computers in 1983 almost matched pantyhose in sales. But if these sales figures seem relatively modest, it is the potential of home computers for bringing about social changes in society that make them so important. And the proportion of U.S. households with microcomputers is estimated to increase from around 12 percent in late 1984, to over 50 percent by 1990.

At present, the main uses of home computers are for playing video games and for word processing (that is, computer-assisted typing and editing.)[4] Electronic games can also be played in arcades (similar to the pinball arcades of a previous era), in the home on video game equipment that attaches to a television set, or on a microcomputer in the home. Only about one-fourth or so of home computer owners have a modem—a computer peripheral that allows one to connect a microcomputer via the telephone system to an information bank like The Source or Compuserve, or to a mainframe computer or a minicomputer at a place of work, or to a network of other microcomputer users.

While it can be argued that most computer uses are of a communicative nature—for example, word-processing—the recent trend toward increasing use of microcomputers as networking devices demonstrates the potential of computers for interactive communication. Such networking illustrates a type of "machine-assisted communication" (Dominick 1983), which differs from interpersonal communication, in which a face-to-face information exchange occurs (without any type of equipment intervening), and mass media communication, in which some mass medium like print, radio, television, or film allows one or several individuals to disseminate messages to a large, geographically dispersed audience. Machine-assisted communication via the telephone, telegraph, or computer networks has certain characteristics of both interpersonal communication (like its interactive nature) and mass media communication (such as that some type of electronic communication equipment is involved). Many of the mass media organizations today use computers as internal communication channels for writing, editing, and typesetting. So computers are both a new medium of communication and a new component in existing mass media institutions.

Computer networking is illustrative of the new type of human communication that will be the hallmark of the "information society" that the United States and several other nations are becoming. The microcomputer can serve two functions in a computer-communication network. It can allow its owner to have an "intelligent" access to networks—emulating the "dumb" terminal but doing a much better job of communicating. It can also serve as a host computer, the node in a network. Microcomputers can be, and are, used as the hubs of an increasing proportion of such networks.

Recently, an explosion has occurred in the growth of applications of computer-communication networks to human communication tasks, needs, and interests. Popular applications are computerized conferencing, electronic shopping, banking, learning, telecommuting, and computerized access to information. Toffler's (1980) futuristic electronic cottage is a reality for many who take part in the activities of "compunicating."[5] The computer also plays an increasing role in other communication activities. Interactive cable television, for example, is highly dependent on a computer that drives the transmission process, but the computer is not the medium in this communication system.

The Evolution of Computer Networks

Computer networks began in comparatively small, localized settings—experiments in management efficiency and office automation and facilitation of academic group research. The U.S. Department of Defense and several major universities followed these initial attempts to form computer networks. Medium to large-scale computer conferences were developed and launched, based on one or more central mainframes serving as "hosts" (Hiltz and Turoff 1978; Kerr and Hiltz 1982).

Due to microcomputer technology, the number of active networks has multiplied in the past three or four years. Hundreds of thousands of new microcomputer owners have been searching for, and finding, communication outlets for their newly-acquired computing power. Such outlets include electronic bulletin boards, online information utilities, dial-in news, banking and shopping services as well as the relatively older conferencing and time-sharing computer facilities, which were established prior to the microcomputer boom.

There are currently several thousand microcomputer-hosted bulletin boards (the most primitive, but very popular, version of computer networks) operating in North America. These electronic bulletin boards compete with more established, commercial utilities, each of whose subscribers number in the tens of thousands. Subscribers to these commercial utilities are also utilizing their own microcomputers (Glossbrenner 1983). While the commercially offered utilities are profit-oriented, many of the publicly accessible electronic bulletin boards are run by individuals for non-economic motives. In northern California, for example, there are currently a dozen bulletin boards devoted to political discussions, none of which turns any profit for its operator or charges its users. These bulletin boards are constantly busy. Some users of these bulletin boards are highly devoted, active members of vibrant networks (Rafaeli 1983).

Asynchronous and Interactive Dimensions of Computerized Communication

The communication environment brought about by the computer as a medium is very different from that inherent to earlier modes of communication. Computerized communication combines certain communication concepts in a novel way. The nature of communication has always been, at least in part, a product of the technology of the medium. This technology-dependency of the communication process has never been more so than today, with computers assuming an increasing role as both media and participants. Changes in information technology have begun to redefine the process of communication itself. One should understand the parameters of this new communication process, prior to proposing the most appropriate ways to study it.

As illustrated by the various applications of computerized conferencing, the communication process among participants in computer networks is typified by several characteristics. This process is asynchronous, it is interactive, and it is relatively fast. Also, communication via computers has become a nonlinear process. Computerized communication is different from mass-mediated communication as we have come to know it in the past, in that the audience is "demassified" in nature, although it is potentially as large as ever. Messages can vary between anonymity and intimacy, as is illustrated by the comparison between computerized CB radio emulations and "matchmaking" electronic bulletin boards.

The asynchronous nature of communicating via computer permits the retention and accessibility of messages and entire, multiparticipant conversations over extended periods of time. For example, a participant in an electronic messaging system may receive a message from another participant whenever he logs on to the computer; thus time can be managed or controlled. The discourse is enriched by the ability to reach back into the automatically compiled minutes of past communication events. The possibility of asynchronous communication also fosters interaction among people or groups who are not on coordinated schedules. In fact, as Toffler (1980) pointed out, asynchronized communication enables society to harbor many asynchronized life rhythms, such as flextime work arrangements.

Interactivity is perhaps the most salient of the characteristics of computerized communication systems. The interactivity of a system could be indexed by the degree of participatory user activity on the system. Users of computerized communication—who are the combined set of what used to be called sources and receivers—have a very high degree of control over their communication process. If the degree of interactivity in a communication system is measured by the ratio of user activity to system activity (Paisley 1983), computerized communication approaches infinite interactivity. In computerized communication there is very little system activity that is independent of the users/participants. Those individuals who use the system determine its content. Computerized communication is, at least potentially, a very democratic process.

Computerized communication is nonlinear because the traditional, unidirectional model of the communication process is demonstrably no longer applicable. During the era of print communication (from the days of Gutenberg until the present), a linear model of the communication process that postulated sender, receiver, and channel was formulated and widely used by communication scholars (Shannon and Weaver 1949). This linear model persisted through the next phase of communication technology, the era of film, radio, and television (from approximately the 1920s until the present). Mass audiences were treated by several generations of communication researchers as the receiving end of a linear process of communication. Some later expressions of this linear model of communication modified it to include a feedback component. This linear model is not appropriate to computer communication. For instance, the distinction between sender and receiver is blurred, as one must speak of users in a computerized communication system. Further, no single message is at the center of the communication process. It is no longer possible, in the context of computerized communication, to distinguish a message from feedback. If "sender" is no different from "receiver," and "message" is indistinguishable from "feedback," perhaps the model based on these concepts must be replaced.

The demassification of the audience inherent in computerized communication is an extended and enlarged version of the market segmentation processes witnessed by the electronic and print media in past decades. The U.S. magazine industry moved toward specialization, especially after television became widespread (about 1960). The radio industry also evolved toward an increasing number of formats, in part due to competition with television. Cable television furthered this demassification process by providing individual themes for any one of its numerous channels. (Newer cable televison systems now have over 100 channels.)

In a similar but more pronounced way, computerized communication focuses on topical information of a specialized kind. Demassification means that information is now available in a much wider variety than previously. The target audience for each individual type of communication content is shrinking. Both the audience and the medium are thus being demassified, although the aggregate effect need not necessarily be a smaller audience on the whole (Smith 1980).

We have shown that communication mediated by computers, at least in the form that it takes today, assumes several new traits. It is an asynchronous and interactive process. Communication can no longer be mistaken to be linear in nature. Through the versatility of the new medium, the audience will be more demassified than previously. The audience may be as large as the audience addressed by the more traditional, older media, but the nature of the medium itself will undergo demassification.

Paradigms in Compunication

The scientific study of computerized communication is still in search of a unifying paradigm. The earliest studies of computer networks that were within the communication studies discipline have taken one of three tracks. These studies were either descriptive, evaluative, or they were studies of effects. In the following, we suggest a fourth, possibly more appropriate, direction.

Computerized communication was first studied in a descriptive context. The driving question was that of finding out what systems existed, and what they could do. As is the case in many other fields of study, a descriptive phase has to precede any deeper foray into the nature of computer communication. Typical research questions answered by this descriptive research were and are: What are the characteristics of the system's hardware and software? Who are the users? How large is the network? How many of these systems are in operation? What are the first-level costs of using the system? Hiltz and Turoff (1978) undertook this challenge rather early. In their book, they describe several (both real and projected) computer communication systems.

The next phase in the communication discipline tradition of the study of compunications were and are evaluation studies. This family of studies is typically conducted in light of certain goals set by the owners or funders of the computerized communication system being evaluated. This type of study can be "likened to the mapping of available knowledge gained from exploratory studies of computer-mediated communication systems" (Kerr and Hiltz 1982).

Finally, a large proportion of the research on computerized communication systems has been centered on the effects of such systems. Individual, group, and social-level effects of these systems are investigated. The focus here is on the actual consequences (intended or unintended) to the users of the computer communication system, in the terms of the systems' sponsors. This research focus on impacts is a direct parallel to the past emphasis in mass communication research on media effects, except that such effects are much more complex to identify and quantify in an interactive communication environment.

Danowski (1982) presents a methodology for analyzing computer-mediated communication systems. Kerr and Hiltz (1982) review studies pertaining to a long list of cognitive, affective, and behavioral impacts of compunication on individuals and groups.

These three research phases—description, evaluation, and assessment of effects—resemble very closely the phases that communication research has undertaken in studying other mass media of communication in the past. The theoretical paradigm dominating these three phases is the linear, source-message-channel-receiver model. Paisley (1984) now calls this model the "dominant non-paradigm." For a variety of reasons, the linear model is not only inappropriate for studying computerized communication, it may no longer be very appropriate for investigating the more traditional mass media as well.

Print communication lent itself to study most easily with such a linear model. For several generations, researchers focused on the process of communication as a flow of information, opinion, and effect from a source to a receiver. The study of film, radio, and television also progressed from description and evaluation through a focus on direct effects. As these media ascended to their current position of prominence in society, communication researchers followed their rise by asking the same general questions: describing, evaluation, and measuring effects. The orientation of the research was predominantly that of the source. Occasionally, however, researchers used the receiver's point of view. Rarely was communication itself anything other than an independent variable.

A Convergence Model

Certain students of mediated communication awakened to the need for a better fit between their paradigm and the mediated communication processes. Proponents of the need for a new paradigm to drive communication research suggest using a convergence model of communication, rather than continuing reliance on the "dominant non-paradigm" of linear effects. The convergence model does away with the mechanistic and linear conceptions of the communication act that is studied. Instead, the cyclical and dynamic characteristics of communication are highlighted. The interdependence of all participants in the communication process is emphasized. Rather than asking, What does communication do to people?, the convergence paradigm advocates the question: What do people do with communication? (Rogers and Kincaid 1981). Thus *communication* is defined as a process in which the participants create and share information with one another in order to reach mutual understanding.

Studying computer-mediated communication with a convergence paradigm amounts to looking at communication behavior as a criterion variable rather than as a predictor of effects. Convergence models should allow a holistic, rather than mechanistic, study of the communication process. The unit of analysis can be the process (such as a network link of information-exchange) rather than individuals. Key concepts within the framework of this paradigm are information and meaning, convergence and divergence, mutual understanding and agreement (Rogers and Kincaid 1981).

We return, at this point, to the special characteristics of computerized communication discussed earlier in this chapter. The interactive nature of this process, its asynchronous and rapid timing, the demassification of medium and audience, and its nonlinear, interactive nature, all demand a new theoretical framework for its study.

The convergence model that is emerging in the study of interactive communication seems to be ready-made for this research purpose. Convergence is a process in which participants create and share information in order to reach mutual understanding. In this convergence model, research is directed at the degree to which such mutual understanding occurs and the degree to which such understanding correlates with agreement.

The active, participatory nature of the highly interactive computerized communication systems can only be studied with such a convergence perspective. The nonlinearity implied by the design of the computer-communication systems and the asynchronous nature of the process both require a framework of study that avoids the mechanistic-causal and psychologistic biases inherent in the previous "non-paradigm."

The technology used to design the components of interactive computer media is heavily reliant on cybernetics, the mathematical study of guidance and control processes (Wiener 1948). Cybernetics has been used as a model for studying human behavior mostly in the context of human-machine or human-nature interactions. We suggest that convergence theory, which was derived from the premises of cybernetics, be used to study the human-to-human communication systems that are today proliferating through the widespread diffusion of microcomputers in society.

Ironically, Wiener's work on cybernetics was introduced just prior to Shannon and Weaver's (1949) information theory and linear model of communication. Both theories had a widespread impact on communication research and thought. Information theory, however, lingered on in the form of linear theories of communication and the S-M-C-R model (Rogers and Kincaid 1981). Is it possible that convergence theories, so closely related to cybernetics, will be the driving force in an epistemological change away from linear, one-way conceptions of communication?

An epistemological change of this sort is urged by Paisley (1983). Such an intellectual shift is called for not only because of the inadequacy of prior theoretical orientations in the discipline of communication studies but because "*technological change* has placed communication scientists on the front lines of a social revolution" (Paisley 1983). A new kind of society, the Information Society, is being formed by the new communication and information technologies. Communication science can not ignore the emergence of these new technologies. Our central position as students of communication in this age of communication provides a rare opportunity. We cannot afford to use old, blunt, and no-longer-appropriate tools in confronting this new research challenge.

Notes

1. The historical sections in the first part of this article are adapted from Rogers and Larsen (1984).
2. A 1982 survey found 3,100 microelectronics manufacturing firms in Silicon Valley, with two-thirds of them having fewer than ten employees (Rogers and Larsen 1984).
3. This material about the invention of the microprocessor comes from a personal interview with Ted Hoff in August 1982 (Rogers and Larsen 1984).

4. As indicated by a survey of home computer users in the Silicon Valley area of northern California (Rogers et al. 1982).
5. Oettinger (1971) coined this term over a decade ago when he observed that "computers and communications have long since become inseparable. It is time to reflect this union in the fusion of their names."

References

Danowski, J.A. A network-based content analysis methodology for computer-mediated communication: An illustration with a computer bulletin board. Paper presented at the International Communication Association, Boston, 1982.

Dominick, J.R. *The dynamics of mass communication*. Reading, Mass.: Addison-Wesley, 1983.

Feigenbaum, E.A. and P. McCorduck. *The fifth generation: Artificial intelligence and Japan's computer challenge to the world*. Reading, Mass.: Addison-Wesley, 1983.

Glossbrenner, A. *The complete handbook of personal computer communications: Everything you need to go online with the world*. New York: St. Martin's, 1983.

Hiltz, H.R., and M. Turoff. *The network nation: Human communication via computer*. Reading, Mass.: Addison-Wesley, 1978.

Kerr, E.B. and S.R. Hiltz. *Computer-mediated communication systems*. New York: Academic, 1982.

Oettinger, A.G. Compunications in the national decision-making process. In M. Greenberger (ed.), *Computers, communications and the public interest*. Baltimore: Johns Hopkins University Press, 1971.

Paisley, W. Computerizing information: Lessons of a videotext trial. *Journal of Communication*, 1983, 33(1), 153-161.

———. Communication in the communication sciences. In B. Dervin and M. Voight (eds.), *Progress in the communication sciences*, Vol. 5. Norwood, N.J.: Ablex, 1984.

Rafaeli, S. The uses and gratifications in communicating by electronic bulletin boards. Paper in review, 1983.

Rogers, E.M. and D.L. Kincaid. *Communication networks: Toward a new paradigm for research*. New York: The Free Press, 1981.

———. and J.K Larsen. *Silicon valley fever: Growth of high-technology culture*. New York: Basic, 1984.

———. et al. *The diffusion of home computers*. Stanford, Calif.: Stanford University, Institute for Communication Research, Report, 1982.

Shannon, C.E. and W. Weaver. *The Mathematical theory of communication*. Urbana: University of Illinois Press, 1949.

Smith, A. *Goodbye Gutenberg: The newspaper revolution of the 1980's*. New York: Oxford University Press, 1980.

Toffler, A. *The third wave*. New York: Bantam, 1980.

Wiener, N. *Cybernetics: Or control and communication in the animal and the machine*. New York: MIT Press and Wiley, 1948.

7

The Person-Computer Interaction: A Unique Source

Robert Cathcart and Gary Gumpert

This article explores the person-computer relationship. It uses the accepted interpersonal communication model as the basis for determining the psychological parameters of person-computer transactions. It is proposed that the computer functions as a proxy in the interpersonal communication dyad and that this, in turn, creates source ambiguity resulting in dissonance. The essay examines the means of resolving this state of dissonance and suggests that the traditional dyadic model is being modified by the person-computer paradigm.

Ithiel de Sola Pool in the foreword to *The Coming Information Age* states that prior to the introduction of the computer, every communication device "took a message that had been composed by a human being and (with some occasional loss) delivered it unchanged to another human being. The computer for the first time provides a communication device by which a person may receive a message quite different from what any human sent. Indeed, machines may talk to each other" (Dizard 1982, xi-xii). With the rapid introduction of computers, both personal and institutional, into everyday life we are witness to one of the most profound changes in communication since the invention of cuneiform and the clay tablet—the technology which first made it possible to imprint symbols of human vocal messages and to transport them physically from sender to receiver.

Each new technological communication innovation has its effect on information transfer and processing. Each new technology has expanded human communication capabilities over time and space, and has resulted in altered interpersonal relationships (Cathcart and Gumpert 1983). Communication technologies themselves have been altered in usage through various combinations of media: for instance, the telephone with radio, the photograph with motion pictures and television, and so on. Today, another radical shift in communication is occuring through the combining of the computer with electronic and telephonic communication technologies.

Types of Computer-Human Functions

Computers, in conjunction with other media, have already made remarkable inroads into our daily activities and have altered certain interpersonal relationships. At present, human-computer interaction takes place in the following three ways:

1. Unobtrusive functions—those in which the utilization of the computer is not evident to the user, for instance, digital recordings and telephone connections (Covvey and McAlister 1982).
2. Computer-facilitated functions—where persons use a computer for the purpose of expediting communication. This function refers to communication *through* a computer rather than *with* a computer. Electronic mail, for example, represents a change of medium (paper to display screen) in which the computer is interposed between sender and receiver. In this case, the computer is a high speed transmitter of what is essentially a written message. The same holds true for computer bulletin boards and other computer networks which allow messages to be stored and retrieved at the convenience of senders and receivers without reliance on intermediaries such as printers, librarians, postal clerks, telephone operators, and so on (Levy 1983).
3. Person-computer interpersonal functions—situations where one party activates a computer which in turn responds appropriately in graphic, alphanumeric, or vocal modes establishing an ongoing sender/receiver relationship (Cathcart and Gumpert 1983). Two basic types of computers are utilized in this activity: the dedicated computer which is fundamentally unifunctional, and where the interaction generally requires no training of the operator; and the multifunctional computer where training is required and the person must accommodate himself or herself to particular forms of computer language such as BASIC, Pascal, LOGO, and so on (Papert 1980).

   Today we have no choice but to interact with a computer rather than with another human being when we wish to locate information stored in data banks, pay monthly bills via telephone connection with a bank's computer, or withdraw or deposit money from a twenty-four-hour bank teller machine. In these cases, there is no way to process the transaction other than by "talking to" or having "dialogue" with a computer. For these transactions, the computer is programmed to interact in a manner replicating face-to-face interpersonal communication (Vail 1980).

It is our intention to discuss the phenomenon of communicating *with* a computer rather than *through* a computer. We will argue that in this situation the computer serves as a proxy for another person, thus establishing an interpersonal communication dyad. For there to be interpersonal dialogue, there must be an exchange of messages. The interpersonal reciprocal act implies the existence of both a sender and a receiver, as well as mutually understood linguistic codes.

Characteristics of Dyadic Communication

The accepted dyadic model of interpersonal communication posits interchangeable sender-receiver roles in situations where immediate and continuous feedback produces communication which is symbolic, non-repeatable, and nonpredictive. That is to say, the message is created through symbolic processing, the meaning of which in turn is determined by the relationship of sender and receiver. In addition, such messages, at least in face-to-face dyads, are evanescent and therefore nonrepeatable. Such interpersonal dialogue is characterized by a maximum sender-receiver control of the exchange and equal responsibility for the outcomes.

Person-Person Communication

It is the dynamic relational quality among sender, receiver, and symbols which makes interpersonal exchanges creative and nonrepeatable (Millar and Rogers 1976). Each interpersonal exchange is situation specific—created on the spot, the outcome not known in advance. There is always a creative element in human communication because of the symbolic nature of language and the fact that meaning is produced through the symbolic process, not by the symbols themselves.

Though there are grammatical rules and norms which govern the use of language, these do not limit the outcomes of communicative exchanges, i.e., the meanings and motives which are the result of the exchange of messages. Because of the creative and nonpredictive qualities of interpersonal communication we are always "at risk" in such situations, in that misunderstanding, rejection, and loss of self-esteem are always possible (Miller and Steinberg 1975).

This is not to say that face-to-face interpersonal communication is the only or even preferred way to have dialogue. Rather, it is to point out how important sender and receiver relationships and symbolic functions are in all human communication.

Mediated Interpersonal Communication

We have suggested elsewhere that when a technological medium is interposed in the interpersonal dialogue, there are produced significant differences in the content of the exchange and the relationship of sender and receiver. We have referred to this activity as mediated interpersonal communication (Cathcart and Gumpert 1983). For the mediated exchange to work as interpersonal communication, there must be tacit agreement that the participants will proceed as though they are communicating face to face. For example, when a television personality communicates directly to an audience *through* a television camera, interpersonal norms are used to simulate face-to-face interaction, and the viewer tacitly agrees to respond as though face to face with the television personality. This "intimacy at a distance" is the basis of parasocial interaction (Horton and Wohl 1956). A telephone conversation is another example of a mediated interaction which proceeds as if it were a face-to-face encounter (Aronson 1971).

It is apparent in these two cases that the interpersonal sender-receiver relationship is maintained even though there is limited feedback and some restrictions on the possible outcomes. We maintain that for dialogue to take place it is important that an interpersonal sender/receiver relationship be established and that the symbolic process be maintained.

Person-Computer Communication

The interaction of person and computer, on the other hand, creates a unique mediated sender-receiver relationship. This idiosyncratic relationship happens because computers, unlike other machines, can as Ithiel de Sola Pool suggests, "respond" and produce messages, apart from any human source. For the first time a technology can not only speed and expand message exchange, but it can also respond with its own message to a human partner. The computer in this mode becomes a proxy for a sender-receiver in the communication dyad.

In this situation we tacitly accept the computer as a substitute for some person with whom we would interact were the computer not available. We give the computer commands, we ask it questions, and it responds. It gives us commands, asks us questions, and we respond. We appear to be reproducing dyadic communication based upon an immediate sender-receiver relationship. But, this dialogue is a strange one in which the human, capable of speech, responds through tactile means of keyboard, touch-tone or touch-screen as the computer, a machine, responds with visual displays and sound reproductions/imitations of the human voice. The computer user agrees to communicate as though the computer were a message source, simulating interpersonal exchanges even though in reality there can be no duplication of the creative, nonpredictive human interpersonal dialogue.

The computer is programmed to respond in certain ways depending on the input it receives from the human user. There appears to be a dyadic exchange which develops according to message input and feedback, but in actuality the exchange is always limited and predictive because of the binary nature of information processing within the computer. This makes the relational and control aspects of the dyadic exchange quite different from that of the human interpersonal dialogue, even where the person-computer interaction has been programmed to replicate person-to-person communication.

The key to the difference between person-to-person communication and person-to-computer communication is the "program." Because computers must be programmed, the outcome of the "dialogue" is predictable, even though the computer branches its responses based on the unprogrammed inputs of the user. What happens, of course, is that the user becomes "programmed" to provide the kinds of input that are compatible with the computer. Thus the relationship shifts to one where the computer is in control and is responsible for the outcome of the communicative exchange.

Impact of Person-Computer Communication

It may seem strange to speak of control and relationship when examining human-computer interaction, but this tract is worth exploring if we are to understand computer effects on intrapersonal and interpersonal communication. In the person-to-person dyadic model, sender-receiver roles are interchangeable, just as message and feedback are reciprocal and creative. A relationship exists based upon the perceptions of sender or receiver and the symbolic content of messages exchanged. This relationship is furthered or altered depending on what is disclosed, what needs are satisfied, and what social and cultural norms prevail. The content of the message and feedback is shaped at the outset and throughout by the relational aspect of sender and receiver interaction. Examples abound and are obvious. The way a parent and child communicate is shaped in part by the familial relationship and evolves depending on what each discloses and what each perceives the relationship to be at any given time. The unique relationship and the limitless possibility of producing meaning symbolically implies that the outcome cannot be predicted—only indicated.

When the human dyadic exchange is *mediated*, the sender-receiver relationship is altered. If we cannot see, but only hear the source, less information is processed and our perception of source is altered. We necessarily compensate for this loss and in so doing the relational quality of the interchange is modified. When the only way we can reach out and touch someone is through the exchange of oral-aural signals, we cannot have the same relationship as when we can physically touch, smell, taste, and visually see the other person. When feedback is delayed as in letter writing or attenuated as it is in radio and television interactions, there is a marked difference in the alliance of sender and receiver. Still, in every mediated communication there is a relationship based on the perception that each person has of the other, and the fact that whatever messages are produced are the result of symbolic processing between two or more indivduals.

The Problem of the Source

What is our perception of, and relationship to, the source when we interact with a computer? Here we face a dilemma. The source of the message/feedback produced is not the hardware, even though we are "talking to" the machine. We might consider that the programmer who creates the software is the source. But, then we are confronted with even more ambiguity. A programmer produces software for a particular computer in suitable computer language. That program can be used in millions of computers. But, when one of us is "talking" with the computer, we are not interacting with the programmer. We have no idea who that person is, nor did the programmer have in mind trying to reach us with a message. The programmer's interaction is with the computer, not with the user.

Conceivably the program itself could be the source of the message and feedback. The user activates the program and once initiated the program begins to guide and direct the form and content of the dialogue. The program, however, provides only a framework or a context in which the exchange takes place. Only certain kinds of inputs and responses will be accepted. The program cannot decide to pursue a new line of inquiry—create new meanings or symbols—nor can it change its role in relationship to the user as the dialogue proceeds.

Computer as Human-Like

The ambiguity regarding the source in the person-computer dialogue necessitates resolution of a dissonant situation. If the user is to accept that communication *with* a machine rather than *through* a machine is possible, then it follows that some adaptation or tacit agreement must occur if that state of ambiguity is to be resolved. One common solution has been to anthropomorphize the computer—to attribute human characteristics to the machine. Examples abound, like "I usually get my way, but it makes me do tricks first." The relationship appears to vary, depending on the program utilized. A numer of relational possibilities exist: tutor, coworker, competitor, student, and master. It is interesting to note that much of "computerphobia" is to be found in the potential master-slave relationship (Turkle 1982). For example, one can buy a software program called ABUSE in which the computer will insult you, make fun of your inputs, or shame you by solving problems that you cannot solve. In any case, anthropomorphizing the computer is required in order to establish the relationship necessary to make dialogue possible. We supply the computer—the machine—with human attributes. Thus, we can react as though it were a proxy—human—source and a receiver.

The desire on the part of the user to reduce ambiguity and dissonance is but one of several factors that result in the anthropomorphizing of the computer. Humanizing the computer is also prompted by the physical and electronic properties of the comptuer and by the nature of programming which uses the interpersonal dyad as an operational model. Interpersonal cues are provided by the programs, and the user agrees to participate in a simulated interpersonal relationship. Such reciprocity makes anthropomorphizing the computer different from that with other machines such as automobiles, airplanes, and boats.

Human-like symbolic responses are programmed into the computer. Complicated and subtle relationships are created where a computer user chooses to allow his or her reaction to be manipulated and controlled in order to fulfill the ideal of interpersonal dialogue. For example, the user relinquishes control of the dialogue to a "humanness" which is programmed into the machine. The result is a symbiotic relationship that permits the user to experience an ongoing dialogue. Instead of viewing himself or herself as a machine which has been programmed to provide the "correct" responses—which is, after all, what interaction with a computer is all about—the user imputes human qualities to the machine in order to make possible the semblance of interpersonal communication.

Additional user accommodations are dictated by the anthropomorphized computer relationship. For example, most computer programs require the fabrication of spontaneity—the impression that what is occurring has not been programmed. Unlike high risk/low predictability face-to-face relationships, person-computer relationships are predetermined, and articulated spontaneity is built in. This pretense of uniqueness is, of course, the basis of any interactive program in which branched responses and combinations are possible. Theoretically, the computer has a response or alternative for every one of our idiosyncratic inputs. It is, however, the spontaneity or nonrepeatability of the person-to-person dialogue that cannot be fulfilled.

Interpersonal communication is creative—unpredictable—precisely because of the symbolic potential of human language. The ambiguity of human language and the relational and situational attributes of the human dialogue allow unique and unpredictable outcomes. That is, human communication is a process of continual creation and re-creation of reality. In short, anthropomorphizing the computer is required because the user agrees to interact as though the dialogue were spontaneous and unique, thus producing the necessary person-to-person relationships so essential to interpersonal communication.

Computer as a Tool

Computer anthropomorphizing, though widespread, is not the only means of reducing the dissonance faced in communicating with a computer. An alternative is to take the position that the computer is "just a machine" (Turkle 1982). In this case the relationship appears to be that of a human being to a tool which extends human capabilities; the computer is an extension of the human brain and a high speed calculator.

If one views the computer as a machine that does nothing but what it is told to do, then the user must think of computer responses as messages to himself or herself from himself or herself. The user must accept the position, tacitly or directly, that he or she is both source and receiver of the produced dialogue. This appears to alter the sender-receiver relationship to one where the user is master of the machine. It does not, in reality, alter computer control of the dialogue. To interact with a computer the user must accept the computer on its own terms. That is, the user must use the computer's language and believe in its computations.

When people insist that their relationship with the computer is that of person-to-machine, it suggests that they are willing to accept what we call "inanimate dialogue." The tacit agreement in this case is that the user knows best and is always in control of the machine, but will go along with computer commands and responses in order to get what he or she wants. There still must be established a satisfactory sender-receiver rationale in the person-computer interaction in order to overcome the dissonance created when a person receives machine-made messages, that is, interacts with a machine which produces messages apart from any other human.

Control and Predictability

Whether one anthropomorphizes the computer or insists that it is only a machine, the computer exercises significant control over the dialogue and its outcomes. This is true because the human is forced to use the computer's language and the language of the computer insists on unambiguous information and predictable outcomes (Phillips 1983).

Despite continued progress in developing computers which can respond to human language in human-like language, and which have artificial intelligence that resembles human intelligence, computers do not engage in symbolic processing. Computers exist because they circumvent the unpredictableness of human symbolization. They reduce all information to on-off signals. They cannot be sidetracked. Once the inputs are received and transformed into binary signals, the outcomes are inevitable. Anything which does not fit is rejected. Anything which is accepted is made to fit into the binary mode. Thus, there can be no new meanings generated, no creative outcomes as the result of chance encounters. The whole purpose of computers is to eliminate such "human error"—creativity—and to speed the process of counting and calculating by eliminating those aspects of human communication which get in the way of reaching predictable conclusions.

Uniqueness of Human Communication

It is the ambiguity of human language and its metaphorical functions which make it possible for humans to act together in unique ways. It is also the ambiguity of language and the randomness of symbolizing which allows for emotional, even capricious and unpredictable interaction. A computer, on the other hand, is never capricious. Its language is digital—intelligible to computer circuitry. Computers reject the disorderly and the ambiguous. Computing is a process of regularization. If the computer is not able to respond to the question asked, the question must be changed. The computer is entirely logical. Even when computers randomly program, their randomness is perfectly random.

When the human chooses to, or is forced to, interact with a computer, the control of the dialogue shifts to the computer even though the human must turn on the machine and give it commands to make it function. The control shifts because the computer demands that the dialogue be carried on with its language and within its programmed parameters. As put by Professors Ducey and Yadon (1983) in describing the need for computer courses in the curriculum:

> Once students have accepted and become familiar with the need to state a problem and its solution in concrete, precise and unambiguous terms, the true creativity of computing is more easily appreciated (p. 6).

Even though humans program computers, it is not possible to program a computer to be reciprocal and spontaneous with its human counterpart.

The human, on the other hand, is more adaptable and can learn to func-tion digitally, can learn to be more like a computer. The implication is that as we continue to computerize facets of our society, interpersonal skills will shift from the management of human relationships and the formulation of intersubjective truth to the management of information and the arriving at mathematical certainty (Carey 1982).

The person-computer dyad resembles the interpersonal communication model, first because it is designed to emulate it by producing a responding mechanism which interacts with the user as a proxy for another human; and second, because it requires the user to implicitly agree that there is an interpersonal sender-receiver relationship functioning. It differs from the two-person dyadic model in that it cannot produce spontaneous and un-predictable outcomes.

The binary nature of the computer and its processing is the antithesis of human symbolic processing where senders and receivers produce meanings and solve problems based on the ambiguities of human language and the evolving dynamic interpersonal relationship. The fact that computers can perform extremely complex operations at high speed and have almost limitless memory makes them very potent tools in the extension of human power. They are, however, of a different order than past machines that have also extended that power. Computers force humans into relationships which have serious implications for the future. Armelle Gauffenic (1983), a specialist in systems analysis at the Ecole Supérieure de Commerce in Paris, has put it very well:

> As systems develop they gradually deprive the individual of interpersonal contact and communication, thus altering his affective psychological and emotional balance and isolating him in a relationship with the machine. While the nature of that relationship depends on the interface potential, the general rules obtaining are those of computer science; efficiency, rationality, the utmost simplification, etc. This new form of communication, this lan-guage of modern times, in which any concession to the superfluous—to mere form—is out of the question forces the mind and the intellect to conform to its mould if they are not to become impotent and maladjusted (p. 137).

Conclusion

The person-computer interaction is a reality. Dialogue with computers will increase as millions of personal computers are purchased and interface with mainframes and databanks. The implications of this mediated inter-personal connection are already altering our society—even before the effects have been realized. The computer brings with it extraordinary bene-fits: relieving people of tedious tasks, making possible massive storage of information and instant retrieval, solving medical and scientific problems in minutes which before would have taken a lifetime. But at the same time, disquieting consequences also loom on the horizon: speed and efficiency become the norm, mathematical processing replaces human symbolizing, individual isolation is increased, the frailities of human behavior become exaggerated.

Every technological innovation in communication alters the status quo and every such development results in the reallocation of communication priorities and values. This paper is not intended to denigrate or disparage the computer and its uses, but rather to stimulate research into the functions and uses of computers as human proxies. It is not the technology of the computer which is in question, but rather the process which occurs when the human and the computer interact, as they must. Social scientists have long been aware of, and have studied, the effects of the human-machine interaction in our postindustrial society. The need now is to study the consequences of a system in which machines and humans converse with each other.

References

Aronson, S. The sociology of the telephone. *International Journal of Comparative Sociology,* 1971 (September), 12, 12-28.

Carey, J. The mass media and critical theory: An American view. In M. Burgoon (ed.), *Communication yearbook* 6. Beverly Hills, Calif.: Sage, 1982.

Cathcart, R. and G. Gumpert. Mediated interpersonal communication: Toward a new typology. *Quarterly Journal of Speech,* 1983 (August), 69, 267-277.

Covvey, H.D. and N.H. McAlister. *Computer choices.* Boston: Addison-Wesley, 1982.

Dizard, W.P. *The coming information age: An overview of technology, economics, and politics.* New York: Longman, 1982.

Ducey, R. and R.E. Yadon. Computers in the media: A new course in the curriculum. *Feedback,* 1983 (Spring), 23, 6.

Gauffenic, A. Nineteen eighty-four: From fiction to reality. *Impact of Science on Society,* 1983, 2, 133-138.

Horton, D. and R. Wohl. Mass communication and para-social interaction: Observation on intimacy at a distance. *Psychiatry,* 1956, 19, 215-229.

Levy, S. Travels in the network nation. *Technology Illustrated,* 1983 (February), 56-61.

Millar, F.E. and L.E. Rogers. A relational approach to interpersonal communication. In G.R. Miller (ed.), *Explorations in interpersonal communication.* Beverly Hills, Calif.: Sage, 1976.

Miller, G.R. and M. Steinberg. *Between people.* Chicago: Science Research Associates, 1975.

Papert, S. *Mindstorms.* New York: Basic, 1980.

Phillips, G.M. *The rhetoric of computers.* Paper presented at the meeting of the Eastern Communication Association, Ocean City, Md., April 1983.

Turkle, S. Computers as rorschach. In G. Gumpert and P. Cathcart (eds.), *Inter/ Media: Interpersonal communication in a media world* (2d ed.). New York: Oxford University Press, 1982.

Vail, H. The home computer terminal—transforming the household of tomorrow. *The Futurist,* 1980 (December), 14(6), 52-58.

8

Mass Communication Theory and the New Media: Major Assumptions in Light of Technological Change

Harvey C. Jassem and Roger Jon Desmond

Mass Communication theories are based on assumptions relating to the structure and operation of media and the interplay between media and consumers. The authors delineate twelve major assumptions that provide the basis for much mass communication theory and examine how these assumptions or premises are affected by the so-called communication revolution. The media, the media institutions, and the environments in which they operate are all changing. It is suggested that while elements of mass communication theory will survive the communication revolution, there are enough changes taking place to make dangerous any reliance on existing media theory.

As we proceed through what is being called the communication revolution, the very nature of mass communication is changing. Were this not the case, there would be no need for labeling this era as revolutionary. It is likely that what we know about communication and mass communication will become dated and obsolete with each significant change in the discipline. This article calls for a reexamination of what we know about the phenomena of mass media and mass communication as delineated in the theories of media and mass communication.

Why Theory

There is an ingrained belief common among scientists, humanists, and perhaps most people we choose to call rational, that while we may understand little of life and the world around us, there are causes, effects, and systems operating. We covet universalities and generalizations that seem to explain or predict things and that appear to be confirmed over a period of time. With the understanding that comes along with such confirmed "truths" comes a reduction in uncertainty and greater opportunity to adapt to, or control, our environment. Such assumed truths are the grist of theory.

While there are many common uses of the term *theory*, Donohew and Palmgreen's (1981) view that "a theory is a tentative explanation to assist in understanding some small or large part of the 'reality' around us," and Kerlinger's (1965, 11) view that "a theory is a set of interrelated constructs (concepts), definitions, and propositions that presents a systematic view of phenomena by specifying relations among variables, with the purpose of explaining and predicting the phenomena," serve as good operational definitions.

For the purposes of this analysis, there are at least two senses of the term *theory*. The first sense is typified by Kerlinger's writing: Formal, "covering law" theories attempt to create predictive generalizations from the perspective of the philosophy of the social sciences. The second sense of *theory* in mass communication is the less formal set of operating assumptions used by media managers and programmers to assist them in the analysis of their professional activities. Such concepts as "least objectionable programming" theory and economic and programming formulae are subsumed under this second category. Both types of theories are employed in the analysis of media, and both will be discussed in light of the changes in the "new media."

Theories are as true and good as they are useful over a period of time and over different situations. While accepted "truths" or theories may change over time, the temporal quality exhibited by theories is not an indication of their insignificance. Not only do theories serve as explanations, they also serve as templates through which phenomena are viewed and interpreted, though the view they help to provide may be as limiting as it is enlightening. It is tempting to define phenomena as the applicable theory suggests while deviant findings may be disregarded as mistaken rather than regarded as worthwhile threats to existing theory. And because we use theory to make sense of the world, we may begin to see only those parts of the world that the theory accounts for. Theory, thus, is not only a result of our thinking and research, it helps focus and shape our thinking and research.

Thus, theories do more than explain and satisfy curiosity. They may be important agenda setters for those doing pure, social, or policy research and activity. They help suggest further questions, explanations, and theories. They purport to have a handle on some "truth." Yet, of course, theories are not inherently "true." They are functional for a finite period of time. That is not a criticism of theory, though the inappropriate assumption that theory and truth are synonymous would warrant criticism. It should matter little to the person who is curious about the appearance of the sun tomorrow whether the sun revolves around the earth or whether it just appears that way when the earth rotates on its axis. Either conceptualization (or theory) works equally well and allows believers to "know" that the sun will appear again tomorrow. Indeed, either theory could be used to explain why it is possible to keep time by the sun. In this example, when the questions became more complex, the two theories were not equally valuable. The latter theory was generally believed to better explain more phenomena, thus it was adopted as being correct and the former was largely abandoned.

Accepted explanations—theories—change. The changes may result from better measurement techniques and devices, more complex questions, changes in the systems or relationships being theorized about, etc. The history of mass communication theory has been fraught with changes and revisions. Whether those changes were the result of better "measurement" or of changes in mass communication itself remains to be seen.

Development of Mass Communication Theory

Mass communication theory arose with an assertion that there was something called mass communication in operation. That so-called mass communication was an outgrowth of mass media. The definitions and development of these terms are somewhat circular. In modern terms, mass media arose with the development of printing and efficient means of distribution. The printing press, improved transportation, and the concentration of people in urban areas made it possible for a person to cost effectively share within a brief period of time his or her ideas or information with people who were beyond earshot.

People had shared ideas with those out of earshot before. The works of artists, writers, playwrights, etc. had long been available for consumption by people not immediately conversant with their creators. It seems, however, that it was the reproduction of multiple copies of the message which allowed many people to receive it at about the same time in different places that, combined with the emergence of new economic and social forms (industrialization and urbanization), led to the "recognition" of mass media and hence, mass communication. Developing as they were at a time of changing economic and social arrangements, the new media and communication forms were subject to the analytical terms, tools, and theories of the day. Golding and Murdock (1980) suggest that the first research or explanatory perspectives of media had roots in "the theory of mass society":

> Social structure was seen as an amorphous mass surmounted by a dominant elite. Inequality was conceived as organized around an uneven distribution of power, not of property, and the mass as inhabiting an unbanished, industrialized world in which primary social ties were weakened and individuals had become susceptible, manipulable, and "atomised." With this picture of urban America researchers concentrated on the vulnerability to elite manipulation of the masses by the new and powerful tools of the modern media (p. 63).

Lowery and DeFleur (1983, 3-4) also point to the "term 'mass,' as used in 'mass society,' as the intellectual source from which we obtain the concepts of 'mass' media and 'mass' communication." Mass society referred to one in which social differentiation increased, informal social controls lost their effectiveness, the use of formal social controls increased along with widening differences, and "open and easy communication as a basis of social solidarity between people [became] more difficult. . . . It was this conceptulization of mass society that dominated the thinking of those intellectuals who were first concerned about the effects of the new mass media" (Lowery and DeFleur 1983, 10-11). And it was the effects that the media had on people that was the focus of most mass communication research and consequently most mass communication theory. Kepplinger (1979, 175) notes that "mass communications research . . . skipped over the first phase of every 'normal' science, which is the description of its phenomenon, and hastily took up the search for effects."

The search for the effects media had on the newly formed masses led to mixed feelings, but by and large confirmed the notion that media did indeed affect its audiences. Rigorous empirical research, borrowing from the tools and methods of more firmly established social sciences, began to be seriously used to examine media effects as early as the 1920s. Early findings (and theory) suggested that "the media had great power to influence every individual, more or less uniformly" (Lowery and DeFleur 1983). Theories such as the "uniform influences," "hypodermic," and "magic bullet" theories of media effects suggested that the media audience was composed of a mass of people who were similarly and directly affected by the media. This notion reinforced the concept that mediated communication, in which a message was constructed by a complex organization and disseminated (with the help of technology) to many people, was in fact a new form of communication—mass communication. Whether one accepts the proposition that mass communication started with the advent of the printing press or one traces its roots to pre-industrial architecture, pictures, stained glass, etc., notions of early, organized message-distribution functions generally acknowledge the presence and definitional importance of large, anonymous audiences being communicated to/with by an elite (or media elite) class.

These mass communication, mass media, and mass society theories were put to the test in such contexts as politics, advertising, and war propaganda. With the arrival of the mid-twentieth century, investigators found that observed and reported effects of mass communication did not conform to the then-existing theories of uniform influences (Lowery and De-Fleur 1983). The repudiation of "magic bullet" type theories was accompanied by the introduction of theories based increasingly on the assumption that audience members were distinct individuals, members of different important social groups, subject to many influences, and used the media selectively to serve their own needs or wants. Hovland et al.'s (1953) "individual differences" theory and Lazarsfeld et al.'s (1944) "two-step flow," both of which were introduced in the 1940s, suggested that audience members were not identical and that in studying media effects one would have to look at backgrounds, compositions, and relationships of and among audience members.

Reinforcement theory, play theory, selectivity theory, and uses and gratifications theory characterized media theory in the 1960s and 1970s. Social scientists shifted their focus from one geared to find the effects of media on their audiences to the ways individuals use media for their own needs.

The newer examinations of mass communication stopped conceiving of the media as monolithic. Not only might people have individual needs and wants, the media differed from each other and thus could not equally satisfy given needs or wants. Media form and content were seen to significantly affect or alter the impact of mediated communication.

As social scientists focused on the importance of the individual audience member and medium, the belief that media cause direct effects on audience members was generally replaced by a "limited effects" perspective. More recently, there appears to have been a shift back towards the media-as-cause-agents theories or models (National Institute of Mental Health 1982). As indicated earlier, such a change in perspective may result from different measurement tools and techniques, different philosophical/theoretical perspectives, different questions or thresholds, or simply the accurate measurement of changing effects (McDermott 1975, 83). There has also been more attention paid to the problem of documenting long-term effects. Speculation is mounting that the failure to find significant effects of short-term media exposure may not negate the possibility that there are effects resulting from cumulative long-term media exposure.

Theories of mass communication, while highly effects-oriented, have also attempted to deal with the other aspects of the system. In a 1979 article, Hans Kepplinger wrote:

> Increasingly in the last years, research on effects has been supplemented by research on causes. Here exist four different approaches: the "Grand Theory" approach examines the historical or economic requirements for the development and organization of the mass media. The supporters of this approach usually agree within the limits of a Marxist or structural-functional theory of society in which their assumptions and statements transcend the structural-functional classifications of the 1940's. The "News Value" approach investigates the requirements which phenomena must fulfill in order to become the subject matter of reporting. The "Uses and Gratifications Approach" inquires into the causes of media use, and not into the effects. The "Diffusions Approach" inquires into the sources and thereby into the causes of information, personal views, and behavior. Thus, the research of effects has extensively lost its formerly exclusive rank and has become, among other things, an equivalently rank approach (p. 167).

Legal research, historical research, and qualitative research have also gained stature in the field of mass communication. Accompanying the increased understanding that results from these newer mass communication research methods are newer noneffects types of theoretical perspectives. Often descriptive in nature, sometimes normative, they tend to focus on the structure and causes of what we call the mass communication process.

Assumptions of Mass Communication Theories

Stephen Littlejohn (1983, 264) has pointed out that "theories of mass communication are exceedingly difficult to integrate and to organize." But as valid as Littlejohn's observation is, there are a number of general organizing schema that appear to reflect the assumptions and perspectives of those doing mass communication research and theory building.

The following list of assumptions characterize the majority of media theories. While they are not exhaustive, they do recur with great regularity throughout most media theories.

The Audience Is Passive

Both explicitly and implicitly, the major theories of mass communication make reference to several dimensions of audience passivity. One sense of the inactive audience emerges from the work of Katz (1957) and others, concerning the tendency of audiences to ignore political and public health messages in newspapers and on electronic media. By positing a "two-step flow" and other mediating constructs, these theorists underline the importance of interpersonal communication processes of persuasion. The majority of theories are concentrated on apathetic audience members who rarely exhibit evidence of behavior change as a function of exposure to news, public affairs, or public service messages. The concepts of opinion leaders, Thayer's (1979) "epistemic communities," are invoked to illustrate the persuasive function of individuals, themselves informed by media, talking to other individuals.

Another dimension of audience passivity concerns the less formal theories of entertainment. Writers who focus on the mass appeal of network television frequently refer to the passivity of audiences in that members "take" what they "get," or that they generally will choose the least objectionable offering from one of three networks. Called the theory of least objectionable programming, the notion that audiences passively accept what they are given pervades the literature concerning entertainment programming. Data from rating services document the general tendency of viewers to stay tuned to one station or network. There are numerous references in the programming literature to programmer's inability to overcome this inertia.

When samples of audience members are asked what they would prefer to the offerings of the networks, they generally are unable to articulate any novel categories of programs which do not closely resemble their favorites. Other theories, such as the formal generalizations of diffusion of innovations, also suggest a passive audience (Rogers and Shoemaker 1971). The most firm conclusion of the majority of speculations is that mass media are sometimes significant sources of information about ideas and products, but are not extremely important as initiators of adoption.

The presumption of a passive audience is also implied in regulatory philosophy that requires each individual broadcast station to operate as if its audience depended solely on it. Each station is expected to provide a wide range of entertainment and information programming, as well as contrasting points of view on controversial issues to insure that listeners of a single station get a potpourri of news and entertainment.

Much media attendance appears to be a function of habit. People watch television or listen to radio at approximately the same time daily, regardless of what is being aired. They decide whether to read a newspaper or magazine, then subscribe to the one(s) they want. They go to the movies during holiday periods or for their weekend dates. Extraordinary news events or film releases may temporarily increase media consumption. Television viewing regularly decreases during the summer months.

Delayed Feedback

Media generally deliver messages one-way to the audience. The responding audience member must turn to the letter, telegram, telephone, pocketbook, or ratings services to voice his or her feedback. The separation between producer and receiver is great. It may involve distance and time separation and is enhanced by the essentially one-way delivery systems media employ. This separation has implications for the content of feedback as well. Mass media feedback is limited in the sense that letters to editors, editorial rebuttals, and audience ratings are vague, abbreviated, and often imprecise in comparison to the instantaneous and multifaceted, emotional and informative dimensions of conversations.

Simultaneous Reach

Most media in America are designed to be distributed throughout their primary reception areas in such a way as to be available to all or most consumers at approximately the same time. Broadcast programming strategies rely on the assumption that programs are available to consumers only as scheduled. Network schedulers do not merely try to compete against the program offerings of the other networks, they compete against specific programs, which is possible only because schedules are set by the media and hence are universal. For marketing and promotional reasons, most other media release their fare simultaneously. While it may well be the case that people in the media businesses set their outlets' schedules to best suit audience wants or needs, ultimately it is the producer, not the consumer, who sets the media's time schedule. In other words, the consumer has to see the film according to when it is playing in the theater, buy the newspaper or magazine when it is on the newstand, buy the book when it is in print, and watch or listen to the broadcast programs when they are aired.

The notion of simultaneous media reach affects more than distribution and consumption strategies. It serves as an important underpinning of such theorizing as McLuhan's (1974) concept of the "global village," where large societies become like traditional villages in that information about many aspects of society is available to everyone. Theoretical constructs like "agenda-setting" and Gerbner's cultivation theory depend partially on the idea that information and entertainment wash over societies in simultaneous waves (Gerbner et al. 1980).

Mediation by Gatekeepers

As early as the 1920s, references were made in the literature on communication to the gatekeeper who edits, selects, or emphasizes certain parts of mass-mediated fare for audiences. In the 1950s, White (1950) and others built theories around the predictability of gatekeeping behaviors, and the social constraints which operate on persons and institutions in the process of dissemination.

The structure of traditional media systems has been such that gatekeeping is inevitably a source function. Gatekeepers control the spigot of content at the source, with no gatekeeping by consumers of messages. An implication of the assumption of mediation is that messages can never be personally tailored for individuals, since gatekeepers necessarily act on stereotypic notions about intended audiences.

Audience Members' Choice of, and Support for, Entertainment and News Fare Is Predicted by Demographic and Psychological Factors.

The category of theories labelled "individual differences" theories of media effects grew out of the work of Janis, Hovland and others in the early 1950s (Janis 1962). Searching for general factors which would predict persuasibility, these psychologists studied sex, age, education, and other demographics to determine their role in the media persuasion process. Many theories of mass communication still include demographic variables.

Early network programmers needed audience descriptors in order to convince advertising clients to buy time on the networks. Audience demographics were convenient devices for describing audiences to advertising agencies and their clients and became the industry standards for targeting the mass audience. Reliable information is vital to programmers, so they can schedule programs for optimum commercial exposure. Radio, television, and to a lesser extent newspaper management personnel inevitably employ age and sex descriptors to talk about their audiences and readers.

Receivers Are Anonymous

The majority of media theories are based upon conceptualizations of an anonymous audience, in the sense that people do not know each other and thus cannot share their impressions of programs or articles with one another in meaningful ways. Their heterogeneity prevents associations which they might form to achieve some end or purpose in reaction to media stimuli.

Eliot Friedson (1953) has pointed out the faults of this conceptualization in his frequently cited critique of the assumptions of the mass audience, by pointing out such collective activity as the social behavior surrounding film attendance. Nevertheless, the image of a heterogeneous, disorganized audience continues to dominate theories of mass communication.

Audience Members Are Able to Articulate the Functions Media Serve for Them

A large body of theoretical literature and research concerning the uses and gratifications surrounding audience behavior emerged in the 1970s. Growing out of the functionalist theories in sociology, the focus of this approach has been on the specification of individuals' motives for selecting certain channels and content, and the extent to which desires and needs are fulfilled by various media. With few exceptions, the theoretical writings in this tradition suggest that people are capable of categorizing media fare in terms of its utility for them. One of the most important implications of this assumption is that people are consciously aware of the ways that media operate in their lives, and that people deliberately choose media to gratify needs they can specify.

Concern with Effects of Receiving Messages, in Terms of Knowledge Gain, Behavioral Change, or Attitude Change

One implication of this aspect of media theory is that a large (perhaps the largest) dimension of media behavior has been nearly ignored by theorists: the inherent, affective pleasure which surrounds the reception of entertainment. From theories of newsreading to the enormous literature concerning television and social behavior, the assumption is that the most salient aspect of mass media for social scientists is their function as information "machines." When entertainment has been relevant for study, theorists have generally ignored affective pleasure or disappointment as important elements in media systems. Two exceptions to this assumption are the work of Stephenson (1967) in play theory, and the research program of Zillman (1982) and his colleagues in the synthesis of Aristotelian and psychological theories of entertainment. Even the inherent pleasure involved in a large range of choices among and within media, and the rewards involved in choice making, have been relatively ignored in theories of mass communication.

Minimizing the Role of Interpersonal Communication in the Selection of Media, Attention, Evaluation, and Subsequent Effects

Although the assumption which pervades the literature is that mass communication is different from interpersonal communication, few theories have addressed the complex relationships among these phenomena. Theories such as observational learning, cultivation, uses and gratifications, and many others have concentrated on the relationships between individuals or aggregates and mass media fare, to the exclusion of the role of conversations between people in these processes. This concentration, like many of the others described earlier, arises from conceptualizations of mass media as institutions which serve large, anonymous audiences. An exception to this assumption is the theoretical work in the diffusion of innovations, where media and individuals are conceptualized as interrelated (Rogers 1973).

Mass Media Fare Is Expensive to Produce

The production of mass communication messages generally requires large institutions and sophisticated equipment. It is a very costly enterprise. As a result, those persons without sufficient financial resources cannot generally get access to the media. Additionally, content of mediated messages must be designed to satisfy the interests of those financing the enterprise.

Media Have Geographic Attributes

Distribution patterns for specific media types have generally had geographic dimensions. Most geographic distribution patterns have been a function of technology but have had content and usage implications. Newspapers and broadcast stations have served as local general-interest media. The localism attribute of radio and television stations has not only technological roots but also is required by FCC statute. Newspapers have been local organs primarily because of the problems they would encounter in distributing their product over a large distance in the brief, timely fashion current news requires. Magazines, books, and films, which needn't be quite as time bound, tend to be distributed nationally and as a result their content has no local focus.

Limited Number of Channels

In television, and to a lesser extent in radio, it has long been accepted that there are a limited number of stations available in any given area. Most American television markets have only three commercial and one noncommercial broadcast television signal available. This situation exists for both technical and economic reasons and has implications for programming, economics, and regulation.

Facing little competition and an audience with few broadcast alternatives, programmers neglect even substantial minority tastes in order to appeal to larger audiences by offering so-called lowest common denominator fare. Programs that garner shares of fifteen are not likely to be scheduled when lower common denominator programs could earn shares of thirty.

The limited number of outlets provides the justification for ownership duopoly regulations; the "no more than one-station-per-market" rule makes sense when the number of stations (and hence potential owners) is limited. Similarly, regulations pertaining to content (including the Fairness Doctrine and the Equal Opportunities provision of the Communications Act) are justified on the basis of spectrum or channel scarcity.

Another implication of this sort of limited market is that in such a closed market, prices for commercial time may also be unnaturally high (U.S. v. N.A.B. 1982).

While the limited-channels attribute is most noticeable in the broadcast media, to a lesser extent limited channels or outlets are present in many other media forms as well. Consider newspapers, for example. Only two percent of American cities and towns have more than one local newspaper. And while cable television supplies more channels to consumers, nearly every American cable system is a local cable monopoly.

The Assumptions Discussed

These general assumptions have characterized theorizing about mass communication and media since the beginning of the so-called electronic age. Textbooks, academic journals, and popular writings are so replete with them that it is difficult to imagine unique treatments of the process of mass communication without the majority of these "truths." The following discussion considers these assumptions about mass communication in light of the new technologies of the "communication revolution."

The Passive Audience

One major change brought about by the new media is sheer diversity; increasing adoption of cable television, for example, means more choices in entertainment than were available before cable. While viewers may not exercise all of their optional channels, the typical cable organization offers them a minimum of three times as many options as they had in broadcast television. While estimates of cable erosion into broadcast channels vary, there is no refuting recent evidence for marked declines in audiences for network television.

If even a minority of novel options are selected, viewing of television becomes a more active process. Viewers may have several choices of programming within a given category, a choice of two sitcoms where only one was available before, and may have to choose in a deliberate manner, rather than accepting the program from an available category.

If passivity is a function of an unwillingness on the audience member's part to get up out of the arm chair every thirty minutes to check the television listings and change the channel, perhaps such a viewer would be willing to sit down once a week and program his or her video recorder to tape an assortment of programs which could later be viewed passively (without having to periodically get up to change programs). One wonders whether such behavior should be categorized as being active or passive.

When we mean by *passive,* the audience member's reluctance to act as a result of a media message, such passivity may be alleviated, at least in part, if acting is made easier through communication advances. The passive viewer, who might not have been willing to rush to the hardware store or even to go to the phone to call a toll-free number in order to purchase said widget, may purchase said widget if to do so all he or she needs to do is press the "buy" key on the hand held remote key pad. Thus, we may find that there is greater physical passivity but more selection in the viewers' future.

Since each cable franchise offers a community access channel, more viewers than before have the option to become program writers and directors. While only a minority of consumers will ever utilize the access channels, their presence offers more possibilities for activity than have been available in television's first three decades. Also, when new media are combined, viewers have geometrically more opportunities to become active and selective. The combination of a video recorder and cable television, for example, offers the possibility of making choices available at any time, allowing individuals to choose when, instead of what, to watch. Such behavior would critically alter the least objectionable program aspects of audience selectivity.

With respect to the tendency of audiences to fail to respond to information-persuasion campaigns, it is unlikely that any of the new media will drastically increase audience behavioral responses. Recently, with the advent of "900" telephone lines and interactive cable systems (QUBE, for example) viewers have, in the millions, voted for favorite comedians and the like, but the extent to which this represents active behavior is arguable.

An extremely important direction for future research will be the investigation of how the use of new media will initiate change in consumer activity levels, especially with respect to the exercising of choices. Will the increased offerings available to cable subscribers precipitate an increase in program sampling, or will the cafeteria of options constitute an unmanageable information overload? Will the use of storage devices facilitate individual scheduling of programming-for-oneself, or will they ultimately reduce viewers' explorations of availabilities? Most significantly, will the use of some of these devices affect users perceptions of activity and involvement, and if so, what might the implications of such altered perceptions be?

Feedback

If we mean by feedback any measure of audience reaction, response, or behavior resulting from a message, expanded interactive media may enhance the opportunities for audience feedback. Audience behavior, as measured in media consumption and ratings, can be more easily monitored with interactive media. Consumers who electronically order particular entertainment or information offer fairly direct feedback. Interactive cable systems have the capability of polling subscribers and tabulating responses instantly. These new methods of feedback appear to put in jeopardy the definitional quality of delayed feedback. Yet closer examination reveals that the type of feedback that lends itself to instantaneous retrieval in the new communication age is of the most limited and primitive type. It would hardly help clarify communication. While interactive cable systems could poll subscribers on an unlimited number of issues and tally the results by individual subscriber, such polling would be constructed by the message sender, not the receiver. Further, subscriber participation would probably require real-time attendance and would thus be inconvenient. Similarly, such polling has not been, nor is it likely to be, extensively used.

For the media audience to respond meaningfully, traditional (and delayed) forms of feedback such as letters, phone calls, or interpersonal communication may prevail through the media revolution. While electronic mail could speed up the letters, more distant program or content sources may make the relevant party to whom feedback is addressed more difficult to locate. Thus, while the most simple forms of feedback may be quickened by the new media, more substantive forms of feedback are likely to remain more delayed and difficult than they would be in interpersonal communication settings.

While this analysis considers interactive media, it excludes fully and equally interactive mediated communication such as video and audio teleconferencing. Such extensively interactive media, where all parties are equally responsible for and capable of sending and receiving messages, are not forms of mass communication despite their use of television technology. It may well be that feedback-eliciting systems have value only for viewers or consumers who gain satisfaction from a feeling of involvement in the communication process. Future research regarding feedback should address these feelings and perceptions: Do digital feedback systems offer substantial returns in user's feelings of involvement, or are they viewed as extraneous and unnecessary by viewers? Is the activity of providing limited feedback intrinsically rewarding, or, as in the case of qualitative audience rating systems, is some type of incentive needed to provide inertia-overcoming audience activity?

Simultaneity

New media allow for both more and less media simultaneity than ever. Through satellite transmissions, the entire United States can receive a plethora of identical media messages simultaneously. Yet, because consumers cannot (or may not want to) attend to more than one medium at a time, recording, storage, and delayed access devices are being developed. Thus, more simultaneity is possible due to the increased choices available and the taping/storage/retrieval options available, and audience members may no longer attend media presentations in a time-joined fashion. Some content may be more prone to viewer-initiated delay than others. Dramas, situation comedies, and other essentially time-free content store well; news does not.

The storage and random retrieval of media messages strains the widely accepted premise that media messages reach audiences simultaneously. That premise is a basic assumption of the notion of reference groups in Thayer's (1979) epistemic communities. These groups are composed of friends, kin, or co-workers, who influence each other by focusing on issues that they learn about though media, and who, in turn, are directed to certain information and entertainment via conversations. These conversations then become components of the agenda-setting function of media. An assumption inherent in this line of thought is that these conversations take place soon after exposure to media messages, and further, that many such groups are having similar conversations following exposure to the same media messages. If simultaneous media message consumption is adversely affected by the media, these theoretical concepts will need modification.

An empirical question which will be important in the investigation of the processes is: How will the "checkerboarding" of simultaneously transmitted programming affect personal agendas? Will people continue to discuss news and information to the extent that they do now, when exposure is individually controlled?

Gatekeeping

The gatekeeping function and the role of the gatekeeper are changing. Traditionally, media content has been selected and arranged by people working for the media. Not only are gatekeepers' jobs affected by issues of greater channel choice and simultaneity, the gatekeeper faces a new world in which many of his or her traditional roles will be taken over by consumers. Audience members have increased ability to record and store information and, hence, can select and schedule such to suit their own needs. While the gatekeeper is still likely to edit content from many sources, now more than ever so too will the consumer. In a sense, the consumer's world of information and entertainment may not be so much limited by gatekeepers as organized by them. Indeed, with the increased potential for consumer overload, the organizing function of gatekeepers may become more critical than ever.

One new development pertaining to gatekeeping is the enhanced possibility of some gatekeeperless forms of mass communication. There exist at least two forms of mass media which require no gatekeepers. Public access television, where persons make their own television programs and air them on a first-come-first-served basis is one. Another involves emerging interactive databases where users deposit and/or call up information on their own.

As consumers enhance their gatekeeping function by taking a more active role in selecting, scheduling, and providing information, the role of the media gatekeeper is not likely to be eliminated. The resultant changes in the roles of gatekeepers and consumers need to be further examined. With more choices available to consumers, some of which may include gatekeeperless options, might audience members' tastes and expectations change in such a way as to affect the standards of the remaining gatekeepers? Might a plethora of unedited choices result in a consumer backlash where more gatekeeping or editing is called for?

Individual Differences

An extremely relevant issue in the segmentation of audiences arises with the new media: As channels multiply, so do audience types with respect to the combinations of media that people will use, and the uses to which they will put them. Adding numerous cable channels, computer networks, and recording devices to individuals' repertoires will inevitably create types such as storers, who frequently record, or music viewer/listeners, who primarily seek music channels and stations, and many other types. It is possible that the *uses* of media will become a more accurate method of describing audience behavior than the demographic groups now employed.

In the past, one major rationale for these demographic descriptors was that they represented target groups of consumers, convenient for advertisers to segment. If various types of media users (many of whom edit out commercials) continue to emerge, they are not likely to be described by age, sex and occupation.

Advertisers and programmers who are interested in reading people will need to identify user types and learn more about how "type" members behave as consumers. This research effort may eventually benefit managers and their marketing staff as much as the application of psychographic profiles have been of use to the magazine industry. There is every reason to believe that these data will also shed light on the questions addressed by scholars who operate from the uses and gratifications paradigm; information will emerge concerning patterns of use across various media.

Anonymous Receivers

The rise of certain new media means that the "to whom it may concern" quality of writing and programming, characteristic of many products, will change, becoming more personalized and localized than ever before. Cable access channels currently offer customized programs to neighborhoods and various minorities. Many of these also offer audience participation, in terms of panels from assorted local groups and program input from viewers with special interests. On many access channels, virtually anyone in a community may participate in program design and production. Low-power television, with it's emphasis on local control, may offer similar opportunities.

While this development will not change one aspect of anonymity—that audiences are still anonymous to large-scale programmers—audiences of access channels will become known to each other in ways that previously were not possible. Through the process of seeing fellow community members on television, viewers will perceive new possibilities for the medium.

With improved feedback mechanisms, the receivers' anonymity will become less certain than it now is. As audience choices are recorded (for billing purposes, and so on) producers/suppliers will know more about their audience than was possible before. As audience members become involved in interactive media services, they may get to know a limited amount about their co-audience members. It is unlikely, however, that the masses will know many more audience members than they currently do. They may know more about their fellow consumers in an aggregate sense (the results of instant telephone polls may be commonly available, for example), but this will not substantially change the anonymous nature of being an audience member. With increased addressability, however, media producers/providers may more effectively reach particular individuals or particular groups of audience members. It will be relevant to examine just how specifically media can be targeted and what the incentives and disincentives are for such targeting strategies. It is also of some importance to determine whether the increased addressability affects the receivers' perceptions of their importance or of their role in the communication situation. What, for example, is the difference to the consumer between getting direct mail addressed to "Dear Friend" and getting the same mail personally addressed?

Functions Can Be Articulated

How the new media will enter into the lives of people will depend upon a number of social and economic conditions which must exist if certain media, currently on the research and development horizon, are to survive. In order to know functions, people would need to be aware of the ways in which such things as low-power television transmitters give them new options for creating programs. Functions will emerge after these media take their place in consumer acceptance.

Consider one classical mass media function: surveillance of the environment. Traditionally, uses and gratifications research has offered generalizations such as the observation that the younger one is, the more one tends to rely on electronic news to find out what is developing in the country, as opposed to newspapers, the surveillance choice for older Americans. These findings are typically based on responses to close-ended survey items originally derived from essays on topics such as "How I find out about recent developments in the state legislature." In the final analysis, the functions of mass communication become the proportion of positive responses to such standardized items from any given sample. No novel, unique, or idiosyncratic functions can possibly emerge from such data.

A new programming concept is embodied in the cable-cast music television (MTV) channel, offered by Warner-Amex Communications and several other companies. Concert clips, fantasy scenes, and skits of three to five minutes in length accompany the audio track of popular music. The channel is being rapidly adopted where it is available, and some observers and critics have suggested that it is altering the manner in which people are listening to recorded music. When they hear a selection on record or radio, their thoughts presumably turn to the video of the song, or to what a video might look like to accompany a song for which no videotape exists.

If music television does, if fact, serve these visualizing music functions, a new, previously unarticulated function of a novel program concept has been born. How could traditional research methods discover and assess this phenomenon using paper and pencil tests? Presumably, many novel functions such as this surround and emerge from the new media.

Music television is a programming concept which emerged from the new channels made possible by cable television. New functions emerged from the program, and it is likely that the future will offer numerous channels in search of such content, with undreamed-of functional possibilities. The specification of emerging functions will be an important research priority in the next decade: What uses will be born from such innovations as holography, home-accessible databases, flat-screen television, pocket television, and numerous other new media?

Emphasis on Knowledge Gain, Behavioral and Attitude Change

Many of the new media have been implemented in order to entertain people. The multiplication of channels by cable television has yielded numerous entertainment options, and a few information options, such as Cable News Network. The question of the inherent pleasure in these options will become a part of communication theory in the next two decades. One reason for this interest is that the gatekeepers for cable systems will need to determine the economic viability of some of the offerings, and in the past, many theoretical issues in mass communication have emerged from such pragmatic concerns.

One channel on some cable systems now features video games, so players are able to use their video game systems or personal computers to play, and possibly store, new games. In the past two years, the commercial video game industry has amassed more leisure money than has the film industry. Entertainment is and should be as important to communication theory as it has been to the economy. Certain services, such as home ticket purchase for concerts and athletic events, are now offered through several teletext systems. The extension of information about entertainment thus becomes a domain of mediated entertainment.

It is not likely that theory and research into information and persuasion will be deemphasized in light of the new media. The point is that, given the enormously increased options for entertainment provided by new media, the play dimension of these activities will become increasingly relevant for theorists.

As the lines separating information and entertainment shift, perspectives on studying effects should also change. While some services are developing as distinct information (CNN) or entertainment (HBO) services, the production values and content which previously made distinctions between information and entertainment reasonably easy to recognize are changing, and such labeling is becoming more difficult. It may be increasingly important to understand how audience members perceive various media choices and what about those media alternatives lends them to be used/perceived as they are.

Interpersonal Communication and the New Media

In many ways, new media will emphasize talk between people as an important mediating variable in the communication process. A recent study found that interpersonal communication processes were important in helping people to become aware of specific options available on cable television (Bezzini and Desmond, 1982). Of course, interpersonal diffusion has always been a channel for information about new products, but the current wave of new media offer many topics and opportunities to "keep up with the Joneses."

Video teleconferencing has opened up opportunities for meetings and committee work. A likely development is that it will be used for two-person conferences as well. Usually, whenever technology has been available, it has been used for other reasons than were originally intended. Although video telephone conversations may not qualify as true interpersonal communication, it is difficult to specify reasons why. There is no inherent reason to argue that technological mediation through video disqualifies communication from being interpersonal, and such extensions as hearing aids and contact lenses have become part of the process for a long time.

Finally, conversations between people will be important in the "wired city" envisioned by futures forecasters such as Toffler. It may become important for people who work in information processing jobs at home to seek conversations that fulfull their basic needs to communicate with others. Because of the use of teleconferencing, the social function of traditional conventions may be a loss that people mourn. How will they chat with colleagues in informal ways? It is possible that the function of this social dimension of networking may be so important that it will inhibit the adoption of teleconferencing for some groups. While some writers have argued that new technologies may threaten interpersonal communication, it is possible that our need for face-to-face interaction will threaten some aspects of the adoption of new technologies. How these channels will accommodate to each other should be of as much interest to future theorists as how the telephone modified interpersonal communication.

Mass Communication Is Expensive

The traditional notion that mass communicating involves substantial expense is not likely to change much as a result of the communication revolution. While there are new economics resulting from technological advances (satellite time-sharing, for example), McLuhan's statement that "whereas Gutenberg made everybody a reader, Xerox makes everybody a publisher" (McLuhan 1974, 52) is overstated; in all but a few cases, reaching a large audience will continue to require a substantial investment. Some technological advances make it economically possible to produce heretofore inordinately uneconomic ventures (such as specialized networks, specialized magazines, etc.). But with the exception perhaps of cable access (which relies on expensive facilities), the generalization that mass communicating is considerably more costly than interpersonally communicating remains valid. The implications of the exceptions do, of course, merit examination.

Geographic Attributes

The characteristic geographic attributes of media are currently experiencing stress. While local media and markets will likely survive, particularly with the use of distance-insensitive satellites for distribution, many media are becoming national (or larger) in scope. Newspapers, for example, which had traditionally been local due to the timely nature of their content and the inverse relationship between time and distant delivery, now have among their ranks national editions of *The Wall Street Journal, The New York Times,* and *U.S.A. Today.* The unleashing of direct broadcast satellites will make simultaneous reception of shared television signals possible throughout North America without the contraints of cable television wires or local coverage patterns of traditional television stations. Satellites will likely encourage more national distribution of media and will make long distance electronic mail and telephony more efficient and affordable.

In the future, distribution of information and entertainment may have little to do with physical distribution or transportation systems or with the cities they built. A person's media access or richness may have more to do with his or her satellite terminal than with his or her proximity to a large city or market. Such changes may significantly affect content, advertiser options, consumer habits, and audience member affiliations or identification. If identical information, entertainment, and commercial resources are available throughout the country, or even internationally, the concept and importance of one's home community may change. It is conceivable that in such an environment the population might feel less defined by its geographic boundaries. Conversely, people may find themselves with a greater need for meaningful distinctive characteristics to turn to as the embodiment of their home.

Limited Number of Channels

In media, the "limited number of channels" notion may be more a measure of perception than of fact. It would be difficult to deny that there has long existed an abundant array of media to choose among in the United States. Some media choices (local newpapers and broadcast stations, for example) are easier to receive than others (distant newspapers or broadcast stations). One change brought about by the communication revolution is that some of the new media technologies such as cable television, satellite-delivered television or newspapers are making it rather simple to receive what had heretofore been difficult.

Accompanying the increased availability of media outlets may be some consumer confusion. Reading television listings that cover thirty or more channels (rather than the more typical half-dozen or so) and learning the more complicated tuning devices cable television brings may affect audience perception of the number of channels or choices available. Objectively measuring the amount of choice available is becoming a serious problem. Until recently, television meant three commercial networks, one public service, and, in large markets, one or more independent stations. Counting the number of television channels available provided a rough measure of the television choices available. Cable television changes that when many of the channels it carries duplicate (sometimes simultaneously, sometimes not) other offerings on the cable.

Similarly, how does one measure the impact video cassette and disc machines have on channel choice? One such machine can be used to exhibit an almost limitless number of movies, programs, information, etc. produced and supplied by sources running the gamut from networks and major film studios to universities or private entrepreneurs. Perhaps it would be expedient to categorize video discs and cassettes apart from television when measuring the number of channels or alternatives. Such a decision seems arbitrary, however. No matter how one categorizes these machines, it will be difficult to assess their precise impact on the "limited number of channels" issue.

As difficult as it may be to know precisely the extent to which choice and diversity are changed by the new media, it is clear that both choice and diversity are increasing and that the audience is taking advantage of many new media alternatives. This has serious implications for theories of audience socialization, media programming, media economics, and so on, all of which were based on a premise of limited channel choice.

Comments

Theories and conceptualizations are based on assumptions. The assumptions and the attendant theories must be subject to challenge and change as our understanding of the studied phenomena changes or as the phenomena themselves change. In the area of mass communication it is clear that many of the traditional "truths" are changing. The media, the media institutions, and the environments in which the media operate are all in need of serious reexamination. This study indicates that while elements of mass communication theory will survive the media revolution, there are enough changes taking place to make dangerous any reliance on existing media theory.

Subject to redefinition are the media practioners' perceptions of their roles, the relationship between audience members and media, and media economics and regulation, to name a few. We may integrate media more fully so they become more a part of both work and play. There will be new uses and new users of media. There will be new producers as well. James Curran (1982) suggests that these changes will, at the very least, change the power structure in society. It is imperative that we keep a vigilant watch on new media developments. Our frame of reference should be a wide one, not limited in scope to the values and templates old mass communication theory provided. Newer and more appropriate theories will assist us in understanding, coping with, and shaping the new media age.

References

Bezzini, J. and R. Desmond. Adoption processes of cable television. Paper presented to the International Communication Association, Houston, 1983.

Curran, J. Communications, power, and social order. In M. Gurevitch, T. Bennett, J. Curran, and J. Woolacott (eds.), *Culture, society, and the media.* New York: Methuen, 1982.

Donohew, L. and P. Palmgreen. Conceptualizing and theory building. In G.H. Stempel and B.H. Westley (eds.), *Research methods in mass communication,* Englewood Cliffs, N.J.: Prentice-Hall, 1981.

Friedson, E. Communications research and the concept of the mass. *American Sociological Review,* 1953, 18, 313-317.

Gerbner, G., Gross, L., Morgan, M., and Signorielli, N. The mainstreaming of America: Violence profile no. 11, *Journal of Communication,* 1980, 30(3), 10-30.

Golding, P. and G. Murdock. Theories of communication and theories of society. In G.C. Wilhoit and H. deBock (eds.), *Mass communication review yearbook,* Vol. 1. Beverly Hills, Calif.: Sage, 1980.

Hovland, C.I., Janis, I.L., and Kelly, H.H. *Communication and persuasion.* New Haven: Yale University Press, 1953.

Janis, I. An experimental study of psychological resistances to fear-arousing communications. *Journal of Abnormal & Social Psychology,* 1962, 65, 403-410.

Katz, E. The two-step flow of communication: An up-to-date report on an hypothesis. *Public Opinion Quarterly,* 1957, 21, 61-78.

Kepplinger, H.M. Paradigm changes in communication research. *Communication,* 1979, 4, 165-181.

Kerlinger, F.N. *Foundations of behavioral research.* New York: Holt, Rinehart, & Winston, 1965.

Lazarsfeld, P.F., Berelson, B., and Gaudet, H. *The people's choice.* New York: Duell, Sloan, & Pearce, 1944.

Littlejohn, S.W. *Theories of human communication.* Belmont, Calif.: Wadsworth, 1983.

Lowery, S. and M. DeFleur. *Milestones in communication research.* New York: Longman, 1983.

McDermott, V. The literature on classical theory construction. *Human Communication research,* 1975, 2, 80-93.

McLuhan, M. At the moment of Sputnik the planet became a global theater in which there are no spectators but only actors. *Journal of Communication,* 1974, 24(1), 48-58.

National Institute of Mental Health. *Television and behavior,* Vol. 2. Washington, D.C.: U.S. Government Printing Office, 1982.

Rogers, E. Mass media and interpersonal communication. In I. de S. Pool and W. Schramm, *Handbook of communication.* Chicago: Rand-McNally, 1973.

Rogers, E.M. and Shoemaker, F.F. *Communication of innovations, a cross-cultural approach.* New York: Free Press, 1971.

Stephenson, W. *The play theory of mass communication.* Chicago: University of Chicago Press, 1967.

Thayer, L. On the mass media and mass communication: Notes toward a theory. In R. Budd and B. Ruben (eds.), *Beyond Media: New approaches to mass communication.* Rochelle Park, N.J.: Hayden, 1979.

U.S. v. National Association of Broadcasters, D.C.D.C., 553 F. Supp. 621 (summary opinion of Judge Harold H. Green, 79-1549), November 23, 1982..

White, D.M. The "Gate Keeper": A case study in the selection of news. *Journalism Quarterly,* 1950, 383-401.

Zillman, D. The anatomy of suspense. In P. Tannenbaum (ed.), *The entertainment functions of television*. Hillsdale, N.J.: Lawrence Erlbaum, 1982.

9

The Political Economy of
Database Technology

Irving Louis Horowitz

This article examines the limits of a databased technology in a period that is beginning to move beyond the euphoria of information computerization. Without minimizing the extraordinary significance of a computer-oriented databasing, extending to the rationalization of available information for the purpose of serving multiple ends, it is clear that the marketplace has become increasingly reflective about databased technology. Every type of issue from the free flow of information, and the high costs of advanced technology to generate sophisticated information for few users is now under review. The political economy of technology makes it plain that advanced techniques or the complex processing of data does not in itself obviate traditional concerns of cost, demand, and affordability; quite the contrary, such considerations have become even more central precisely because start-ups are fiscally costly. Thus for some time to come there will be a multitiered informational environment; increasingly subject to fiscal scrutiny. In addition to market forces the author asserts the existence of behavioral patterns that tend to slow down the rapid implementation of a computer-based informational environment. Neither a crude empiricism nor a negative moralism is a realistic position. The pattern is one of absorption of new procedures by a constant reassessment of the costs and continued utility of the old technology.

The rapid evolution of the new information technology has brought about a revolution in the way we perceive data. We have acquired a positive perspective on the actual and potential uses of computer-manipulation of information to reach higher standards in the assessment, evaluation, and use of social science products and activities. Such a perspective generates, if

149

only indirectly, high expectations for improved control of decision making and evaluation of the consequences of policymaking. How realistic these expectations are, given the state of the art in new information technology, is the subject of this article.[1]

Technological developments have served to highlight central questions about information in general: What data are critical? How accurate do they need to be if we are to draw conclusions? How can limited resources be used efficiently? How should data be collected and disseminated? Who should pay for the gathering and use of statistics? (Montagnes 1981). While these are pressing matters, my focus here is on the specific issues generated by a socioeconomic environment in which computerized databases have become common. It is sufficiently challenging to identify the central contradictions between a databased statistical environment and a socioeconomic marketplace which is still evolving in its responses.

In the publishing industry, electronic databases were initially seen in a *unilinear economic fashion*, that is as a product transferred from seller to buyer, and as a mechanism for moving beyond hard copy to diskette or cassette rendering of basic information. The revolution in software came with *optimal scenarios* extended to include a shift from a "hard copy" society to the "automated office." When such ultimate expectations turned out to be, more or less, exercises in hyperbole, foundering on the limited market demands for such a total overhaul, the thrust of databasing shifted to a more imaginative framework: how to use high information availability to develop better policy, management, and evaluation procedures.

Nature of Databases

There is nothing qualitatively new in the creation of databases as such. Indeed, databases and online searching have come to be viewed as analogous to scientific inquiry in general (Herter 1984). Excellent files on a wide variety of subjects have been collected and maintained for many years in manual, hard copy form. Even the storage and updating of materials maintained in hard copy and disk form cannot be viewed as particularly revolutionary; microfilming has been with us for some time. What the computer gives us is multitiered and multitracked access to information maintained in databases: the ability to rapidly process and select large quantities of information from a given database in various ways. From a single online database, a business organization can produce printed items such as catalogues and product announcements, specialized newsletters, editorial and planning lists, checking documents, order lists and create advertising materials, and output magnetic tape for both in-house use and external systems. Additionally, dial-up, online searches of central databases can be performed from communicating terminals using logical search commands similar to those of online host database systems. This networking is no small step in the integration of rationalization of business functions formerly thought of as separate and discrete (Singleton 1981).

In the publishing area, for example, databasing can be used to rationalize production and marketing functions, providing coordination with the full cycle of reviews, promotions, advertisements, etc. Databasing rationalizes the following prototypical functions: selection of book titles in production, logging of appearances of books in newsletters and exhibits, retrieval of lists and copy for other uses, production of planning lists for direct mail efforts, production of output in specialized formats to serve as input to other systems, and finally, generation of formatted output to substitute for forms previously completed clerically.

The technical foundation of any system is a well-defined database. The engineering principle behind the new vision of the database is that nothing is recorded more than once, and the purpose of databasing is to bring the entire processing function into a unified field or frame of reference. But, again, it must be emphasized that what is new is not the notion of the database but rather the concepts of multiple access and use coupled with database comprehensiveness.

Impacts of Computerization

Computer manipulation of databases has profound consequences for the academic workplace, no less than for knowledge systems in general. Social scientific information systems will change traditional hierarchical arrangements within academic culture. Traditional distinctions between top "theorists" and middle level data gatherers are becoming blurred as information experts perform decision-making tasks. The manner in which databases are set up may define the kinds of questions one can ask and the information they will yield. In some fields, especially economics and psychology, top scholars have helped develop principal databases to ensure control of the information they contain (Franklin 1982). The behavior of social scientists to one another in all probability will still be dictated by personal power relationships; but just what constitutes power may itself be largely determined by the ability to access and utilize databases.

Decision Making

Electronic access to databases also permits use of data to back up long-range economic planning in a systematic logical way, displacing seat-of-the-pants intuition. Better control of statistics may reduce creativity and risk taking. But whether databasing increases or decreases creativity, it will lead to more control of numbers, ease modelling, and permit use of "what if" scenarios and give us correspondingly more confidence in decision making. In business, complete data on impact of changes in sales volume, inventory, or commissions are easily maintained with databases, which is why economists have so readily adapted to the new technology. Today the researcher can access information from commercial databases like Standard & Poor, Dow Jones, and Dun & Bradstreet, which means better ability to predict both specific and general economic trends. While the ability to draw policy implications from such databases is far from self-evident, they do support the rationale of such decisions and clarify who implements them.

Individuals seeking policymaking roles will increasingly be required to understand how information is gathered as well as how it is used. No longer residing in bound folders or manuals, data in computers is accessed easier and can be used by decision makers. Different spending levels for marketing activities can be modelled against variations in sales and prices. The validation is done by use of computer-programmed historical data from a wide variety of sources. But again, the same problems remain. In validating common sense with computer analysis, does one maximize profits or minimize risks? Historical trends may or may not be an accurate gauge of future trends. Unaccounted for larger environmental variables, which are seemingly endless, like Brownian movement in molecules, limit the predictive potentials of any database, however thoroughgoing. While theoretical generalizations will sometimes be informed by a data-rich environment, the place of specialization and the intellectual gambler will hardly dissolve (Benjamin 1980, 86-87).

Demographers work with population sizes and shifts, political scientists with voting patterns, sociologists with stratification networks, and, of course, economists have the monetary system as a touchstone to guide them. Whether databases in each of these areas will be structured with sufficient flexibility to be "plugged into" other bodies of data, gathered and filtered in different ways, will become a major challenge for those working with the new technology. If we only multiply databases as self-contained monads without "windows" linking databases to other databases, resulting in a larger world, then we will simply compound the problem of determining what constitutes knowledge by providing a systems overload of information, without the synthetic framework to make the new data broadly useful. Indeed, the risk is that theory construction will itself be dismissed as idle chatter.

Information: "Old" and "New"

Although there continue to be significant technical developments taking place within software development (notably, the ability to systematically and rapidly retrieve great amounts of information formerly thought to be disparate and unrelated), the major changes affecting computerized databases will largely depend upon developments in technology as such. The era in which activities involved in the creation, reproduction, storage, and retrieval of data each stood alone as distinct functions is now past. New trends of technological convergence have led to a more encompassing field of information management. The convergence of databases is really an aspect of the new technology, which in turn is part and parcel of revolutionary changes now customarily referred to as postindustrial society.

Computerized databases as a specific subfield of information services have added to, rather than displaced, traditional reference works. Once again, we witness the phenomenon previously described elsewhere as multitiering, that is, tracking information in a variety of forms that manage to coexist much more neatly than the advocates of hard copy might have predicted. To be sure, databases have introduced new wrinkles to the knowledge world, i.e., the value of information is estimated by the time spent linked to the database, or by whether one orders an offline print; but such matters must be seen within a larger context in which the expense of storage and retrieval is still far greater than in print media. Even when hard-copy, inexpensive formats exist, the speed, convenience, and timeliness of databased information delivered electronically may overcome price resistance; this of course will depend on the value ascribed to rapid access to the information.

Databased publication has raised demands for standardization of information presentation, rationalization of the way such information is delivered, and above all consistency in what one receives. If a reader depends on Dow-Jones coverage of financial performance of public corporations, statistics on location, sales, product lines, number of employees and such coverage is adequate for only fifty percent of U.S. public corporations. The information purchased is subject to the same criticisms of incompleteness or shoddiness as information delivered in hard-copy book form. The difference is that information is corrected more rapidly in the online form than in hard-copy form. One keeps returning to the obvious but elusive truth: The form in which information is delivered does not absolve one from analyzing the substance of what is delivered. One must evaluate the cost of electronic information services in terms of return on investment. And these are empirical, not a priori matters.

The evolution of databases is highly instructive on a number of counts: First, technological push can operate in the absence of economic pull. That is to say, even without broad market demands for electronic databases, the innovative impulses of high technology have plunged forward. Second, the costs of this plunging forward may bear little relationship to cost-benefit efficiencies. The economic costs of databased products are more excessive than one might have either predicted or expected to the point where wide-band databases, such as newspapers, are so inefficient as to just about be priced out of the market. Third, the current situation of excessive supply for inadequate demands is not likely to be easily reversed; quite the contrary, new technological pushes are such that computer databases are moving into areas where economic inefficiencies are simply taken for granted rather than reduced.

Political Economy

The political economy of the new information technology is such as to maintain rather than replace older modes of delivering systematic information in hard-copy form. The increasing costs of transmitting information over telephone lines, the relatively small amount of databased information actually used with respect to the amounts created, and the institutional rather than personal uses of computer databases will add new labor costs rather than reduce old costs. All of this adds up to a political economy of information that is going to remain multitiered, i.e., to be delivered in both print and electronic forms for many years to come, much as radio and television coexist. For, while it is correct to note that the economic constraints on high technology may not stem innovation, such constraints may be quite significant in containing levels of use.

While electronic publishing as a whole is not a natural monopoly, specific software or databases may be. But such databased forms are "monopolies" in such a narrow band of information, that the ever present fears of a master database seem remote and exaggerated. Ithiel de Sola Pool has provided a realistic appraisal of the actual economies of databased publishing: "Once an organization has compiled a bibliography of all the chemical journal articles of the past twenty years, no other sane entrepreneur will attempt to duplicate that massive effort. . . . The chemical bibliography may have both too much and too little information for biochemical engineers; there may be room for a specialized bibliography for them" (Pool 1982, 21-22). Depending on the cost of creating the database, the same may be true in a wide variety of areas: Real estate agents may not cover rooming houses or hunting lodges, consumer compilations may be good on prices, but weak on quality tests.

Issues of Regulation

Once the question of monopoly is introduced, then the role of governments also becomes evident. For just as the Sherman antitrust legislation aimed at preventing an undue concentration of industrial power in one firm, so too are mechanisms required to inhibit similar forces of technological concentration in the area of information gathering and diffusion. For the moment, the government, that is, the state, seems more intent on the rational-bureaucratic potentials of such databases, rather than their potential for mischief and/or malice.

What makes this a matter of importance is the need to see databased information as providing an opportunity and a challenge, and not just a mechanism of conspiracy or repression by the all-powerful forces of the state. Once again, it becomes apparent that changes in technology do not necessarily represent an automatic change in the marketplace. For the present and immediately foreseeable future, databased delivery of information is an additional resource, but while it may alter it will not eliminate the system of publishing and exchanging ideas, much less fundamentally transform the social order in any generalized sense.

The purpose of these remarks is not to disparage or discredit innovation in information technology. Such an exercise would be futile at best and idiosyncratic at worst. The forms of technological innovation in this area must continue to amaze even the most hardbitten skeptic; were that the power of the market be viewed with equal appreciation by technologists for whom the brave new world carries few tremors or doubts.

Behavioral Dynamics

The rapid acceptance electronic formats inspire is in part a function of a behavioral transformation in which the information comes to the user rather than the user going to the information (as is the case in conventional hard-copy formats). There is no need to move from drawer to drawer or search from volume to volume of an index, no need for physical movement about a library. The cost of this behavioral *immobilisme* may be the reduction of serendipitous findings, and sometimes it may not occur to the user to consult alternative sources if they are not entered into the database. On the other hand, sometimes, desired information may *only* be retrievable electronically. We are in a transitional psychological mood, one in which we take for granted a multitiered information environment, instead of approaching knowledge needs in a unidimensional mood. Electronic databases are value-added media, rather than alien intrusions aiming at the destruction of other, more conventional forms of delivering information.

It is entirely understandable that technical scientists would emphasize what computerized databases can do for potential users rather than what these human beings can do for databasing. Nonetheless, it must be appreciated that some quite real imbalances need to be redressed. There is now an entire negative literature on databasing, expressing everything from generalized fears of the computerized society to particularized fears about violations of the privacy of individuals and new sources of class polarization. The more sophisticated the information technology becomes, the louder will become the concerns expressed by a variety of critics—often drawn from the social sciences and humanities—until we have a virtual crescendo of assault. It would be risky to think that simply because databasing is more benign an activity than genetic engineering, we will be able to entirely avoid the same impulses to foreclose aspects of the new technology.

Information and Knowledge

The multiple relationships among those who create, diffuse, and use information are entirely germane to databasing. The need in the next phase of the new technology is to search out the psychological and sociological dynamics underwriting the assumptions of information technology. We must acknowledge the profound differences between information and knowledge, and no less significantly, how potential for the misutilization of information and misunderstanding of what constitutes knowledge comes about because of inaccurate evaluation and dissemination of unwarranted interpretations of computer-derived research results.

Databases have logical components and using them sometimes requires specialized training. Many databases are highly sophisticated, including controlled vocabularies, concept codes, and hierarchical tree structures. Beyond that, user guides and database manuals must also be collected and updated. If one takes as a task the study of social stress, one is often led to data on divorces, abortions, illegitimate births, infant deaths, fetal deaths, disaster assistance, school dropouts, etc. But of course such easily quantifiable information assumes rather than proves the existence of social stress. Beyond that, such indicators do not properly indicate, as any good social scientific approach would, that certain stress levels are functional, and not dysfunctional, with respect to economic development and industrial expansion. And while it might be pleasant to speak of integrating databases and systems analysis, the raw fact remains that the sophisticated rendering of scientific ideas extends far beyond databasing; and at times, can even be falsified by the arbitrary and premature narrowing down of research fields (Straus 1983).

Knowing and Questioning

Some of the differences between raw data and refined knowledge can be further illuminated by a small-scale example rather than a large-scale metaphor. If one wants to determine who publishes the largest number of scholarly journals, that data is available through an online search of *Ulrich's Guide*. A first count shows that Pergamon is first and Elsevier a distant second. But this does violence to what we know about the publishing world in common sense terms; namely, that Elsevier is the largest publisher of journals worldwide. The reason our search tells us differently is that all Pergamon's journals are published under the Pergamon imprint, whereas Elsevier publishes its journals under six to eight different imprints in accordance with general research areas and in several countries. As a result, unless we know this a priori, we may ask the question in such a way that we derive false results at the holistic level. It takes a certain qualitative type of knowledge to ask the right question, or to ask it in such a way that we get results in keeping with what we know to be the general contours of real world experience. The answers one gets from electronic databases may be limited by the requirement that users define their terms and ask their questions properly. Every user has been frustrated by not knowing how to ask the question that will yield the evidence common sense dictates is there.

Quantitative measurement of scholarly output has become a widely employed application of databasing. Coding data in terms of the names of key actors in the world of scientific and quasiscientific productivity is both a natural consequence of the systematic storage of information and presumably a technique for assessing the quality of research. The hidden assumption of citation analyses nationally or worldwide in a given area of research is that they yield a true sense of the worth of a particular publication, and since citations may be disproportionate to the actual levels of individual productivity, the assumption would seem to have some credibility. However, here too one must exercise the greatest caution: Citations rarely, if ever, disaggregate approbation or disapprobation with a point of view taken; citations are often a function of extrascholarly values; for example, high numbers of citations are given to departmental chairmen who have the power of appointment and dismissal. At times too, citations are subject to a contagion effect; once an article is established as important or seminal it continues to be cited long after such seminal values have been absorbed into the common wisdom. The electronic database phenomenon has changed our notions of scientific and scholarly value, but too rarely has it given us an appreciation of the distinction between the quantitative and the qualitative, or even the negative and the positive.

To illustrate this point further, even closer to home, the same database inquiry which inverts the relative size of Elsevier and Pergamon also reports only 19 journals within the Transaction Periodicals Consortium. But Transaction has 32 journals in its Consortium. The discrepancy results from the assumptions made by the inquirer who defined a journal as a serial publication which appears regularly at least four times or more per annum. Transaction has a number of journals which appear only three times annually, others that appear semiannually, and still others, irregularly. The way the question was asked distorted the results. It takes a peculiar kind of special knowledge to know how to structure criteria for inclusion and exclusion.

Database Coverage

One may also get inaccurate information by questioning the wrong database. If one inquires of the United Airlines flight information system what connecting flights there are from Newark to Jacksonville, the information derived from the Apollo system will not include flights scheduled by People Express (and other regional carriers as well) between these two terminals. There may be valid or at least understandable commercial reasons for failure to report all connecting flights within a computerized system that has the capacity to do so. Nonetheless, one must again have a certain amount of a priori knowledge of the limitations of a database to ask the right questions of the system: Specifically, what are the criteria of inclusion and exclusion of information. Problems in the use of computerized databases for scholarly research use deserve and are receiving serious attention (Falk 1983). Still, we do not yet have reviewing mechanisms for databases. These perplexing and frustrating microscopic examples of limitations in the new informational environment should not be lost sight of in our search for larger macroscopic issues.

Conclusion

We must broaden our understanding of information and its retrieval. At the same time, we must also be in a position to extend the economic infrastructure of knowledge. If we fail in that critical task we will indeed end up with what the late C. Wright Mills referred to as "crackpot research," a form of research in which the only questions permitted are those which the hardware and software as currently defined can logically handle. Rather than function as a handmaiden for economic innovation, such limitations can frustrate research and innovation. In a worst-case scenario, broad-ranging questions may not be permitted; further, no social actions will be permitted that are not sanctioned by existing data or information.

We must broaden the conception of political economy to include those aspects of the new technology that involve considerably more than market factors and their expansion. Indeed, the least malleable and most difficult to pinpoint aspects involve social benefit rather than economic cost items. Computer intelligence is, after all, part of organized intelligence as such, and hence involves educational questions: What should be the appropriate community expenditures in this area? How should appropriations be made—from new taxes or taking away from other ongoing activities? To what extent, if any, should considerations of reducing existing racial, gender, ethnic, etc., inequalities be invoked? We can see quite dramatically that the political economy of databasing is nothing short of questions of political economy at a time of technical innovation.

What we have then is a need to respond to dangers from quite opposite sources: a fast moving technology which threatens to leave many behind and create even new varieties of uneven social stratification, and a slow moving normative foundation which is severely challenged by the imposition of anomie and atomization of the valuational base of society as a whole, i.e., the reduction of ethical and aesthetic issues to possibilities of gain and loss on one side and sound and noise on the other.

Computerized databases are an enormous resource, a significant development heralding the twenty-first century. They finally shift emphasis from the means of production as a critical variable in the social order to one based on means of communication (Rosenfeld 1980). At the same time, they fail to do away with longstanding problems of good and evil, right and wrong, beauty and ugliness. One can hope that knowledge, and here we mean specifically the knowledge gained from hard work in the sciences, social sciences, and humanities, will inform the revolution in information technology. If information and knowledge move in tandem, this will comprise a revolutionary step towards solving larger normative and distributive concerns. However, if either is permitted to subvert the other, in the name of a crude empiricism or an even cruder moralism, we have the capacity for an equally gigantic reaction: a darkness born of modernity.

Beyond the issue of inverting economic costs and educational opportunities through databased technology is whether a constituency exists to move beyond considerations of market demand and consumer supply. Innovation in databased technology has been so rapid that in many areas informational needs have not caught up. In many fields of scholarship, traditional indexing and abstracting services may be quite ample to meet current needs. If the sales of conventional informational sources is any indication, this narrow view is fully confirmed. Yet, while the new technology is focussing upon the finer points of informational retrieval, with sophisticated techniques of term weighing and relevance judgment, many fields of learning still operate in a parochial and gross fashion (Brittain 1984). Nothing could prove more dangerous than a dizzy-with-success attitude to technical developments in computer-related technology. There is little evidence that technological innovation marches lock-step with economic change into a brave new world. The present databasing environment confirms the need to study the past and plan for the future as we examine the present.

Note

1. This article should be viewed as part and parcel of an ongoing effort to connect the social structure and scholarly publishing. In this regard, I would suggest the following three pieces in particular, which I authored, to be read in conjunction with "The Political Economy of Database Technology": Horowitz and Curtis 1984; Horowitz 1983; Horowitz 1982.

References

Benjamin, R. *The limits of politics: Collective goods and political change in postindustrial societies.* Chicago and London: University of Chicago Press, 1980.

Brittain, J.M. Internationality of the social sciences: Implications for information transfer. *Journal of the American Society for Information Science,* 1984 (Winter), 35(1), 11-18.

Falk, J.D. America: History and life online: History and much more. *Data Base,* 1983 (June), 6(2), 114-125.

Franklin, J. Primary information on-line in biomedicine: An appraisal. *Scholarly Publishing,* 1982 (July), 13(4), 317-325.

Herter, S.P. Scientific inquiry: A model for online searching. *Journal of the American Society for Information Science,* 1984 (March), 35(2), 110-117.

Horowitz, I.L. The impact of technology on scholarly publishing. *Scholarly Publishing,* 1982 (April), 13(3), 211-228.

_____ . New technology, scientific information and democratic choices. *Information Age,* 1983 (March), 5(2), 67-73.

_____ and M.E. Curtis. Fair use versus fair return: Copyright legislation and its consequences. *Journal of the American Society for Information Science,* 1984 (March), 35(2), 67-74.

Lacy, D. Culture and the media of communication. *Scholarly Publishing,* 1982 (April), 13(3), 195-210.

Montagnes, I. Perspectives on the new technology. *Scholarly Publishing,* 1981 (April), 12(3), 219-223.

Pool, I. de S. The culture of electronic print. *Daedalus: Journal of the American Academy of Arts and Sciences,* 1982 (Fall), 111(4), 21-22.

Rosenfeld, H.N. The American constitution, free inquiry, and the law. In J.S. Lawrence and B. Timberg (eds.), *Fair use and free inquiry: Copyright law and the new media.* Norwood, N.J.: Ablex, 1980, pp. 288-309.

Singleton, A. The electronic journal and its relatives. *Scholarly Publishing,* 1981 (October), 13(1) 3-18.

Straus, M.A. *Social stress in American states and regions.* Paper presented at the International Conference on Data Bases in the Humanities and Social Sciences, Rutgers University, June 10-12, 1983.

10

Behavioral Impacts in the Information Age

Frederick Williams, Ronald E. Rice,
and Herbert S. Dordick

As we look to the behavioral impacts of the communications technologies, the question arises as to general areas of inquiry. Are there impacts that tend to transcend individual technologies and reflect changes in our attitudes and behaviors in communication? Examples of such loci of impact are described, including: attitudes, perception of time and space, connectivity, mobility, increased choice, and socialization. These are intended to illustrate the concept of generalized impacts rather than to be an exclusive list. Moreover, they carry strong implications regarding the direction of needed research into the new technologies, research that of necessity will be markedly interdisciplinary.

The Behavioral Perspective

In the midst of the burgeoning body of research literature on the new communication technologies, it is important to ask whether such inquiry should not be given more visible direction. Much current literature is more practically than theoretically oriented—focusing, for example, on problems of implementation, productivity, or most pragmatically, market analyses. Also, research tends to be technology specific—that is, often concerned with a single medium, such as cable, word processing, or electronic mail. The present chapter, by contrast, proposes a concentration more upon behavioral and theoretical lines of inquiry, topics more reflective of human impacts than of the technologies themselves.

This article is speculative and prescriptive, rather than a review of research. In essence, it is argued that there are certain generalized types of behavioral impact regarding human reaction to, interaction with, and management of the new communications technologies. They reflect such topics as attitude, perception of time and space, a phenomenon we have chosen to call connectivity, mobility, increased choice, and child socialization. These are but a sampling, offered mainly to illustrate a general point about the importance of behavioral impacts. Nevertheless, our main point is to stress the importance of thinking more behaviorally and theoretically about technological impacts and of devoting at least some of our research to these aspects.

Six Areas of Behavioral Impact

Attitudes

That *attitude* would be an important behavioral correlate of the impact of new communication technologies might seem at first overly obvious. Yet, as one analyzes the components of attitudes, layer upon layer of inter-related and increasing complexities are apparent. For example, there are the attitudes involved in the diffusion of innovations (Rogers 1983). Among the components in this theory is the concept that the attitudes of individuals vary as adoption and use of an innovation becomes increasingly imminent—for example, from initial knowledge to the desire to reaffirm that a choice was correct. The attitude cycle ranges from fears or uncertainty associated with encountering change to feelings of self-satisfaction about the decision made—even if it was one of rejection. The notion that decisions require confirmation and reconfirmation is an important component of innovation theory.

Although most of the development of theories in this area involved such innovations as agricultural methods, public health, or family planning, the premises remain very relevant to communication technologies, and indeed, have been applied to such innovations as videotex (Rice and Paisley, 1982), word processing (Rice and associates 1984), home computing (Rogers, Daley, and Wu 1982), and computing in the schools (Williams and Williams 1984). These and other studies tend to show that prior attitudes about the medium or the utility of the information obtained through the medium greatly influence actual adoption and use of the medium.

Other attitudinal dimensions of media use were explored in the monograph *The Social Psychology of Telecommunications,* by Short, Williams, and Christie (1976). Among their contributions was an examination of the concept of the "social presence" of a communication technology. One major question concerned the consequences of technological constraints upon interpersonal perception. For example, because a telephone link cannot easily transmit nonverbal information, we may be restricted from many of the subtle cues that support a highly personalized interpersonal exchange, as in face-to-face communication. Or, in another example, a business letter, because it also lacked paralinguistic cues, could be even more impersonal.

Yet, as emphasized by Short and colleagues, it is not simply the technological restrictions that affect our attitudes about the personalness or social presence of communication via a particular medium. There are further complexities relating to the choice among alternatives as well as stylistic variations within the limitations of a given medium.

As for choice, the idea is that if one has selected a medium of less social presence over another one, this is itself a message of impersonalness. One example would be the proverbial "Dear John" letter in place of a face-to-face conversation or telephone call. As for style, every medium, regardless of its technological constraints, provides some latitude for textual variation which can affect actual and perceived degrees of personalness. That is to say, a telephone call or a telex can vary in personalness, depending upon what is said. Even a computer message exchange can be worded personally and have a quality of informality—in fact, this is encouraged under some conditions. In essence, choice of medium and style are complex attitudinal considerations, and although this idea has been with us for nearly all of the history of commentaries upon communication, it is still being overlooked in many contemporary studies of the new technologies.

A similar line of research has been applied to assess individual's experiences in teleconferencing, and attitudes loom large as an explanation of the impact. For example, in an analysis of computer teleconferencing in Phillips's (1983) study, the transcripts of three computer conferences were studied and their content analyzed, with emphasis placed on whether (and if so, how) the medium's special characteristics enhance or diminish the emotional dimension thought to exist in such conferences. Results indicated that the ability to communicate asynchronously, while alone at a terminal, in a written mode, and without the expectation of immediate feedback, encouraged "stream of consciousness" communication of thoughts that exhibited spontaneity and creativity. Once users passed the initial period of frustrations associated with learning new procedures there was a tendency to be very open, conversational and personal in writing style, with numerous instances of emotional expressiveness, humor, metaphorical language, and overt sociability.

The necessity of having to write down thoughts did seem to negatively affect the amount of participation by some users, and several people expressed discomfort with the literacy and computer skills required for the computer conference. Several others mentioned that lack of immediate feedback was somewhat disturbing. In general, however, the study revealed a predisposition towards acceptance of and positive attitudes towards computer-mediated communication.

The attitudes associated with a medium—*media stereotyping*—is itself an intriguing theoretical topic. How do such attitudes arise? Do they have some type of general nature? Some thoughts on this were earlier advanced in a paper by Williams and Rice (1983; see also Rice and associates 1984, chap. 3) in which it was argued that such stereotypes are often multidimensionally complex (see also Phillips 1982). For example, results of an ongoing media-attitude survey of university students had yielded a multidimensional factor structure showing differentiation of newspapers, broadcast, television, cable television, videocassette machines, telephones, radio, and computers, relative to attitudinal factors of familiarity, importance, and personalness (Williams, Phillips, and Lum 1982).

That stereotyping studies might even be applied in a "developmental" sense to children was advanced in one of the author's projects involving investigation of youngsters in computer camps (Williams, Coulombe, and Lievrouw 1983). Children, roughly 9 to 12 years old, who had just several hours of initial experience with small computers, related complex attitudes about them. Factoral dimensions corresponded to general evaluation, quality, ease of use, and expense. Moreover, by use of these factors, it was possible to determine certain differences by age and sex of the subjects. Generally, younger children as well as girls found the machines a bit more complex.

The notion that stereotype attitudes might be used to predict the probability of adoption of a technology is another avenue of attitudinal research. Examples of this are the studies conducted in conjunction with the establishment of a teleconferencing system by the Atlantic Richfield Company. Ruchinskas and Svenning (1981) examined individuals' intentions to use video-conferencing, as well as factors predicting use of current communications options, in an 800-employee study spanning ten operating companies of a large U.S. corporation. Use of available communication alternatives (telephone, face-to-face, writing) is predicted primarily by current work activities and cross-locational communication needs (Ruchinskas 1982). When considering use of a new communication option such as video-conferencing, employee attitudes and beliefs assume a much more prominent role (Svenning 1982). Beliefs about teleconferencing's attributes and expected benefits appear to be particularly critical to whether employees intend to use this new technology (Svenning 1983). Cross-locational communication needs become more influential in explaining the frequency of projected video-conferencing use. Svenning and Ruchinskas (1984) suggest that although beliefs and expectations about new communication technologies will influence initial or trial usage—that is, communication behavior—the cross-locational communication requirements and work activities will be more influential in predicting routinized use.

However interesting researchers might find the relationships between attitudes toward and behavior with new media, let us conclude this first section with a caveat. Although attitudes can be readily associated with behaviors such as the adoption of a communication technology, experience involving it, or even with message effects, there is nevertheless no comprehensive theory of the matter. One is still hard pressed to argue theoretically as to the fundamental structures or processes that link attitudes to use of new technologies. Are we researching the "fringes" of somewhat interesting, yet superficial correlates, or are we moving toward some type of comprehensive modeling? To date, we have few results or even little thinking to help us answer this question.

Time and Space

Another area in which behavior is influenced by new media is communication through time and space. Marshall McLuhan drew much attention to this issue in the 1960s. McLuhan claimed that "pre-alphabet people integrated time and space as one and thus lived in acoustic, horizonless, boundless, olfactory space" (McLuhan and Fiore 1967, 56). The arrival of print media led to the disintegration of this boundlessness. The length of time through which communication content existed could now be varied, and the duration between transmission and response could be extended past the originator's lifetime. Communication could also be transmitted through geographical space. Speaker and listener did not have to be co-located in either time or space. But the linear, sequential nature of printed text constituted a severe limitation compared to the simultaneity and holism of pre-text communication. McLuhan and others have argued that this constraint has led to everything from poor memory to the Enlightenment, though there are convincing counterarguments (see Pattison 1982).

Electronic media such as radio and television supposedly reduce the constraints of print-mediated communication by removing the linear form of text and by reintroducing sound and visual dimensions. Yet each medium has temporal and spatial limitations. The listener/viewer does not control the content or timing of the programming, although he or she may control, to some extent, decisions about whether and where to receive the communication. Thus, the time element is constrained while spatial and aural elements are expanded.

Now, with computer-based communication systems, these elements have been further released from constraints. Admittedly, the visual and aural elements have not reached life-like quality, but for the purposes of this argument, they will soon become satisfactory. Technologies such as electronic mail, computer conferencing, voice mail, personal computers and the like, facilitated by value-added networks, local area networks, videodiscs, and communication satellites, remove many limitations on communication. Content can be sent independently of the location and timing of the intended recipient; the recipient does not have to receive the message at a specific time or location—or even at a specific terminal. Communication may be simultaneous but may also be sent and received at the convenience of any participant. The technical devices themselves are now small and light enough to be carried around. This release of communication from temporal, spatial, and technical constraints leaves perhaps only the most intractable constraints—economics, politics, and education. Only a small segment of society may have the resources, access, and training necessary to use these media.

Let us now consider some of these concepts relative to social organization. For one, the new technologies may cause many of the inherent economies of scale in information exchange and evaluation to be lost as communication becomes temporally and spatially unconstrained. Specialized users will be better able to focus on specific information unencumbered by indexing, queuing, scheduling, and physical and cost obstacles. However, as these individuals become involved in their own world of electronic communication, they may also become overloaded not only by the amount of information and communication they receive—a common enough concern—but also by the speed of information processing demanded from them.

The Japanese are particularly concerned about this; their term for information overload is "information speed." This increased speed and amount will not only burden the cognitive and logistical abilities of users, but could change the nature of intraorganizational relations. For example, physical or cultural aspects of the office will have less affect on electronic communication between participants, so the shared meanings and social behaviors necessary to understanding each other may be harder to identify. Thus, misunderstandings may develop because of the incomplete nature of the communication in spite of increased speed and amount of information. Further, because the regulation and immediate adjustments inherent in physically or temporally simultaneous communication may be lessened, exchanges may become uncoordinated, lagged, overlapped, or misinterpreted. For example, if one person has a communicative style developed from familiarity with rapid feedback from electronic messages, while the other has less demanding expectations because of the natural delays involved with meeting people face-to-face, the first may interpret small delays in response as unreciprocated messages, leading to negative evaluations of the relationship.

Related to this potential behavioral consequence of changed information patterns is that individuals may be less able to use organizational filtering and evaluation processes to control the content, timing, and amount of information, seen by many as the primary function of organizations (see Galbraith 1977; Rogers and Agarwala-Rogers 1976; Weick 1969). That is, the electronically mediated communication may displace organizationally mediated communication.

This is particularly paradoxical, because one of the primary rationales for office communication technologies is to reduce, by filtering and evaluating, the amount of information managers must process. Contrary to the popular belief that managers need more rapid information, managerial attention may be the scarcest organizational resource (Simon 1973). For example, within an organization, individual managers may at once gain control of information processing technologies (personal computers, electronic mail) while losing control over what information they must process.

Results from some electronic mail studies show that the introduction and use of such systems leads to increased upward and diagonal organizational communication. Perhaps higher-level managers do not want (or need) this increased communication and will then decide to instruct their secretaries to filter these messages. This implies some loss of control over the system. People who have slightly overlapping interests and information sources will be less likely to "bump into" others with whom they might gain from unplanned communications. These coincidental communication activities may prompt fruitful exchange of rumor, contextual information, quickly diffused crisis information, insider tips, warnings about subsequent activities, insights into organizational culture, first impressions, and the like.

Or, consider an area such as supervisory evaluation. Office information systems can increase managerial span of control—how many subordinates are supervised by the single superior (Rice 1980). At the organizational level of analysis, this is efficient, may be effective, and clearly reduces costly managerial overhead. From the point of view of supervisors and workers, however, the amount of interaction between them declines. Supervisors will not be able to devote as much time in training, socialization, or evaluation of each subordinate. The strategy then taken may be to use the technology to aid in performance evaluation—such as using line-counts as an indicator of performance by word processing operators, or using the number of transactions processed by insurance clerks. These indicators may be largely unrelated to the organization's mission or the performance of the employees, however.

Forms of social integration that allow the development of commitment, quality control, and adaptation may be lost as temporal and spatial co-location lessens. One alternative, of course, is to simultaneously relocate the supervisory functions—allow the subordinates to evaluate their work, develop new procedures, and create self-designing work groups. The lesson here is to transform the control mechanisms in ways appropriate to transformations in spatial and physical communication interactions.

Often, the behavioral strategy taken is not appropriate, however. At the interface between organizations and their customers or clients, there is a growing tension between strategies taken to optimize internal information systems and strategies taken to provide customers the information they need. Singer is a particularly astute critic of the more common approach by organizations—optimize the system at the cost of clients and customers with atypical needs (Singer 1977, 1980a, 1983). This strategy involves altering the human-system interfaces to "buffer" the system from exceptions, complex information searches, communication cycles that do not match processing cycles, and requests for information that are not natural byproducts of transaction processing. This choice results in the creation of "crazy systems" and "Kafka circuits." Individuals become caught up in delays, system-generated errors, apathy, form letters, unlisted phone numbers, unidentified service personnel, or an inability to establish a wider context for the information. Thus, the efforts put into rationalizing and streamlining the organization's information processing create economic externalities which are passed on to nearby powerless clients and customers.

Connectivity

One of the most obvious and consequential implications of new communication technologies is that people are able to expand their personal networks (see Rice 1980, 1982; Rice and associates 1984). This capacity can be termed *connectivity*. One indication of the steady growth in the awareness of the importance of communication networks is the burst of network analysis research (see, for example, Rice and Richards 1984; Rogers and Kincaid 1981). Personal networks are crucial in getting a job, meeting people, maintaining psychological and social health and exerting power.

Until recently, becoming part of networks, maintaining them, and expanding them, has been limited by the temporal and spatial constraints of communication channels—telephone calls and letters, exchange of research through journals, or transportation requirements. However, computer conferencing, electronic mail, electronic publishing, community computer-based bulletin boards, online databases and computer-based private packet-switched networks have created the technological potential to communicate with others throughout the world at any time of the day or night (see Glossbrenner 1983; Hiltz and Turoff 1978; Rice and associates 1984). Again, however, economic, political and regulatory constraints still overwhelm this potential.

Those newcomers who do gain access to such systems immediately expand their network of communication partners. The obvious benefits of this ability include better access to needed information, social reinforcement from those with common interests, reduction of social status and appearance as criteria for interaction, or greater exposure to the social activities either in one's nearby community or nationally.

The potential disadvantages are less obvious. For example, the freedom from temporal and spatial constraints leads to increased transience and mobility. If one can work from a terminal that only requires an accessible telephone line and respond to messages at convenient times, there are fewer constraints on mobility, and fewer ties between one's workplace and a residential area. Indeed, one need not work at any specific place—such as a company building—at all. Or one can easily choose a new set of personal or casual contacts based upon interests of other users listed in a directory maintained in computer storage as part of the communication system. Thus, local social structure may disintegrate while "electronic" social structure become more integrated. Another consequence of the loosened constraints of time and space, with concomitant increases in the potential for connectivity, is that as the number of linkages increases, the duration of specific linkages may decrease due to constraints on individuals' information processing capabilities. So, we may have wider, but less substantial, networks.

Perhaps historical, temporal, and spatial constraints on communication established baselines for cognitive and social thresholds. If those baseline constraints are now becoming relaxed, will our typical communication behaviors become insufficient? In a communication environment freed from temporal and physical constraints, information resources are likely to become more important than—or a means of obtaining—material and social status resources. Further, individuals may need to acquire their information wealth early on in order to survive in this environment. Information, freed from temporal constraints, has less tangibility and longevity than some material goods. This need creates pressure to quickly search the communication system for rewarding contacts.

Further, as previously discussed, these systems facilitate contacts that are unfiltered by physical or status differences or control over material goods. But, because humans have limitations on their processing abilities, unlimited contacts thoughout a system cannot be long maintained. Sources of information must be evaluated rigorously, and only the useful ones can be maintained; further, they must be reciprocated or the individual's resources invested will be lost. This hypothesis for the growth and development of networks in an information environment has been tested and generally supported by Rice (1982), who analyzed two years' worth of computer-monitored data from a nationwide computer-conferencing system. Studies of people's recreation of their direct personal communication networks identify numerous constraints:

1. That a network node needs, and indeed can handle, only about seven first-step choices to reach its systems' "others" (Bernard and Killworth 1977).
2. That about seven strong links within a "small world" network are involved before the weak link to a target from a different social structure is found (Korte and Milgram 1970).
3. That knowing a respondent's 7th- and 8th-ranked sociometric choices provides no significant additional information about subsequent choices, while knowing the 1st through 6th-ranked choices provide decreasing, but significant, information (Rapoport 1979).
4. That intensity of relationships involving other nodes becomes asymptomatic around 12 to 15 nodes as, presumably, the time needed to maintain network contacts reaches its limits (Cowell and Wigand 1980). Further, this level of communication activity and recall may be sufficient.

Computer simulations of the efficiency of networks for allocation of job tips indicated that ninety percent of job vacancies were filled with three direct contacts and associated two-step links or with only eight direct links (Delaney 1980). The point here is that considerable evidence implies that there are upper limits on the number of communication links we can instantly remember, or which we need in order to function in society. An implication of the foregoing is that an extreme emphasis on increased connectivity may be misplaced effort. Becoming fully connected may not be necessary, both because of natural human processing constraints and because of the nature efficiency of even sparse networks.

All this discussion of information environments is not to deny the role of political and material wealth in the creation and maintenance of social stratification. However, information wealth—increased connectivity and control over production, licensing, and retrieval of information—will interact with material wealth to create different kinds of social stratification.

One aspect of this interaction has to do with increased information wealth—specifically, personal contacts—as a product of social and material wealth. Greater social wealth and status leads to increased likelihood of shared traditions and social means, attendance at certain schools, access to certain job categories. These situations lead to increased density within the social category and greater cohesion, and increased attractiveness as objects of contact. That is, such individuals are seen as leaders or information carriers and are valued as contacts (Rice 1982; Rytina and Morgan 1982). Thus, a minority group with information and material wealth (such as a technocratic elite) can develop high cohesion and considerable contacts with the majority and other minorities, becoming a dominating minority. However, intellectuals without social traditions and material wealth will not be as attractive as potential contacts due to lack of connectedness within their own "class." This is because their ability to focus on specific content and specific contacts rather than on a class-bound group (due to the loosening of temporal and spatial constraints) paradoxically prevents the development of contextual and institutional norms within that group. Until information wealth is converted into social and material wealth, or until it becomes a central social norm itself, purely information-rich social members will continue to be excluded from social and political power.

Change in Mobility

Exchanging communication for transportation has been a dream of telecommunication enthusiasts from the earliest days of our desire to communicate at a distance. But it is not at all clear that these proposed substitutes for travel have indeed reduced travel. On the contrary, the history of the telephone, at least in the United States, provides considerable evidence that travel increased as telephone use increased (Gottman 1977). Contrary to the technology assessments of those who believed there is a trade-off of communication for travel (for example, Nilles et al. 1976), a better argument can perhaps be made that the telegraph and the telephone were the first in a long line of information technologies that have stimulated increases in mobility. This is not to say that some travel has not been displaced by communication but rather that the telephone, the computer, and the emerging information technologies offer society a nontravel choice as well as more reasons and needs for travel—in other words, more mobility.

We are a mobile society; mobility is imbedded in the American psyche. The great historian of the West, Frederick Jackson Turner, viewed the history of the United States in terms of the inexorable westward movement of the new Americans and speculated that once the nation had been traversed we would no longer have the energy and drive of our early years. Census data reveal that at least 50 percent of the American population moves every five years and of these, about 20 percent move to another state (Long and Boertlein 1976; Parsons 1949; Stein 1966; Whyte 1955). Other sociologists have frequently commented on the rootlessness of the American and the decline of family and community. Bennis and Slater (1968), in

particular, have studied the alienation and tensions produced by this transience and rootlessness.

As we have exhausted our geographical mobility—coast to coast, Alaska,
Hawaii—we are now seeking the mobility that electronic communication
provides. Society will use the information technologies in support of more
travel. Links to family and community will be maintained through communication. As the preceding section suggested, there will be fewer reasons
to stay in one place because communicating on the move—while driving,
while travelling in air or on the high seas—will provide the connection to
relatives, community, and perhaps even more important, to jobs. The fragmentation and alienation feared by the sociologists may be avoided by
connections provided by the information and communication technologies. *Community* will take on new meaning, as Bell argued. Communities will no longer be bounded by geography but encouraged by common
interests on communication networks (Bell 1976).

The citizen's band (CB) radio "mass event," shortlived as it was, was
nevertheless significant in its illustration of what MacCannell (1976) saw as
evidence of a "new species of commodities (do-it-yourself programs) that
reflect the modern fragmentation and mutual displacement of work and
leisure and the emergence of new synthetic structures as yet unanalyzed,"
and the blurring of distinctions between work and leisure. A unique opportunity to study the intersection of mobility and human communication
was missed when the CB radio boom died, rather precipitously. For a brief
period in the seventies, the Federal Communications Commission was
receiving more than 500,000 requests for CB licenses a month. In 1977 the
FCC estimated that there were more than 25 million CB radio users in the
nation (Bowers 1978). What did this suggest about the nature of our societal need for communications? From the few studies of CB users, Bowers
found that the expressive uses of the medium ("gossip, messages containing
little or no direction") amounted to almost 70 percent of all message content. Only 1.8 percent were classified as instrumental or seeking to achieve
some useful end and 29 percent were classified as intrinsic or consciously
about the maintenance of the CB community. This seems to point to the
ease with which different interactive communication media are adapted to
a very wide variety of human needs. It raises questions about the doubts
expressed by researchers who have examined teleconferencing, computer
conferencing, and other of the new information/communication technologies—for example Williams (1977) on psychological distance and self-
confidence, Reid (1977) on evaluation of the person and the substance of
the conversation, and Short et al. (1976) on various limitations to teleconferencing as compared to face-to-face meetings.

Increased Choice

There is little doubt that the wide and varied number of new media now becoming available will create new opportunities for societal and individual choice. As mentioned earlier, communication scholars have long been concerned with the processes of choice.

Missing from much of these analyses, however, is the recognition of the human as an economic factor. Theories of the adoption of innovation make only a casual reference to the cost of the adoption process, referring simply to the notion that adoption is more likely if "a comparative advantage" is evident. Whether the resulting comparative advantage is worth the cost of adoption requires measuring the relative value of that advantage. For example, with the rapid emergence of many information technologies and communication media, advantages of one over the other may be revealed by taking time into account—time to change practices and procedures, time to train or retrain staff, or time to install. It is evident that there should be increasing interest in the economics of time.

Time is a scarce resource, but in a rather special way. Time itself is in a sense inexhaustible, but each individual's time is not! As Sharp (1981) so rightly points out, Psalm 90 tells us rather optimistically that "seventy years is the span of our life, eighty if our strength holds." Given that this scarce resource must be spent—over that we have no choice—then it is necessary that it be spent well. Even if we narrow this problem down to the selection of one medium or information technology over another, the value of time, or its cost, depending on how the decision maker chooses to view time, cannot be overlooked. The economics of time has always been important to the processes by which people choose one medium over another. However, it has not generally been made explicit in analyses of such decision making, nor has it been integrated with the attitudinal factors normally used to explore the adoption process and the choices made in that process.

Economics traditionally deals with the allocation of resources where the value of an allocation is a material one. Definitions of economics, indeed, stress this measure of value, usually money. Robbins' (1945, 16) definition of economics as "the science which studies human behavior as a relationship between ends and scarce means which have alternative uses" suggests that time is a proper subject for economists and, further, that its allocation among alternatives may play an important role in making choices.

Because economics has traditionally dealt with allocations of material goods, time allocation studies have focused on time at work, where such measures are important for measuring productivity improvements following the adoption of new production techniques. Ghez and Becker (1975) broadened these studies of time economics to include time spent for leisure and for "creating intellectual capital"—education. These important studies led to the large time allocation research of Robinson and Converse (1972) which sought to show just how time was being spent. They did not, however, deal with how choices were made. Further, these analyses assumed very sharp distinctions among human activities; few, if any, crossovers or joint allocations of time were considered.

With the advent of new information technologies, distinctions that seek to define work as that which takes place solely in the office are becoming blurred by work that can be performed at home, in a car, or while flying. Communication technologies that allow trading off time in one kind of activity for another kind of activity can easily allow for extremely efficient utilization of time and, indeed, are supposed to do just that. As noted earlier, one can choose to displace time with comparative ease, but with potential attitudinal and behavioral constraints and costs. Clearly, this may raise the question of the value of time saved or shifted versus the cost of human or behavioral efficiency.

Another factor for which time becomes an important economic variable in choice concerns the extent and degree to which these information technologies create the network marketplace (Dordick et al. 1981) and the self-service economy (Gershuny 1978). For the bank or supermarket, will the investment in information technology match the customer's desire and willingness to alter traditional habits? How long will it take? Is the value of the customer's time saved worth the customer's personal investment in learning a new pattern of behavior? Increasingly, labor-saving devices are seen as time-saving devices. Visiting the bank, shopping, and increasing worker productivity are all factors that require an appreciation of the economics of time and how this affects attitudes and behavior.

In his study of the social impact of the telephone, Pool (1977) makes the point that the telephone is a purposive technology—that is, we can do what we want with the technology. Unlike the example of Robinson Crusoe (Von Neumann and Morgenstern 1964), who was acting as a rational economic man in a "dead" environment (there were no responses to his decisions, at least before meeting with Friday), we make choices among information media, expecting and getting feedback. Moreover, feedback determines our next action. We choose a medium for communication and not merely for consumption. We select a data base for purposes of using the response for some activity that will influence our next action. In other words, there is a sort of strategic game going on in our choice behavior. This leads us to consider new tools for the analyses of choice behavior among information technologies, such as game-theoretic models of behavior.

Information technologies are creating many more opportunities for choosing how we live and work, how we spend our leisure, and how we select among ever growing information and communication-based alternatives. We can choose to do or not do our banking on the network; we can shop or not shop on the network; we can create our own newspaper electronically or purchase one that can be used to wrap fish; we can decide to travel or not travel for a meeting. All of this makes choice behavior an increasingly complex phenomenon. Analyses that do not relate attitudes and behavior to the economics of time and to the passage of time will not yield a comprehensive understanding of how choices are made. The "snapshot" nature of the demographic-style research on which communication science has been constructed must be broadened to include choice as a variable in the analysis, as well as an understanding of time as an economic factor in choice.

Socialization

All that has been discussed in this article may carry significant consequences for the most impressionable segment of our population—our children. Our concluding topic concerns the consequences of information technologies for the process of socialization.

Today's children have access to a broad variety of communication products and services never before as widely available to the nonadult population. Television is an example. According to the 1982 Nielsen summary, youngsters between two and five years old averaged a total of twenty-five and one-half hours of television viewing per week. This was more than their preteen and teenaged counterparts and only about three hours less than the population at large. Young children were nearly as likely as adults to be watching during the prime time hours, and an estimated seven million children under five were still viewing in the 10:30 to 11:00 P.M. period. It is noteworthy, too, that this same Nielsen summary comments on a continued increase in overall hours of television viewing by the general population as presumably an effect of new viewing alternatives (e.g., cable), since network viewing has decreased slightly.

Our children's massive exposure to television, and some of the worrisome consequences thereof, are already well known through surveys of literature by Comstock and his associates (1978), or by the give-and-take between Gerbner's (1979) cultivation theory and its critics (e.g., Hirsh 1980). Now, all evidence points to greater exposure, and especially to adult themes, because of children's increasing access to programs via vid eocassette, videodisk, cable, pay-cable, and soon, direct broadcast satellite. Many children grow up encountering more life "experience" through media-conveyed depictions of values, ethnic and personality stereotypes, and sex-role characterizations, than they witness in the real world around them.

When children are not viewing television, they now have increased opportunities for exposure to technologically-conveyed experiences via videogames (either in arcades or in the home), electronic toys, and wide use of records, tapes, and entertainment radio. Many children are also avid users of small computers for purposes other than videogame playing. This exposure is expanding rapidly as schools attempt to join the computer age.

Against the context of the world's total child population, those born into this specialized environment are few in number—perhaps 5 percent or less. However, because their environment is so visibly distinctive, children socialized by the media could be the catalyst for shifts in the values and expectations of work and leisure, family, morality, wealth, land, goods, ideas, and even of deity. Because technology is becoming so pervasive, we may be rearing a new generation more likely to ask how rather than why, or when rather than if.

Then there is the possibility of change in the nature of childhood itself. In lay terms, *childhood* simply refers to the period in one's life between infancy and puberty. But in more reflective uses, the suffix -*hood* is critical. The stress is upon the "condition" or "state of being" of that age range. For example, beyond a contrast in chronological age, how is childhood different from adulthood?

Looking to the past, researchers such as deMause (1974) have argued that in the history of Western civilization, until the nineteenth century, there is little evidence that adults associated any separate sort of experience with childhood. In our literary heritage, there are references to children but few to the state of childhood. Art, expecially of the medieval period, has often depicted children as miniature adults.

Our society may be returning to a lack of distinction between childhood and adulthood. Some researchers argue that we are experiencing a "disappearance" of childhood (Elkind 1983; Packard 1983; Postman 1982; Winn 1983). The main evidence is the fading contrast between many experiences of adulthood and childhood. On one hand are the problems of children whose parental circumstances, attitudes, or behaviors deny them a full chance to be "child-like." Of course, this includes the consequences of today's changing family conditions, from single parenting to the attempt to "hurry" a child's development (Elkind 1983). But changes are also a consequence of the new access to adult experiences afforded by changes in the media.

Media ecologist Neil Postman (1982) has been the leading proponent of this view. In *The Disappearance of Childhood*, he makes two critical points. First, as others have reflected, prime-time television delivers the adult world of violence, sex, and nearly every deviant behavior directly to the child viewer. Children are left to find meaning in adult themes of murder, rape, and adultery. The loss of the "adults only" barrier diminishes the contrast between adulthood and childhood.

Postman's second point—and one that relatively fewer observers have raised—is that as our culture increasingly moves its messages from the printed page to the audio/visual media, the barrier of literacy is lost between adults and children. Those of us who experienced World War II as children could not follow the gory details of battle very well in newspaper accounts. We were shocked mainly by pictures in *Life* magazine, or by an occasional (and sanitized) newsreel preceding the feature at our local movie house. Compare this to television coverage of modern wars, from Vietnam to Lebanon, where battles are almost literally fought on our television screens.

Surely, Postman's observations on changes in adult-child media accessibility are clear arguments in support of his "fading contrast" thesis. But, there are further considerations. First, if we have learned anything from more than two decades of serious social scientific research into the effects of mass media, we have learned that there are few so-called magic bullet effects. That is, most messages in the media, particularly on television, do not have a direct effect on individuals. Moreover, there are often vast differences in how individuals interpret what they see. Nonetheless, there is an important communication research challenge implied in the "disappearance" thesis: How do children perceive presumably adult themes in modern media? What exactly are the consequences of the steadily diminishing barriers between child and adult media experiences?

The foregoing are but a few symptoms of the transformation of the socializing environment of the modern child. The gap between children who have access to the technologies, services, or experiences of this environment and their parents may be comparable to the gap between agrarian and industrial societies that developed in the last two centuries.

Yet, only a minority of world cultures are rearing these "technological children." The world context is a highly heterogeneous mix consisting mostly of agrarian and industrial cultures, and including a few tribal ones. The growing tension between the "haves" and "have-nots" presents an ominous challenge to ongoing social change in our near future. This tension appears between developed and developing countries and as stratification within contemporary societies. Some countries (and groups) hope to bypass the historical stages of socioeconomic change by modernization plans—for example, countries like Saudi Arabia or the attempts in our country to retrain industrial workers to compete for high technology jobs. Yet rapidly imposed technological change can create revolutionary responses, such as in Iran.

In the discontinuities between today's generations—parents and offspring, teachers and students—we may be directly witnessing the processes of broad social change. Indeed, the most significant behavioral impact of the information age may ultimately be upon our children. Many of our children's problems—and opportunities—are based on adaptations and responses to this rapidly changing environment. This may ultimately be the most important area of behavioral impact in the information age.

A New Agenda for Research

The topics discussed in this article are meant to illustrate a concept more than to serve as an inventory of specific technological impacts. In most general terms, this concept emphasizes that many of our traditional contexts of communication behavior are changing—some visibly, some subtly. The research questions raised by his proposition are both practical and theoretical. How, for example, can the human benefits of communication be maximized by taking advantage of changes in perception of time and space, new opportunities for connectivity, mobility, and choice? In the closer view, how can current information-related behavior be used to our advantage in the implementation of new communication technologies? Or in the long-range view, how can we maximize the probabilities that our children (and thus future generations) will make the best of these changes?

Such questions, of course, suggest the need for theory, concepts, and research to guide our thinking. But while we might initially see most innovations of the information age as technological extensions of existing communication media, we cannot so easily extend existing communication theories to understand them. Certainly, we can ponder transactional theories of interpersonal communication or group behavior in order to interpret the consequences, for instance, of individuals interacting through computer networks, but none of these theories quite leads in directions that assist the interpretation of many of the areas of impact described in this article. Nor are mass communication theories of particular advantage, for although agenda-setting, the two-step flow, or social dependency theories do aid in interpreting the broad flow of information through large social groups; the context of the new information utilities describes a process many are calling "demassification." Perhaps there are opportunities in the applications of such holistic theories as uses and gratifications or critical theories, yet these approaches may still suffer from the inherent short-coming of creating self-fulfilling predictions. A most profound implication of our behavioral impacts discussion, we believe, is that new concepts and theories are required.

Moreover, they must be markedly interdisciplinary. For example, the understanding of computer-mediated communication systems requires the joint expertise of technical designers who create appropriate systems, or-ganizational development experts who implement them, communication researchers who attempt to identify how content and interactions will change, information scientists who investigate optimum system interfaces and user strategies, and cognitive psychologists who research the correlates of successful information processing. In all, we can no longer afford to maintain strict disciplinary boundaries if we are to understand the rela-tionships of information and behavior, a concern expressed in the title and spirit of this publication.

References

Bell, D. *The coming of post-industrial society*. New York: Basic Books, 1976.

Bennis, W. and P. Slater. *The temporary society*. New York: Harper & Row, 1968.

Bernard, H. and P. Killworth. Informant accuracy in social network data II. *Human Communication Research*, 1977, 4, 3-18.

Bowers, B. *Communications for a mobile society: An assessment of new technology.* Beverly Hills, Calif.: Sage, 1978.

Comstock, G., S. Chaffee, N. Katzman, M. McCombs, and D. Roberts. *Television and human behavior*. New York: Columbia University Press, 1978.

Cowell, R. and R. Wigand. Communication interaction patterns among members of two international agricultural research institutes. Paper presented at the an-nual conference of the International Communication Association, Acapulco, Mexico, 1980.

Delaney, J. The efficiency of sparse personal contact networks for donative transfer of job vacancy information. University of Minnesota, Department of Sociology, Working Paper 80-03, 1980.

deMause, L. The evolution of childhood. In L. deMause (ed.), *The history of child-hood*. New York: The Psychohistory Press, 1974.

Dordick, H.S., H.G. Bradley, and Nanus, B. *The emerging network marketplace.* Norwood, N.J.: Ablex, 1981.

Dordick, H.S., P. Lum, and A. Phillips. Social uses for the telephone. *InterMedia*, 1983, 11(3), 31-34.

Elkind, D. *The hurried child*. Reading, Mass.: Addison-Wesley, 1983.

Gailbraith, J. *Organization design*. Menlo Park, Calif.: Addison-Wesley, 1977.

Gerbner, G., L. Gross, N. Signorielli, M. Morgan, and Jackson-Beeck. The demon-stration of power: Violence profile number 10. *Journal of Communication*, 1979, 26, 173-199.

Gershuny J. *After industrial society: The emerging self-service economy*. Atlantic Highlands, N.J.: Humanities Press, 1978.

Ghez, G. and G.S. Becker. *The allocation of time and goods over the life cycle.* New York: Columbia University Press, 1975.

Glossbrenner, A. *The complete handbook of personal computer communications.* New York: St. Martin's Press, 1983.

Gottman, J. Megalopolis and antipolis: The telephone and the structure of the city. In I. Pool (ed.), *The social impact of the telephone.* Cambridge, Mass.: MIT Press, 1977.

Hiltz, S.R. and M. Turoff. *The network nation.* Menlo Park, Calif.: Addison-Wesley, 1978.

Hirsh, P.M. The 'scary world' of the nonviewer and other anomalies: The reanalysis of Gerbner et al.'s findings on cultivation analyses, part I. *Communication Research,* 1980, 7, 403-456.

Korte, C. and S. Milgram. Acquaintance networks between racial groups: Applications to the small world problem. *Journal of Personality and Social Psychology,* 1970, 15, 101-108.

Long, L. and Boertlin. *The geographical mobility of Americans.* Current Population Reports, Special Studies Series, P-23, No. 64. U.S. Bureau of the Census. Washington, D.C.: U.S. Government Printing Office, 1976.

MacCannell, D. *The tourist: A new theory of the leisure class.* New York: Schocken, 1976.

McLuhan, M. and Q. Fiore. *The medium is the message.* New York: Bantam, 1967.

Nilles, J. et al. *The telecommunications-transportation trade off.* New York: Wiley, 1976.

Packard, V. *Our endangered children: Growing up in a changing world.* New York: Little, Brown, 1983.

Parsons, T. The social structure of the family. In R. Ansker (ed.). *The family: Its functions and destiny.* New York: Harper & Row, 1949.

Pattison, R. *On Literacy.* Oxford: Oxford University Press, 1982.

Phillips, A. Attitude correlates of selected media technologies: A pilot study. Los Angeles: Annenberg School of Communications, 1982.

———. Computer conferences: Success or failure? In R.N. Bostrom (ed.), *Communication yearbook 7,* Beverly Hills, Calif.: Sage, 1983, 837-856.

Phillips, A., P. Lum, and D. Lawrence. A conceptual framework for the cross-cultural study of telephone use. Paper presented at the Conference on Communications and Culture, Temple University, Philadelphia, March 1983.

Pool, I., ed. *The social impact of the telephone.* Cambridge, Mass.: MIT Press, 1977.

Postman, N. *The disappearance of childhood.* New York: Delacorte Press, 1982.

Rapoport, A. A probabilistic approach to networks. *Social Networks,* 1979 2(1), 1-18.

Reid, A.L. Comparing telephone with face-to-face contact. In I. Pool (ed.), *The social impact of the telephone.* Cambridge, Mass.: MIT Press, 1977.

Rice, R.E. Impacts of organizational and interpersonal computer-mediated communication. In M. Williams (ed.), *Annual review of information science and technology,* Vol. 16. White Plains, N.Y.: Knowledge Industry Productions, 1980, 221-249.

———.Communication networking in computer-conferencing systems: A longitudinal study of group roles and system structure. In M. Burgoon (ed.), *Communication yearbook 6.* Beverly Hills, Calif.: Sage, 1982, 925-944.

——— and Associates. *The new media: Communication, research and technology.* Beverly Hills, Calif.: Sage, 1984.

_____ and W. Paisley. The green thumb videotext project: Evaluation and policy implications. *Telecommunications Policy*, 1982, 6(3), 223-236.

_____ and W. Richards, Jr. An overview of communication network analysis methods. In B. Dervin and M. Voigt (eds.), *Progress in communication sciences*, Vol. 6. Norwood, N.J.: Ablex, 1984.

Robbins, L. *An essay on the nature and significance of economic science.* London: Macmillan, 1945.

Robinson, J. and P. Converse, The impact of television on mass media uses: A cross national comparison. In A. Syalai (ed.), *The use of time.* The Hague: Mouton, 1972.

Rogers, E.M. *Diffusion of innovations.* New York: Free Press, 1983.

_____ and R. Agarwala-Rogers. *Communication in organizations.* New York: Free Press, 1976.

_____, H.M. Daley and T.D. Wu. *The diffusion of home computers.* Stanford, Calif.: Institute for Communication Research, Stanford University, 1982.

_____ and L. Kincaid. *Communication networks.* New York: Free Press, 1981.

Ruchinskas, J. Communicating in organizations: The influence of context, job, task, and channel. PhD. dissertation, University of Southern California, 1982.

_____ and L. Svenning. Formative evaluation for designing and implementing organizational communication technologies: The case of video-conferencing. Paper presented at the annual conference of the International Communication Association, Minneapolis, May 1981.

Rytina, S. and D. Morgan. The arithmetic of social relations: The interplay of category and network. *American Journal of Sociology*, 1982, 88, (1), 88-113.

Sharp, C. *The economics of time.* New York: Wiley, 1981.

Short, J., Williams, E. and B. Christie. *The social psychology of telecommunications.* New York: Wiley, 1976.

Simon, H. Applying information technology to organizational design. *Public Administration Review*, 1973, 33, (3), 268-278.

Singer, B. Incommunicado social machines. *Social Policy*, 1977, 8, (3), 88-93.

_____. Crazy systems and Kafka circuits. *Social Policy*, 1980a, 11, (2), 46-54.

_____. *Social functions of the telephone.* Palo Alto, Calif.: R & E Associates, 1980b.

_____. Organizational communication and social disassembly. In L. Thayer (ed.), *Understanding organizations,* 1983.

Stein, M.R. *The eclipse of community: An interpretation of American studies.* New York: Harper & Row, 1966.

Svenning, L. Predispositions toward a telecommunication innovation: The influence of individual, contextual, and innovation factors on attitudes, intentions, and projections toward video-conferencing. Ph.D. dissertion, University of Southern California, 1982.

_____. Individual response to an organizationally adopted telecommunications innovation: The difference among attitudes, intentions, and projections. Paper presented at the annual conference of the International Communication Association, Dallas, May 1983.

_____ and J. Ruchinskas. Organizational teleconferencing. In R. Rice (ed.), *The new media: Uses and impacts.* Beverly Hills, Calif.: Sage, 1984.

Von Neumann, I and O. Morgenstern. *Theory of games and economic behavior.* New York: Wiley, 1964.

Weick, K. *The social psychology of organizing.* Menlo Park, Calif.: Addison-Wesley, 1969.

Whyte, W.F. *Street corner society.* Chicago: University of Chicago Press, 1955.

Wiener, N. *Cybernetics: Or control and communication in the animal and the machine.* New York: Wiley, 1948.

Williams, E. Experimental comparison of face-to-face and mediated communications: A review. *Psychological Bulletin,* 1977, 963-976.

Williams, F., J. Coulombe and L. Lievrouw. Children's attitudes toward small computers: A preliminary study. *Educational Communication and Technology,* 1983, 31, (1).

_____ and H. Dordick. *The executive's guide to information technology.* New York: Wiley, 1983.

_____, A. Phillips and P. Lum. *Some extensions of uses and gratifications research.* Los Angeles: Annenberg School of Communications, 1982.

_____ and R.E. Rice. Communication research and the new media technologies. In R. Bostrom (ed.), *Communication yearbook 6.* Beverly Hills, Calif.: Sage, 1983, pp. 200-224.

_____ and V. Williams. *Microcomputers in elementary education.* Belmont, Calif.: Wadsworth, 1984.

Winn, M. *Children without childhood.* New York: Pantheon, 1983.

11

The Impact of New Technology on the Acquisition, Processing, and Distribution of News

Jerome Aumente

Newer technologies have historically found fertile ground within the newspaper publishing industry from the days of movable type through the most recent advances in computerized editing and production and the satellite delivery of text and pictures worldwide. The latest technologies of videotex and teletext open yet newer dimensions. But electronic publishing also brings with it still uncharted implications for the mass communications industry. The linkage of the computer with the television or display screen for home delivery of text and graphics raises new questions involving the traditional role of the journalist in an age of high technology. The business alliances, matters of privacy and First Amendment protection, and a rewriting of the relationship between passive consumer and information supplier in an age of interactive and transactional exchange via the computer keyboard are just a few of the intriguing areas needing careful attention by specialists in journalism, mass communication, and information studies.

An Overview of Electronic Publishing

Print publishers, and especially the newspaper industry, have been quick to adopt technological advances and each level of new resources has brought about significant change. However, there is an unusual convergence of existing and emerging technologies today that could bring about gradual but major transformations in the way both the print and electronic mass media gather, process, and disseminate news and information.

The convergence of technologies—the personal computer, broadcast television, the rapid growth of cable television, satellite transmission, new initiatives by the phone industry to upgrade data transmission and cultivate new information services—have contributed to the birth of a fledgling industry, electronic publishing, with videotex and teletext offering new avenues for the distribution of text and graphics directly to the home or office on television screens and computer display screens.

Even in the very early stages, the new technologies are affecting traditional assumptions about the relationship between reader and editor or publisher, largely because of the shift in role from passive consumer to active user of the medium. Interactivity has come into play in ways that both the traditional print and electronic mass media have not experienced before in their relationships with readers, viewers, or listeners. Media consumers have shifted the way they approach and value the entire mix of news media resources based on their first, tentative exposure to videotex and teletext.

And all of this is taking place in an uncertain, if exciting atmosphere, in which the print publisher must confront new interpretations of First Amendment protections: when news is suddenly disseminated through a government regulated broadcast or cablecast industry; when sophisticated computer power suddenly offers the publisher great opportunities for measuring use of news and information, but with the concomitant risk of privacy invasion never before at issue; and when the print publisher must confront complex technical and business alliances that are without precedent.

For academic researchers, educators in journalism and mass communication studies, and policymakers and regulatory specialists in the mass media field, the technological trends in electronic publishing create new issues needing attention.

The focal point for this is bundled together under the rubric of electronic publishing. Some consider it an entirely new medium, while others see it as a hybrid of existing media and improved technologies seeking media market space to survive. Both videotex and teletext offer text and graphics displayed on a television set or video display screen in a wide range of colors, animation, and sophistication of image depending upon the level of hardware and software employed.

Videotex is interactive, with two-way exchange possible between home or office users and the electronic publisher's mainframe computers. A stand-alone videotex terminal or a personal computer with appropriate communication equipment allows for access and interaction, and the telephone line or two-way cable television (relatively rare today with only three percent of cable systems two-way) provides the linkage.

Teletext offers one-way selection of text on the screen, transmitted as part of the broadcast signal in the vertical blanking interval (VBI), the unused lines of the TV screen that appear as a black bar when the picture rolls, or delivered as full-channel service over cable television. The latter may have as many as 5,000 frames or screens of information, whereas the broadcast teletext usually involves a 100-frame teletext magazine that is constantly broadcast along with the regular TV signal. Videotex may have up to several hundred thousand frames in the database.

Using a hand-held keypad resembling a pocket calculator, or a more elaborate alphanumeric computer keyboard for videotex, home and office subscribers are being offered a wide range of news and information, games, message transmissions, entertainment, and with videotex, interactive services including home banking, reservation services, and shopping at home. With button selection, we are witnessing the evolution of the usually passive reader of print materials suddenly making an active range of choices on the sequence and amount of news and information selected. Videotex also offers the means to develop community bulletin boards or individual electronic mailboxes within the system.

Before reviewing trends in electronic publishing and the implications for future news gathering and dissemination, it is important to note several factors which put the developments in context:

- Technological innovation is not new to the newspaper publishing industry or the news services. Historically, they have been at the forefront in adapting new technology whether it was the telegraphic delivery of high speed news in the last century, the introduction of computerized technology during the 1960s to the present, or satellite transmission of data to remote printing plants for the creation of better distributed regional or national newspapers. This supportive atmosphere has made the introduction of videotex or teletext very much a reality.
- While electronic publishing represents a somewhat linear progression in technological development, it also brings newspapers and magazines to a significant cross-road. They are redefining their traditional roles, assuming new identities as broader gauge information and transactional service providers, and making long-range commitments that are far different from past expansion that simply meant an improved printed news product (perhaps produced faster, in more geographically dispersed regions, or with improved color or graphics).
- The changes brought about by the new technologies in electronic publishing will have major consequences for the training of journalists. These innovations will affect the reporters and editors, how they process news and information, and even how they define the type and amount

of editorial content of news and information services offered as a database.

- Electronic publishing may have an impact on the way news and information are provided by the electronic media—today's broadcast television, cable television, and radio. We have seen the convulsive changes that came about in newspapers when radio, and then television, recast the media habits of entire nations. In the United States, we have already seen erosion of the once dominant television network viewing as competition from cable television, videotapes, and videodisks all intrude on viewer time. Early data from electronic publishing tests tell us there will be further significant tradeoffs in how a household budgets its media/reading time when videotex or teletext services are available.

- Unlike previous technological advances in the mass media, electronic publishing demands—and may succeed in winning—fundamental changes in the behavior of those who use its services. There are commitments to hardware and supporting communication devices that are more complex. There is a need for action and interaction on the part of the user that is far different from the passive "read and browse" through a printed publication.

- There is a new community of interaction and information exchange among users of videotex through electronic community bulletin boards that binds them together in ways that no previous reader exchange through letters-to-the-editor or op-ed pieces could ever accomplish. There are also new attitudes emerging as to what constitutes news and information when the databases can provide a grand sweep of national and international details or the most local bit of minutia of school lunches or Little League scores.

The Outside Momentum

Unlike previous advances in the mass media, the momentum that is building for videotex and teletext involves many variables outside the control of the media organizations promoting the service. With radio and television, the content provided by the media organizations gave primary life and validity to the development, marketing, and improvement of the crystal sets, radios, primitive black and white TV sets, and then the most advanced digital, high fidelity color television sets. But with the medium of electronic publishing, the explosion of the personal computer market has created an environment of user acceptance for screen images and text information that is related but distinct from the growth of electronic publishing, though the latter has taken steps to benefit from the personal computer market.

In fact, the thrust of long-term planning is to wed the services of the electronic publisher with the memory of the personal computer, letting subscribers download data for later use that is accessed in the videotex database.

It comes as little surprise that the early subscribers to the news and information database services such as CompuServe show a high percentage of personal computer owners and those who use computers in the course of their daily work (Blasko 1982; RMH 1982). There are also market linkages that the electronic publishers want to forge with video game users who may be likely candidates for downloaded games transmitted to the home. This becomes significant as we see the video game manufacturers and computer manufacturers upgrading their products and appealing to the concern of parents with advertisements that assert that dumb terminals lead to dumb students.

In this highly volatile computer market, values are still being defined and electronic publishers talk at conventions of their peers of reforming the social contours of the media landscape to find an entry point for videotex. It is a far cry from the traditional print publisher designing a new publication and selling it primarily on its merits, without such concern for outside technological forces.

The Trends

There have been enough electronic publishing trials and commercial ventures worldwide (an estimated 80 projects involving over $500 million by 1983) to suggest that: 1) the attempts to find space in a very crowded mass media arena will not be easy; and 2) despite the difficulty, enough major publishing and communication organizations are now involved so that something will happen, with inevitable consequences for the mass media as we know them today. The most optimistic predictions are for a $10 billion electronic publishing industry by the early 1990s.

Perhaps one of the most exciting longer-range implications is to be found in the door to interactivity that has been opened, and that will not be easily closed as more individuals discover the sense of involvement and personal control over access to news and information, which the early tests have shown a very positive factor among users. They can retrieve a wide range of news and information in the amount and sequence they wish, building their own daily menu, and doing so in time patterns that meet their schedules rather than the conveniences of newspaper publishing deadlines or fixed-time broadcasting schedules.

The numbers of users are still small, but the early results suggest interesting attitudinal changes about what constitutes timely news, and this will inevitably have an impact on user attitudes toward the timeliness of printed news and broadcast news. Users also like the potential for interacting and communicating with fellow subscribers in the videotex system through community bulletin boards and individual electronic mailboxes. Some systems are encouraging direct dialogue and feedback between editors and readers of the database with information geared to user interests. *Consumer Reports* is involved in several videotex experiments and hopes to create a network of users reporting via videotex feedback, on their consumer experiences. *Better Homes and Gardens* has operated a *Cooks Underground* of videotex users communicating with each other and editors through CompuServe.

Segmenting and Pricing News

An important, and still uncertain, element in screen delivery of text and graphics is the segmenting of database components into a basket of services, with news, sports, weather, traffic, entertainment, consumer services, shopping, banking, stock quotes, financial investment guidance, reviews, and so forth, and games, puzzles, and other transactional services, each with a distinct subindex. Electronic publishers are trying to find the right mix of services and, with Prestel in England, are pricing individual items on the basis of use.

In the United States, videotex publishers have kept to a single package concept with a monthly subscription fee, but specialized services might be charged for on the basis of use. The Viewtron subscriber in South Florida can get the basic monthly service and then pay extra for services like the Dow Jones News-Retrieval.

Electronic publishing offers publishers the opportunity to deliver and price individual items, letting narrow interest services pay their own way. But there is the danger of an "accountant mentality" setting in, with some news items eliminated because they do not register high interest or use, although on grounds of good journalistic and editorial judgment they would be included. Coverage of foreign affairs is an example, where it is essential to get a full and balanced view of international events, even in those countries that might be less popular, or produce less colorful headlines.

There is a real danger that the new technology will remove readers from the broader landscape of news and information as they summon desired information frame by frame, but lose the power to browse through newspapers or magazines where unexpected items may catch their fancy. Videotex and teletext editors are trying various ways to prompt readers into related categories, or new discovery areas, but these techniques will take time to perfect and the single frame limitations of the medium will always be there.

On the other hand, videotex systems such as the CBS-ATT Venture One experiment have tried innovations that allow a user to program the service so that each day a personalized portfolio is ready and waiting with news, information, stock quotes, weather, and traffic reports, for instance.

This personalized service may soon improve. Electronic publishers are watching very closely as research into literal inferences per second (LIPS) proceeds. With the proper software and computer power, it might be possible for a service to monitor a subscriber's choices over time, and from them form a pattern of likely interests. Conceivably, the computer might suggest news and information that a subscriber ought to consider viewing because of interests shown previously in related material in the system.

This videotex interactivity may someday offer the value-added element electronic publishers need to gain entry into the very crowded print and electronic media marketplace.

Technological Advances

The publishing industry received its first push toward large readership when the machine replaced the human hand which did much of the lettering and illuminating of manuscripts until Johann Gutenberg married movable type to the wine press technology and came up with a speedier method of reproducing words on paper. As Ernest C. Hynds (1980) notes, however, there was little real breakthrough initially. Raised letters, inked and pressed to the page, and with a wooden frame holding the type on a flat bed of wood, required a minute to handset a line. But in the 19th century, linotypes allowed the machine setting of letters and the formation of lines of type arranged in columns.

The steam engine fostered rapid printing, and electrically driven presses pushed speed and production even further. Continuous webs of paper from giant rolls and with pages printed from stereotype plates offered speed, double-sided printing, high volume, and the use of pictures in half-tones and multi-column headlines. Each technological breakthrough required another, and as Hynds notes, a major advance came with paper produced from pulp rather than expensive cloth, and newsprint in the 19th century became the magic carpet for mass circulation newspapers to travel upon.

Meanwhile, also in the 19th century, the telegraph transformed the speed with which news and information could travel from the source to the editor and publisher. Paul Julius Reuter founded his news service in 1850, and used 40 well-trained carrier pigeons to span a 100-mile gap in the telegraph network between the Belgian capital of Brussels and Aachen in Prussia where brokers waited for stock information and news from Paris. Today,

Reuters News Service uses telephone, cable, video, and satellite with its data network and can provide five million bits of data per second, and millions of words, rapidly delivered in an information file that can be retrieved from a video display system hooked to a coaxial cable.

The telephone and the typewriter were early technological advances that gave editors and reporters a more rapid means of preparing readable information for typesetting and editing. And in the 1930s, the teletypewriter accelerated the setting of linotype with perforated paper. The wire services then offered their news over perforated tapes that could be fed directly into typesetting machines. But the increments were slow—by the 1960s, according to Hynds, the speed of a line of type setting was fourteen per minute, up from one line per minute with hand-punched keys, and up from four a minute with perforated tape.

Cold type in the decades of the 1960s and the 1970s allowed offset printing, with photocomposition of pages allowing for more display ads, headline choice, graphics, and freedom from the burdens of hot-lead typesetting. Hynds estimates that by 1978, 94 percent of all weeklies and 72 percent of dailies were using offset printing.

The Computer Revolution

Of all the technological transformations in the publishing field, the computer has had the most profound recent effect. In *Goodbye Gutenberg*, Anthony Smith (1980) notes that the term *revolution* is too often applied with indecent haste to what is mere innovation, but alongside two previous transformations worthy of the term in the storage and handling of information—writing and printing—he calls the marriage of printing and computerization, "the third revolution."

Indeed, the computer impact on the newspaper industry has been of major significance in its daily operations, and the stage is now set for the latest phase with electronic publishing. Newsrooms went from handwritten stories to manual then electric typewriters, and today many have fully computerized writing, editing, and composition. For a time, the optical character scanner took typed copy and transformed it into perforated tape or electronic signals for computer composition. But the microcomputer allowed direct writing, editing, and transfer of editorial matter to the composition room.

Today, most daily papers are participating in the computer revolution. Front end systems let screen copy go from the editor and reporter to final page layout electronically, and copy can also be fed in from other high speed wire services as data pumped along at the rate of 1200 words or more a minute (Hynds 1980).

As computers have become smaller, compact, and portable, reporters in the field are able to hook up their units to a phone and feed stories directly into computers at home offices for editing. The Associated Press has begun using the portable units everywhere, from Kentucky (where a judge let a reporter type silently away during the trial with the data fed to a bureau by phone later) to coverage of President Reagan's overseas trip to Japan (Gerstenzang 1983).

There is also increased reverse flow of stories from member newspapers to the Associated Press since nearly 500 AP member newspapers with computer capability went online with the wire service, offering dozens of stories via computerized electronic carbons, where before a newspaper might offer one or two stories per day. Nebraska, the first state to have all of its sixteen member papers on line, found that it offered 16,000 stories to the AP in 1982 without the need for individual dictation of stories (Schmahl 1983).

The current shift in newsrooms is from dumb terminals to computers that can be used not only for word processing and editing but also for storing, retrieving and accessing databanks and research archives. This level of sophistication gives reporters and editors added power to quickly check facts, pull from a variety of sources, and add depth to their stories.

Robotics has also entered the newsroom, and some newspapers are experimenting with computer software that can take the basic facts of a baseball game or wedding announcement, let the editor choose from a variety of leads and formats, and these templates with the added facts become the automatically written story. It has the potential to remove some of the drudgery and dog work from formula writing and leave room for more enterprise reporting.

Satellites and National Newspapers

The satellite has introduced another major technological change in the newspaper industry, and by the late 1970s, the wire services were delivering news to subscriber papers via satellite to earth receiving stations. It was during this period that the *Wall Street Journal* and the *New York Times* also began publishing satellite editions, and other major papers began using microwave transmissions from central city editorial offices to printing plants in the outlying suburban ring of the metropolis to ease newspaper delivery.

Technology sometimes springs from unexpected problems. During the oil crisis in the 1970s, the *Wall Street Journal* found its air freighted papers being left at the airline's loading docks as planes cut back on weight. The problem became a major impetus for the *Journal* to revamp its whole production and distribution approach according to William J. Dunn (1983), vice president and general manager of Dow Jones & Co. Inc., and president of its information services division. The facsimile process of transmission over microwave was expensive and only a few of its plants could afford the cost. The company was air freighting 25 percent of its papers, but this was halted by the oil crisis. It was under this pressure that the company examined satellite transmission and first leased and bought its own satellite transponder (paid out of the savings from the air freight cost reductions). The net result is that today seventeen regionalized printing plants take the satellite feed for local printing of the paper, no copies are air freighted, and with over 2 million copies a day, the *Journal* is the largest circulation paper in the United States, and with national distribution.

The trend toward national newspapers transmitted via satellite with regional printing of the airborne data continues dramatically today with *USA Today,* the flagship experiment of the Gannett Co. In January 1984, it took out a full page newspaper ad to announce "a winter heat wave"—a thermometer showed a December circulation figure of 1.3 million, less than two years after the paper was launched in September 1982. In a short time span, this nationally distributed paper became the country's number three paper in circulation, behind the *Wall Street Journal* and the *New York Daily News* (*New York Times* 1984).

The paper announced a goal of 2.3 million circulation by 1987, with satellite transmission at the heart of this ambition. Printed in Virginia, its pages are transmitted via satellite to printing plants across the nation. It stresses ample color to exacting standards and numerous short pieces aimed at a peripatetic readership; some have suggested its conciseness might make it a ready-made product for future videotex transmission as well. Its large weather map with a splash of colors has been widely imitated by other papers, and the blue and white modernistic vending machines, resembling a computer terminal on a pedestal, began appearing in major markets around the nation.

But *USA Today* has been losing money (costing Gannett a reported $100 million the first year and $70 million in 1983), and it has been fighting gamely to convince major advertising agencies that its circulation constitutes a market worthy of quality ads (Jones 1984).

In a more modest way, *The New York Times* has been slowly expanding its national edition, and by September 1983 had printing plants at four locations in Florida, Illinois, and California. The national edition is a scaled down version of the paper edited in New York and is transmitted via satellite with facsimile pages then made into plates for printing (*New York Times* 1983).

Electronic Publishing: The Inevitable Next Step?

Now that many publishers have computerized their editorial and compositional operations at great expense so that the gathering, processing, and preparation of the print product can be done with efficiency, speed, and reduction in human error, many wonder if the next step, delivery of the product as computerized data, is not inevitable. They have created a network of land lines, microwave, and satellite transmission far advanced from the days of Mr. Reuter, his telegraph and trained pigeons. Publishers also face growing costs for paper pulp, inks, postage, and delivery by truck and human carrier . They also realize that news has a short shelf-life and are looking for ways to recycle and reuse the product they produce.

It is in this context that a market for videotex and teletext seems feasible, not as a substitution for the newspaper or magazine, any more than print was replaced by broadcast news. It also comes in the midst of an international boom into personal computers which is making text on the screen a commonplace for many in our society. The remainder of this article will summarize some of the trends that relate to this, and suggest some of the areas that need further research and policymaking attention in the coming years.

New Media Structures and Services

Within the world of the mass media and communication industries, the corporate budget sheets are being jumbled as new alliances are made to create and deliver news, information, and transactional services through electronic publishing.

Witness the announcement that IBM, CBS Inc., and Sears, Roebuck & Co. plan a videotex venture in the next several years. Or the alliance between Field Publications in Chicago, Honeywell, a computer manufacturer, and Cintel, a major telephone company, to form KEYCOM for teletext and videotex ventures in electronic publishing. Or the alliance between AT&T and Knight-Ridder in the Viewtron videotex commercial venture in South Florida, or the AT&T videotex experiment with CBS Inc. in Ridgewood, N.J., called Venture One. Similar alliances are taking place between cable operators and newspapers, between broadcasters and publishers, and between print or broadcast media in the United States with electronic publishers in England or Canada.

It appears that the usual familial patterns of media conglomerates are changing in response to a new medium that requires different genus and species when it comes to cable, computers, telephone, broadcast, satellite, and the expertise of news organizations accustomed to gathering and editing news. These alliances produce capital and technological expertise in areas of computerization and telecommunication unfamiliar to print publishers.

The newspaper publishing industry is not, however, unfamiliar with media mixing of its holdings. Witness the Times Mirror Co.—owner of the *Los Angeles Times* and *Newsday,* cable television, radio and television, book and magazine publishing. It has also joined with Infomart of Canada (itself the partnership of major Canadian publishers, Southam and Torstar) to create Videotex America. Times Mirror has launched some of the more carefully structured tests of videotex over cable and telephone and is signing up other print publishers across the country to explore the videotex potential.

Worldwide, viewdata has shown its greatest activity in England where the British government developed videotex in the early 1970s, and through its BBC developed teletext services (Ceefax). Independent television developed the teletext service, Oracle. Today, there are over 1.1 million teletext sets in England and approximately 35,000 videotex users in the British Prestel system (Hooper 1983). France has pushed ahead with videotex and tied it into an electronic phone directory and information services program. Canada's Department of Communications developed the Telidon software and hardware with its superior graphics and system that helped set North American standards for videotex. West Germany, Japan, Australia and other European countries have also been involved in electronic publishing in recent years.

In the United States, we see significant experiments and commercial development with Knight-Ridder's Viewtron videotex venture in South Florida in 1983; the Dow Jones News-Retrieval Service with 100,000 subscribers accessing the system through personal computers; and teletext and videotex ventures with KEYCOM based in Chicago, and having launched a teletext service to cable subscribers via the satellite-delivered, superstation WTBS in Atlanta.

CBS Inc. and AT&T have joined together for a recent videotex experiment in Ridgewood, N.J., and AT&T has also developed a standalone videotex terminal for the Viewtron service in Florida. On the West Coast, Times Mirror Co. has tested videotex in the Los Angeles area and is moving toward a commercial rollout. Both Knight-Ridder, through its Viewdata Corp. of America, and Times Mirror, through Videotex America, have been signing up publishers around the country to explore joint ventures.

Time Inc. has invested nearly $30 million in an attempt to develop a satellite-delivered teletext service to cable systems, and in partnership with local newspapers for a two-tiered service of national and local videotex. But Time announced in 1983 that it was suspending the project after three years and was assessing its future moves. It was plagued by the cost of teletext decoders, too high in price and in insufficient numbers, and withdrew from the market.

In Canada, Teleguide has been used as a public access videotex terminal service in Toronto and has been attracting hundreds of thousands of users each month who want news, shopping, entertainment and dining service information, and similar ventures are being developed in the United States. Highly targeted speciality markets for videotex are also being developed such as Grass Roots, aimed at farmers and agribusiness, with similar services planned in the United States.

In short, there is no lack of activity in this crucial first research and development stage of electronic publishing. CBS and NBC are developing national teletext services; the public television stations are involved in research projects; others such as Taft Broadcasting have launched teletext, and the time-sharing operations such as CompuServe and The Source are also supplying news and wire service news and information via videotex and personal computer/telephone access.

User Reaction

Based on observation of these systems, interviews, and review of various reports, it is possible to generalize about some of the early use of electronic publishing. News is rated high, if not at the top of the list, in popularity among users of videotex and teletext (*Viewtron Newsletter*, 1982; Sider 1984; Goodall 1983; Blasko 1982; RMH 1982). News supplied by the Associated Press in videotex experiments (the Viewtron trial in Coral Gables, Florida, and the AP/CompuServe experiment) was accessed most often of all the news sources because the service was perceived as most timely (Blasko 1982; Fuller 1981; RMH 1982).

Electronic publishers have found that a cluster of services will attract potential subscribers, and that no single element such as news will be the reason to sign on. The convenience of transactional services and timely news, along with useful entertainment and reviews, for instance, will be a better draw than news or banking alone (Videotex America). Parents value the educational potential of the system, with their young children reading off the screen and accessing other news and information (Goodall 1983; Woolley 1982, 1983).

Opportunities for the user to become involved in choice, sequence of information, and control of the time frame, register high as a positive factor in most tests (Woolley 1982, 1983; Heilbrunn 1983; Smylie 1982; Gingras 1983; Sider 1982, 1983). The degree of interest in community interactivity and messaging through the use of electronic bulletin boards and individualized messages to electronic mailboxes scored high and caused some publishers such as Viewtron to devote more attention to the gathering of community news and "micronews" submitted by the subscribers themselves (Woolley 1982, 1983; DeJean 1982, 1983).

Users everywhere are impatient, signalling for a screen of information and then waiting several seconds for the information. Waits of ten or more seconds, which are possible with broadcast teletext, for the requested frame to come around in the cycle are burdensome and have led to editorial decisions to reduce the number of frames in the 100-page magazine or to repeat frames in the batch to cut waiting time. There are few or no waiting problems with phone and cable delivered videotex (Sider 1983 and 1984; Goodall 1983; Carey and Siegeltuch, 1982; Elton, Irving, and Siegeltuch 1982).

The various tests also show that age and gender are factors in use. The young and the male are more likely to use the system (and specific categories in the system) then women and older users (Videotex America 1983; RMH 1982; Carey and Siegeltuch 1982). But even over the course of shorter tests, a leveling out of these differences could be observed as the personal computer market increased and more people became familiar with keyboard access (RMH 1982). In a study of public access teletext terminals, it was found that while men and women passed by the terminal in equal numbers, men used the terminals in a ratio of 4 to 1 over women (Carey and Siegeltuch 1982).

In most of the tests, there was an immediate novelty burst of higher use, a leveling off of access of videotex and then a steady use of the systems. It was clear that a minimum of several months of testing—ideally six months—were needed to measure such usage dips and leveling off (Elton, Irving, and Siegeltuch 1982; RMH 1982; Fuller 1981).

In general, users did not have difficulty with the teletext decoders or the videotex terminals or personal computer accessing, and found the systems user-friendly and not inhospitable (Sider 1984; Carey and Siegeltuch 1982; Videotex America 1983). All of the tests involved data that for proprietary reasons is usually quite closely guarded by the commercial ventures.

Privacy Concerns

The computer is giving editors and publishers new tools that raise questions of both opportunity and concern. Electronic publishers can use their computer readouts to get instant updates, keystroke by keystroke, of what their subscribers are choosing that day, and what items are being ignored or less heavily accessed. This provides an instant marketing tool that allows the editors to measure usership/readership and make adjustments in content. It is something that print publishers require weeks, even months, to acquire through interviews and reader feedback.

But this new found power brings with it several concerns. Primary among them is that editorial content might too strongly reflect what is popular, rather than providing a balanced range of news and information. Readers may develop new tastes and interest over time if they are given an opportunity to read unfamiliar items of news and information. We do not know yet whether or how pressures resulting from instant computer read-outs will affect editorial content.

More seriously, the computer readouts could represent a danger of privacy invasion if an individual or family profile of choice of news items, banking, reservations, shopping, and other transactional fell into the wrong hands. The Videotex Industry Association has written a model privacy code for systems operators to voluntarily adopt. It sharply limits use of market data, gives the subscriber control over use, and may offset growing concerns that could lead to governmental regulatory action if the privacy issue is not resolved voluntarily.

The Changing Role of the Journalist

Electronic publishing is making many new demands on the traditional print journalists. Numerous interviews with videotex and teletext editors indicate that a new language of the screen is evolving. It involves a disciplined process, getting information from many sources, quickly condensing it under strenuous deadline conditions (a deadline-a-minute when there is the opportunity for instant updates), and writing for the eye and the screen. It is sometimes called "printed radio," but is far different because it is written for the eye, and not the ear.

Videotex writers must find grace and elegance in a screen of eighty words. They must integrate graphics into the storytelling, and where possible let the word step aside so that a graphic or animated drawing can tell a complex budget story quickly. This involves a far different role for the writer, and it is a product of the technology.

Teletext must compensate for its limited number of frames (100 in a typical magazine) by a process of constant updates, as many as 2,000 a day with KEYCOM, for instance. It can also offer a breaking story within minutes, with fast updates; the Falkland Islands invasion is said to have propelled teletext into the British user consciousness and showed the particular strengths of teletext as a news medium.

Journalists are being trained to think in clusters of news and information, to design stories that will prompt the viewdata user to related stories in the database or related information services (an online encyclopedia, for instance), and to write in the menu/index and tree structure of information that is peculiar to database publishing.

First Amendment Issues

Broadcasting has always been closely regulated by the Federal Communications Commission because of the scarcity of channel space. The government has defined elements of fairness and equal time which set forth guidelines on how a station must act in presenting controversial issues (all sides must be presented) or letting political candidates for certain office have equal time on the air.

Newspapers and magazines are immune from such government dictates under the First Amendment provisions of a free and unregulated press. What happens then, when a newspaper story appears on the screen as videotex or teletext? Is it subject to FCC regulation? The commission has taken the position that text on the screen is entitled to the same protections and noninterference as text in the paper. But this position is constantly subject to review and will only become an issue if and when electronic publishing becomes a significant new medium. It is replete with concerns that should be closely reviewed for policy and regulatory implications in the coming years.

Timely and Useful News

The early experiences with videotex and teletext show that users expect the timeliest news from these systems. In some cases, they have an unrealistic sense of the computer power and must be reminded that humans must still gather, process, and disseminate the electronic news.

Still, the electronic publishers realize that they must constantly freshen and update their news, and teletext editors were particularly aware of this. The technology, indeed, does allow a writer or editor to complete the story and instantly feed it to the database so that it can be accessed immediately by the user. No other print or electronic medium can offer such a close margin between story completion and delivery (assuming that radio and television broadcasting, except with rare major stories, is confined to the regular hourly broadcast schedule).

If electronic publishing becomes pervasive by the year 2000 or beyond, the teletext and videotex users may well be impatient with the slowness of news from conventional print and broadcast media. This might mean a restructuring and repositioning of the perceived news role of radio and television, just as newspapers had to adjust to the broadcast age.

Teletext and videotex, in competent editorial hands, may be useful for stimulating interest in the fuller newspaper report, or the visuals of the television report. And, in fact, some electronic publishers are experimenting with the interplay of screen text services to stimulate interest in their print product or television news program with references in the various media to the services of the counterpart report. Some publishers are also considering ways of eventually removing useful, but little read, items from their papers and offering them on the screen. Stock quotes, for instance, take up pages of paper, but a small percentage may read only many of the more esoteric stocks. These could be better transferred to the screen.

Cabletext and Video News

The interplay of media is apparent in the use of cable television channels leased by newspapers for screen text services. Recent surveys by the American Newspaper Publishers Association and the Newspaper Advertising Bureau (1983) showed that a majority of publishers in the United States and Canada are very interested in such ventures, as well as electronic publishing through videotex and teletext and low power television.

There are numerous ventures in the United States today where the newspapers put cabletext on a screen that is read in a scroll, without needing decoding devices, and is sometimes supplemented by live news video commentary and reports from the staff of the newspaper's newsroom. Readers are encouraged to turn to the paper for full details. Additional revenue bases are being created, as well as new uses for the news gathered by the paper.

This approach to cabletext and news video is seen as a way of letting newspaper publishers learn how to put up a text database, assemble a staff, and build a market for text on the screen. It positions them for the day when television sets will have built-in decoders, and when there is sufficient, inexpensive videotex and teletext hardware for accessing and justifying a fullscale service.

Summary and Conclusions

The newspaper publishing field, and increasingly the magazine field, are turning toward electronic publishing with videotex and teletext as possible areas for new development. It is the latest step in a natural progression that we have seen in publishing where technological advances have been absorbed, often affecting the way news and information are gathered, processed, and disseminated.

The telephone, telegraph, typewriter, printing advances, computers, satellite transmission, cable television, radio and television broadcasting have all affected print publishing in the past. The way reporters and editors function, what is reported, and how it is presented, have always been the interplay of standards and traditions and the resources and technology that were at hand.

The emerging field of electronic publishing, with videotex and teletext, is very much technology-bound. It is affecting the corporate business alliances that the publishers are designing. It is touching on the news and editorial product, and the relationship to other services. It is concerned about transactional services and interactivity between reader/user and editor in ways that are both new, and startling, to the traditional publishing field.

No one is certain of the outcome. There are still too many marketing variables, technological problems needing debugging, and an evolving and uncertain mass media consumer market to be defined.

What is clear today is that something very real and significant is happening, both in electronic publishing, and in the likely impact it will have on other segments of the print and electronic mass media. It is a fruitful area for teaching and research within the academic community, and for practitioners in the mass media, an intriguing and troublesome world without easy answers.

Note

During the last two years, I have had the opportunity to review some of the principal videotex and teletext experiments in the United States in connection with research for my forthcoming book *Words on Paper, Words on the Screen: Publishing and the New Electronic Pathways* which provides much of the background for this chapter.

References

American Newspaper Publishers Association and Newspaper Advertising Bureau. *Survey of newspaper involvement in new telecommunications modes.* Washington, D.C., June 1983, report supplied to the author.

Blasko, L. Summary remarks in Associated Press/CompuServe report (unpublished) released October 1, 1982 to the author and undated.

Carey, J. and M. Siegeltuch. *Research on broadcast teletext working paper number eight—Teletext usage in public places.* New York: Alternate Media Center, New York University, November 1982.

Compaine, B. *The newspaper industry in the 1980's.* White Plains, N.Y.: Knowledge Industries, 1980.

DeJean, D. Director Videotex Services, Times-Mirror, Co., personal interview, February 18, 1982 and July 18, 1983.

Dunn, W. Vice-president and general manager, Dow Jones & Co., personal interview, October 3, 1983.

Elton, M., R. Irving, and M. Siegeltuch. *Research on broadcast teletext working paper number six—The first six months of a pilot teletext service: Interim results on utilization and attitude.* New York: Alternate Media Center, New York University, August 1982.

Fuller, K. *Report on Associated Press's participation in Knight-Ridder Viewtron project.* New York: Associated Press, Fall 1981, report to membership supplied to the author.

Gerstenzang, J. Portable computers cover Reagan's trip. *Associated Press log.* New York: Associated Press, November 21, 1983.

Gingras, R. Consultant and former director, KCET teletext, personal interview, June 28, 1983.

Goodall, H. Managing editor, Taft teletext, personal interview, June 27, 1983.

Heilbrunn, H., editorial director, CBS Venture One videotex, personal interview, March 23, 1983.

Hooper, R. Building purposeful products. *Videotex '83 proceedings.* London: LondonOnline Inc., 1983, pp. 317-323.

Hynds, E.C. *American newspapers in the 1980's.* New York: Hastings House, 1980.

Jones, A.S. The ad woes at USA Today. *The New York Times,* March 3, 1984, pp. D1-ff.

Kagan, P. Associates. *Electronic Publisher.* No. 4, October 20, 1981, Carmel, Calif.

New York Times. Times will print at 2nd coast plant. September 9, 1983.

———. Advertisement for *USA Today,* January 30, 1984, p. A18.

Reuters. General information supplied to the author by Reuters.

RMH Research Associates, Inc. *Synthesis of findings for AP/Newspaper/ CompuServe program of marketing research,* Fairlawn, N.J., report to AP supplied to the author by AP and dated September 1982.

Schmahl, A. "Electronic Carbons: Editors use Systems: AP Gets Great Story Choice." *Associated Press Managing Editors News,* September, 1983, pp. 12-13.

Sider, D. Vice-president, Time Teletext, personal interviews, February 17, 1982, June 22, 1983, and March 1984.

Smith, A. *Goodbye Gutenberg: The newspaper revolution in the 1980s.* New York: Oxford University Press, 1980.

Smylie, B. Vice-president, KEYCOM, personal interview, November 10, 1982.

Stevens, P. Tiny terminals mean 'portable bureaus.' *Associated Press log.* New York: Associated Press, March 5, 1984, p. 1.

Variety. Results released on LA experiment with teletext. March 22, 1983, p. 25.

Videotex America. Preliminary research findings. Undated with supporting press releases supplied to the author by Videotex America, June 1983.

Viewtron Newsletter. Vol. 2, No. 1, June 28, 1982.

Viewtron Newsletter. Vol. 3, No. 2, June 27, 1983.

Woolley, J. Editor, *Viewtron Newsletter,* personal interview, February 4, 1982 and July 14, 1983.

12

Computer-Mediated Interpersonal Communication

James W. Chesebro

Treating local computer bulletin board systems as a medium of communication, this article assesses the degree to which computer bulletin board systems create, function, and influence interpersonal relationships. As a point of departure, the existence of "computer friendships" are recognized and examined as a mode of behavior which can potentially alter traditional theories of interpersonal communication. Toward this end, five differences between face-to-face and computer-mediated communication are initially identified. These differences provide a context for describing the process of accessing computer bulletin board systems, the geographic demographics of these systems, and the stability, availability, and content of computer bulletin board systems. Among several other observations, it is concluded that some thirty percent of the codable message units contained on these computer bulletin board systems are interpersonal.

The "computer revolution" has the potential of directly affecting all dimensions of our personal lives. Computers have already created totally new options for carrying out daily activities. Electronic shopping, telebanking, video games, word processing, telework, teletext, videotex, computerized educational programs, computer art, electronic yellow pages, home security monitoring, travel and entertainment reservations, and medical and psychological diagnoses can be executed by a computer user within the home. All of these services are now commercially available to the home computer user. Yet, the computer revolution is only at its inception. For example, computer-directed home robots and expert computer decision makers have only recently made their appearance and been advertised in popular computer magazines. At this point, the innovations to be introduced and the long-term social outcomes of this computer revolution are unpredictable, if not literally unimaginable.

This revolution has been with us for less than forty years. In terms of home computer use, the revolution is less than ten years old. Altair introduced the first mass-produced home computer for under $700 in 1975. Nonetheless, the latest available Gallup poll (Pollack 1983a) indicates that some three to five percent of the population or some twelve to fifteen million American households now own and operate a personal computer in their homes. An additional thirty-eight million people are "very interested" in purchasing a home computer. A quiet yet pervasive revolution has begun to occur in the American household.

Of the many possible uses of a home computer, computer-mediated person-to-person communication is now technologically feasible; fully operational and functional telecommunication capabilities can be executed with a home computer. With the proper peripheral equipment such as a modem and telecommunications program, a home computer system allows a user to make contact with virtually any other computer user similarly equipped.

This article deals with this kind of computer-mediated interpersonal communication, offering a six-fold analysis. First, computerized bulletin board systems are described as a medium in which interpersonal relationships can occur. Second, the reported interpersonal experiences on these networks are noted as well as the theoretical issues embedded in such claims. Third, some differences between face-to-face and computer-mediated interpersonal communication are described. Fourth and fifth, the demographics and current state of computerized bulletin board systems are characterized as a preliminary to the sixth and final objective of this essay, a description of the content of computer-mediated interpersonal communication.

Before turning to these substantive concerns, it should be noted that this article is exploratory. While quantifications are offered, more precise measurements and statistical analyses are required than could be carried out here or than the subject matter itself currently allows. As a result, this article should be viewed as a probe, a critical essay, or an exploration designed to reveal rather than to resolve issues.

Computerized Bulletin Board Systems

This analysis focuses on home computer uses of computer bulletin board systems (CBBSs). Capron and Williams (1982, 504) define bulletin board systems as "telephone-linked networks formed by users of personal computers that constitute free public-access message systems." In greater detail, Cambron (1983, 3) describes a bulletin board system as:

> A software package that sets up and operates a computer (usually a micro-computer/personal computer) as an unattended host system accepting tele-phone data calls through an auto-answer modem. There is usually no charge for accessing a bulletin board system other than the cost of the telephone call to it.
>
> A bulletin board system may rigidly control the user's activity or allow almost unlimited access to the computer. In most cases, only one person, or user, can access a bulletin board system at a time.
>
> BBSs usually contain a message database that is most commonly compared to the bulletin board at the local grocery store or laundromat. Anyone can leave a message on it and that message can be read by anyone else accessing the BBS. Facilities also exist for private "electronic mail" exchange between two or more parties.
>
> Bulletin board systems can exchange (upload and download) large blocks of information or programs with the user.

From a user's perspective, these computer bulletin board systems are voluntarily employed, typically operated free-of-charge to the user, have no direct relationship to the user's source or place of employment, and the time spent on these systems generally constitutes a portion of the user's leisure time home activities.

The first computer bulletin board system was established in Chicago in 1978. Within the last five years, hundreds of such systems have come into existence. The latest issue of *The On-Line Computer Telephone Directory* (1983) indicates that some 600 such computer bulletin board systems exist which have been "checked for a valid carrier answer" (p. 6). While this *Directory* is carefully and painstakingly updated every three months, one of the first things one encounters on a computer bulletin board system is the multitude of references to new telephone numbers not listed in the *Directory,* simply because fully functional computer bulletin boards can be and are easily created by any home computer user for a few hundred dollars. Accordingly, the number of computer bulletin board systems has been increasing geometrically.

For many home computer users, these computer bulletin board systems constitute a powerful motive for operating a home computer system. Alfred Glossbrenner (1983, 195), in his book *The Complete Handbook of Personal Computer Communications: Everything You Need to Go Online with the World,* reflects the enthusiasm these systems can generate for the home computer user:

> If you do nothing else with your communicating personal computer and modem, access a computer bulletin board system. In addition to providing

an inexpensive (or cost-free) introduction to life online, CBBSs offer you a front row seat at the "Revolution." The impact of the CBBS phenomenon has yet to be felt, but it's coming. When it arrives, it will sweep the nation.

But there's no need to wait for the number of personal computers to reach the critical mass necessary for this to happen. You can sign on to a CBBS right now and begin enjoying the fruits of the Revolution before it actually takes place.

At present, only ten to fifteen percent of home computer users have the equipment necessary to access these systems. Yet, virtually all households with such capacities use these systems. At present, Gallup estimates that between 945,000 and 1,575,000 home computer users actually employ computer bulletin board systems (Pollack 1983a). Noting a consistent increase in the sale of personal computer modems, Dataquest, a market research firm, has estimated that the yearly sales of personal computer modems will reach 2.2 million by 1987 (Pollack 1983b).

The Interpersonal Dimension

Beyond such enthusiastic use, these computer bulletin board systems also appear to affect the personal lifestyle of the users. Interpersonal relationships and friendships can apparently be readily formed through these systems. In the popular press, these relationships have been described as both easily formed and as powerfully intense experiences for the users involved. For example, Carpenter (1983) has argued that "people are not crying for airline schedules and biorhythms or even stock quotations, but to talk to one another. The one truly revolutionary thing that telecommunication offers is the ability to transform time and space between human voices" (p. 9). In her view, these computer bulletin board systems constitute nothing less than "an entire society on-line . . . a town, a club, a clique, a fantasy world, a dating service . . . or anything one wants it to be" (p. 11). As she concludes, "when you build a computer system, you're building a social system" (p. 11). More typically, users have expressed satisfaction with the quality and intensity of the interpersonal relationships formed through these systems. One user reports, for example, that "I have talked to some people for years without knowing where they live or their real names. Yet they are as much a presence in my life as if they were right in the room. They are my friends" (P. Kerr 1982, C1, C7).

In this context, Williams and Rice (1983) have argued that electronic media and the interpersonal realm are increasingly merging and blurring the distinctions among interpersonal, private, group, and public communication. As they put it, "the traditional categorized distinctions among dif-

ferent types of communication are being reduced by technological change"
(p. 200) which is "blurring lines between interpersonal and mass-mediated
contexts. We must increasingly account for the coalescence of personal,
organizational, and public contexts of human communication" (p. 201).
They specifically suggest that "sources of satisfaction for personal communication needs are no longer limited to face-to-face contexts, traditional
mail, or telephone" (p.220). Moreover, while Phillips (1982) has argued that
computers are perceived as possessing particularly low "familiarity" and
"personal" attributes when compared to other media, Williams and Rice
(1983) report that such reactions are not inherent to the medium, for the
degree of "social presence" (or the personal and social quality attributed to
a medium) can be affected by one's "willingness to overcome those restrictions by persuasive and stylistic strategies" (p. 204). Cathcart and Gumpert
(1983) have thus argued that "the *person-computer interpersonal encounter*
represents another facet of 'mediated interpersonal communication'" (p.
275).

Given such views, in terms of interpersonal communication theory, a
host of intriguing issues emerges. The definition of *friendship* itself becomes an issue. What must people know about each other before they can
say that they are friends? Can friendships be formed if the age, race, nationality, and gender of the significant other are unknown? While friends
may always self-disclose, are there minimum requirements regarding the
physical nature of another that we must know if we are to call another a
friend? Are close physical proximities and physical intimacy essential for
every "close" friendship?

These questions are not easily answered at this point. Existing computer-
mediated communication research has largely bypassed such issues. For
example, while the Winter 1983 issue of the *Journal of Communication*
explores telecommunications, only corporate, commercial, and educational systems are examined. Quite appropriately, the findings reported
deal solely with the effects of computer-mediated communication in business, consumer, and educational contexts. Similarly, while Kerr and Hiltz
(1982) report the results of twelve separate research projects, all of the
computer bulletin board systems examined focus only upon the communication styles of managers, administrators, and professionals participating
in educational, research, and corporate systems. Such systems are task-
oriented; they are not directly designed to encourage personal friendships
nor do existing research reports indicate that interpersonal relationships
are created in such systems. Indeed, Hiemstra (1982) indicates that after
users employ these task-oriented computer systems, they then initiate non-
computerized media (such as typewritten letters or telephone calls) to re-
establish and to reaffirm more personal contacts with the recipient of the
computerized message.

Theories and research findings regarding interpersonal communication have yet to recognize the existence of high technology as an interpersonal variable. Accordingly, computer friendships have yet to be explored systematically by interpersonal communication theorists and researchers. Moreover, we do not expect the issue of computer-mediated interpersonal communication to be easily resolved within the discipline.

McCroskey and McCain (1974) have argued that interpersonal relationships must always involve some kind of judgment about the physical attractiveness of the other. For others, this physical dimension is not apparently a feature of an interpersonal relationship. DeVito (1983) defines a friendship only in terms of the kind of psychological support the relationship provides: "Friendship may be defined as an interpersonal relationship between two persons that is mutually productive, established and maintained through perceived mutual free choice, and characterized by mutual positive regard" (p. 373).

If one takes DeVito literally, friendships can exist without a response to—or even a knowledge of—the physical characteristics of the other. Yet, at this point, we are second-guessing how these authors might react to the notion of "computer friendships." None of these authors has ever mentioned or reacted to the notion of "computer friendships" in print. When mentioned, it seems likely that a good deal of discussion will turn on whether or not a friendship can ever exist if it occurs only through an electronic medium and never as a result of face-to-face contact. Stages in the development of friendship typically include physical intimacy as one of the most important features of the "close friendship" and the "intimate friendship." Thus, an interesting situation emerges. If computer users continue to claim that they have developed friendships through solely electronic connections, will interpersonal specialists readjust their conceptions of friendships, begin to make distinctions among types of friendships, or simply deny that computer relationships exist as friendships? It is unclear how this issue will be resolved. At this point, one can only hope that the issue will be debated. For purposes here, it shall be assumed that computer-mediated interpersonal communication exists and that it is important to distinguish the ways in which face-to-face and computer-mediated communication systems differ.

Distinguishing Face-to-Face and Computer-Mediated Interpersonal Communication

Face-to-face and computer-mediated communication systems differ dramatically. The differences to be highlighted at this point have to do solely with the different modes involved in each system. These modes are so radically different in kind that significant contrasts between the two systems can be immediately noted even before the content of these two systems are investigated. While not exhaustive, a list of five of the differences between face-to-face and computer-mediated communication systems is identified here. These differences relate to the channels used, the type of discursive modes used in each system, the unique feedback systems built into each system, the different kinds of social roles involved in each system, and the use of time embedded in each system. Each of these variables deserves particular attention.

The Channel

In face-to-face exchanges, both verbal and nonverbal channels are employed. Mehrabian (1981) argues that nonverbal communication in face-to-face communication accounts for 93 percent of the social meanings conveyed.

In computer-mediated exchanges, only a verbal mode is employed. The nonverbal channel is eliminated. All social meanings must be translated into and conveyed by a verbal mode. At the same time, a higher degree of control exists in these computer-mediated exchanges simply because the "accidental" information conveyed by the nonverbal channel is eliminated.

The Discursive Mode

In the face-to-face exchange, verbal communication is typically oral. Vocal quality, pitch, and tone are as important as the content or ideas of the message itself. These vocalistic features allow face-to-face communicators to convey a great deal of information efficiently and without the kind of commitment involved in solely written message exchanges.

In the computer-mediated exchange, verbal communication must always be in written form. Beyond the required typing skills if messages are to be conveyed at any kind of reasonable rate, messages must always be input line-by-line. Moreover, asides are not easily incorporated in such systems except by way of parenthetical comments. In addition, modes of response that are typically conveyed only in an oral form must be translated into a written code. For example, a snicker must be included in a transmission in a written form such as "hee hee." Such asides, parenthetical comments, and "oral additions" require greater concentration and consciousness in creating computer messages. Accordingly, one might speculate that those who are more literate in the written mode will be more satisfied and confident in a computer-mediated system (Phillips 1983).

Feedback

In the face-to-face relationship, feedback is synchronistic. While verbal communication in a face-to-face exchange typically involves some form of turn-taking, nonverbal exchanges are constant and immediate. Even the absence of any expressive nonverbal response constitutes a form of immediate feedback to a verbal speaker in a face-to-face exchange. In his classic conception of human communication, Berlo (1960) presumed the existence of a face-to-face interaction and confidently noted that these exchanges were "dynamic, on-going, ever-changing, continuous" (p. 24).

In a computer-mediated relationship, the physical structure and physical requirements of the technologies involved necessarily require that all feedback be asynchronistic. In the computer bulletin board system, one user sends a message. The transmission must be completed before the receiver is able to respond to the sender's message. Only one-way transmissions are technically possible at any given moment. Obviously, because nonverbal variables are completely eliminated in such exchanges, only these one-way verbal transmissions define the feedback system. In particular, while most home computer telecommunication allows for both half duplex (only one-way transmissions) and full duplex (two-way exchanges of information) settings, the full duplex setting only allows for one-way transmission at any given time. A message must be transmitted and received before a subsequent response can be made. Thus, even with the full duplex setting, only one-way transmissions are technologically possible.

Computer users do attempt to compensate for this asynchronistic feedback system. Forms of computer etiquette exist. A user will typically send a relatively short block of copy which allows the responder to deal with a single issue. Moreover, standard abbreviations are used (e.g., "oic" for "Oh, I see!"). In addition, a sender will typically transmit a single message in groups of forty or so characters so that the receiver is able to read the message line-by-line rather than receiving entire blocks of information all at once. This method also decreases boredom for the receiver who would otherwise have to sit staring at a blank screen waiting for a response while the sender is entering data. The transmission of these smaller units of data simulates the more rapid kinds of information exchanges which occur in face-to-face interactions. When the entire message is completed, the sender will typically enter several blank lines (by hitting the "enter" key) to let the receiver know that he or she may now respond to the message. The use of such blank lines eliminates the need for such explicit and formal statements such as "message completed." Although all of these compensation strategies may be employed, the feedback remains essentially a one-way, asynchronistic mode of responding.

Social Roles of the Participants

In face-to-face exchanges, participants necessarily provide a complete and immediate sociological composite of the self. A face-to-face exchange necessarily includes a presentation or "statement" of one's age, race, nationality, and gender. In addition, a host of other messages regarding one's place in society are provided by one's physical characteristics. Occupation, economic income, and even social preferences may be (correctly or incorrectly) inferred by one's physical features. The decision to convey information regarding all of these sociological variables is simply not possible in face-to-face contacts. Our physical presence automatically conveys these social characteristics to others.

In the computer-mediated exchange, a person exerts discretionary control over what kinds of and when sociological information is to be conveyed to others. In addition, the ways in which these sociological factors are to be characterized is determined by the individual to be described. Accordingly, particular attitudes, ideas, and beliefs of a computer user can initially establish a definition of a person for other computer users rather than the apparent sociological classes with which the user would normally be associated.

The Use of Time

Face-to-face interactions always occur in real time. The moment at which a verbal utterance is made, it is conveyed to another. Likewise, the moment at which a nonverbal signal is initiated, it is received by another. Time itself cannot be manipulated in a face-to-face exchange. When something occurs, it is automatically transmitted to others. While we may prepare for social exchanges before they occur, the exchange itself is defined by what we do at each moment during the exchange itself. Even a pause during an exchange may convey meanings to others. Others may, for example, associate a pause in a conversation with slow wittedness, indecisiveness, or perhaps even dishonesty. Thus, during a face-to-face exchange, every moment counts and every moment has a particular characteristic and quality which affects the social relationship.

In computer-mediated interactions, time can be more directly controlled and manipulated. Preparation time can be employed during message construction. Virtually all computer bulletin board systems allow a user to access the system and read messages at their own reading speed. In addition, private mail can be directed to anyone on the system. The construction of the letter to be sent can be as carefully worded as desired. The receiver is unaware of the length of time or precision given to any message he or she receives. Moreover, such mail systems allow two people to stay in contact even if they work at very different hours, live in different time zones, or function during different times of the day.

Even if the users are engaged in on-line chatting in real time, time must be provided for message construction. If nothing else, users must necessarily allow for those who type slower, for transmission problems, for differences in transmission systems, and so forth. Delays, a consistent feature of all computer-mediated interactions may create boredom during an online chat if excessive, but this delayed time provides the user with more time to prepare and to transmit messages. Accordingly, greater attention can be given to the construction of (and hopefully a more accurate statement of one's meanings in) a message transmitted during a computer-mediated exchange. Time thus becomes a variable which can be more directly controlled and manipulated toward personal ends.

In Sum

The mode or method employed to create an interaction dramatically affects the kind of relationship established. Face-to-face message constructions are characterized by the use of a complex, spontaneous, simultaneous, and immediate collage of verbal, nonverbal, and oral symbols. In contrast, computer-mediated message constructions are characterized by the use of written, critical, deliberate, and delayed symbol using. In this context, it is useful to highlight the main difference between these two message styles in terms of the tradition of the discipline. In this view, face-to-face messages function as a form of epideictic discourse, an inherent form of praising and dispraising, and a kind of celebration in the complexity of the "here and now." In contrast, computer-mediated messages more clearly resemble deliberative and forensic discourses in which a delayed and critical style characterizes its mode. In sum, the dominant metaphor of face-to-face communication is the anecdote with its attendant concern for the dramatic, while the dominant metaphor of computer-mediated communication is science with its attendant concern for regulation, precision, and delayed response.

Stages in Accessing Computer Bulletin Board Systems

Beyond the formal features characterizing message construction during computer-mediated interactions, the process of accessing a computer bulletin board system is sufficiently structured to warrant description. The formatting effects of the accessing process generate three particular stages of participation. Rather than propose a formal theory of phases, these stages suggest the overall structural process built into the use of computer bulletin board systems.

Those new to a computer bulletin board system initially experience a kind of "fish bowl" effect. This initial stage is characterized by a kind of random searching of all of the options of the system, the commands used to enter and exit options of the system, in nothing less than a kind of bug-eyed fashion. The command and input of the computer system itself dominates at this stage, creating a conscious awareness of the system itself as a formal grid rather than as a mode for human contact. Phillips (1983) reports that at this stage, users are extremely likely to report a sense of "aloneness."

As other persons on a system are contacted, a second stage of participation emerges. During this early stage of human interaction, exchanges are controlled by the established parameters of the system. If a computer bulletin board system is established for the exchange of computer information, conversations are dominated by such concerns. If the system is established as a dating service, the parameters of the system govern the content of these early conversations. This system-governed stage provides, of course, a convenient set of commonly shared topics during the initial stages of a relationship. At this stage, then, the computer system functions preeminently as a group medium (Hiltz and Turoff 1978) and "a person may participate in widening circles of interaction—programmed access facilitated by the computer" (Williams and Rice 1983, 208).

Finally, as both the system and those functioning on the system become familiar to a user, the constraints of the system can be transcended. Communication no longer needs to be governed by the stated purposes of the computer bulletin board system. Likewise, as more information is gained about other users on the system, new topics—natural extensions—become appropriate. In this sense, the computer bulletin board system begins to function as a more open communication system.

Ultimately, then, the computer bulletin board system begins to function as a vehicle for new explorations. Experience with one computer bulletin board system "naturally" leads to the exploration of other systems devoted to other kinds of topics. These new systems contain an element of "computer shock," which brings us full circle to the first stage in which a "fish bowl" effect occurs.

Geographic Demographics of Computer Bulletin Board Systems

The geographic location of these computer bulletin board systems provides an initial description of the basic demographics of the nature of these systems. A complete demographic analysis of computer bulletin board systems is not possible, for commonly shared data regarding each system is not available; the systems have simply not been in existence long enough for such detailed analyses. In addition, because users may intentionally remain anonymous, a traditional sociological composite of users is likewise impossible. As we shall suggest later, however, even a preliminary geographic analysis does generate a host of important interpersonal considerations which may reveal the motives for establishing and using these systems.

In order to determine the geographic location of computer bulletin board systems, an exploratory analysis was performed. All telephone numbers listed in the *The On-Line Computer Telephone Directory* constituted the database for this examination. Beyond providing the most comprehensive list of actual telephone numbers, this *Directory* is updated by a computer which automatically dials each number to verify a computer connection. If a valid carrier signal is not received, the number is automatically removed from the *Directory*. In terms of both its comprehensiveness and accuracy, *The On-Line Computer Telephone Directory* is "*the* directory of available CBBSs" (Glossbrenner 1983, 191).

The *Directory* lists 603 local telephone numbers, established by home computer operators, which are typically provided to users at no charge. To facilitate use of these systems, all telephone numbers are classified in the *Directory* by the type of operating system that controls each computer bulletin board system rather than by geographic area. To employ the information contained in this *Directory,* all telephone numbers were intially grouped by area code, classified into one of the four major geographic regions in the United States, classified by the state in which these area codes existed, and finally classified, if possible, by the city in which the area codes existed.

The results of the classification are provided on Table 12.1.

Five conclusions emerge from Table 1: First, 71 percent of all computer bulletin board systems are located in the Eastern and Pacific time zones of the United States; second, 59 percent of all computer bulletin board systems are concentrated in eight of the ten most populated states; third, 30 percent of all computer bulletin board systems are located in eight of the fifteen most populated cities in the United States; fourth, 35 percent of computer bulletin board systems are located in ten of the fifteen most densely populated cities in the United States; fifth, the fifteen largest concentrations of computer bulletin board systems are all concentrated in major American cities.

While a host of other variables must certainly be considered, computer bulletin board systems are concentrated in the most populated and most densely populated urban centers in the United States. It may well be that computer bulletin board systems are concentrated in these urban centers to deal with the rather dramatic communication problems which exist in urban centers. Computer bulletin board systems constitute a new kind of social order in which the computer is employed to create a community with new communication options and networks to resolve several of the problems embedded in urban centers.

TABLE 12.1
Geographic Concentrations of Computer Bulletin Board Systems

Regional Concentrations

	Number of CBBSs	Percentage of Total CBBSs
Eastern time zone	289	48
Central time zone	148	25
Mountain time zone	22	4
Pacific time zone	138	23
Canada	6	1
Total	603	101[1]

State Concentrations

State	Number of CBBSs	Percentage of Total CBBSs	National Population Rank of Each State[2]
California	100	17	1
New York	50	8	2
Michigan	42	7	8
Texas	41	7	3
Florida	39	6	7
Illinois	37	6	5
New Jersey	28	5	9
Virginia	22	4	14
Massachusetts	20	3	11
Ohio	19	3	6
Total	398	66	

City Concentrations

City	Number of CBBSs	Percentage of Total CBBSs	National Population Rank of Each City[2]	National Density Rank of Each City[2]
Detroit	39	6	6	5
Los Angeles	30	5	3	2
Chicago	28	5	2	3
Miami	25	4	41	21
New York	22	4	1	1
Newark	22	4	46	17
San Francisco	22	4	13	6
Boston	18	3	20	10
San Diego	18	3	8	20
Seattle	17	3	23	23
Arlington	16	3	—[3]	7
Houston	15	2	5	9
Baltimore	15	2	9	14
Minneapolis	14	2	34	15
Portland	14	2	35	32
Total	315	52		

[1]Rounding off accounts for differences above and below 100 percent.
[2]All national population and population density figures are provided by the 1980 Bureau of the Census Report as provided in the *Information Please Almanac 1982* (New York: Simon and Schuster, 1981), 755, 677-86, 751-54.
[3]Not ranked within the top fifty cities by total population.

Insofar as computer bulletin board systems do function as important interpersonal mechanisms, these systems should be located in geographic areas where interpersonal needs are greatest. Urban settings constitute the areas with the most outstanding interpersonal problems and generally constitute paradoxical contexts which hamper rather than foster the development of meaningful interpersonal relationships. While a larger number of people exist from which more meaningful interpersonal relationships might be created, the urban center itself requires that an increasing number of "blocking devices" be used to prevent "information overloads" (Knapp 1978, 120-31). Thus, the shift to urban centers since the turn of the century has required that an increasing number of people live in ever-decreasing geographic areas without a concomitant increase in the number and quality of interpersonal relationships. Accordingly, population size and population density merge to create overcrowding which functions as a relevant factor affecting communication. Rather than promoting communication, large and densely populated centers are therefore presumed to decrease the total amount of meaningful interpersonal communication.

This preliminary examination of the content of computer bulletin board systems suggests that these systems may well be created to overcome the problematic effects of urban life. While perhaps overstated, Carpenter (1983) may be correct when she argues, "The one truly revolutionary thing that telecommunications offers is the ability to transform time and space between human voices" (p. 9), and that these computer bulletin board systems function predominantly as a new kind of "social system" (p. 11).

The limits of this hypothesis must also be recognized. In some cases, an area code may include both a large city as well as a rural population. The use of only area codes prevents us from identifying the exact location of a computer bulletin board system under such circumstances. In addition, nothing in the data provides a warrant for the conclusion being offered here; evidence for a causal link does not exist, though a systematic pattern is evident in the data.

The Stability and Availability of Computer Bulletin Board Systems

As noted at the outset in this article, computer bulletin board systems have been cast as readily accessible and available to all. In fact, Capron and Williams (1982) define computer bulletin board systems as "free public-access message systems" (p. 504). As a result of the apparent ease of using these networks, some have even claimed that these computer bulletin board systems constitute a new mode for the achievement of participatory democracy. In this view, computer networks are frequently cast as easily created, reasonable to maintain, available to all, readily accessible, and typically operated at no charge to home computer users. Carpenter (1983) has argued, for example, that these systems provide "a place where thoughts are exchanged easily and democratically" and where "minorities and women compete on equal terms with white males, and the elderly and handicapped are released from the confines of their infirmities to skim the electronic terrain as swiftly as anyone else (p. 8). Carpenter concludes that these systems are available to "anyone with a home computer, one inexpensive piece of software, and a 'modem'" (p. 8). In a similar vein, Glossbrenner (1983) argues that these networks "exist today" and are "open to anyone with a personal computer" (p. 1). He observes, "As long as you have a general idea of how to operate your equipment, you can do it right now and obtain the phone numbers of *several hundred* free computer bulletin board systems (CBBSs) across the United States and Canada" (p. 5).

To determine the accessibility and availability of these networks, a 10 percent computer-generated sample (*User's Reference Guide* 1981, 95-96) of 60 computer bulletin board systems was drawn from the 603 computer bulletin board systems listed in the latest issue of *The On-Line Computer Telephone Directory*. Such a sample does not represent all computer bulletin board systems in the United States. Certainly, a host of institutional systems are to be found in corporate, educational, and institutional environments; designed as task-oriented systems, these systems have already been the subject of extensive examination (Hiemstra 1982; Kerr and Hiltz 1982; Williams and Rice 1983). In addition, the nationwide commercial linkages established by CompuServe and The Source are excluded here to highlight the activities of local, grass roots boards created in areas where face-to-face contacts are theoretically possible yet interactions are carried out through computer-mediated systems.

The sample generated roughly approximated and paralleled the national distribution of computer bulletin board systems reported earlier in this study. Specifically, twenty-four states were represented in the sample, with six of the eight most frequently represented in this sample also being ranked in the top ten states as reported in Table 12.1. In particular, the eight states most frequently represented in this sample include California with 12 (20 percent of the sample), New York with 6 (10 percent of the sample), Minnesota with 4 (7 percent of the sample), and Florida, Illinois, Washington, Texas, and Michigan each with 3 (5 percent of the sample). The remaining 38 percent of the computer bulletin board systems were rather evenly distributed (1 or 2 networks per state) among the remaining sixteen states represented in the random sample.

During a one week period at the end of October in 1983, each of the 60 computer bulletin board systems in the sample was contacted by way of a computer-modem connection.

Three findings emerge. First, 9 (15 percent) of the contacts could not be made because the telephone number was no longer in service. Second, 26 (43 percent) of these numbers did not have a valid carrier signal, indicating that the computer bulletin board system was no longer operative, or in one case, generated only a scrambled data transmission signal. Third, of the 25 systems remaining, 11 (18 percent) required a special password for entry into the system or to gain access to the messages on the system. Thus, of the total number of systems contacted, only 14 (23 percent) of the computer bulletin board systems were actually functional and available for public access.

Based on this data, one would conclude that computer bulletin board systems are not readily accessible nor are they available to all. While easily established, the turn-over rate in these publicly accessible and available systems is phenomenally high; the majority of systems shut down within three months due to the time and money involved in maintaining them. Of the remaining systems, nearly 40 percent are operated at a direct cost to the home computer user. Thus, less than 25 percent of the systems currently listed are, in fact, readily accessible and available to all.

The Content of Publicly Available Computer Bulletin Board Systems

The 14 publicly available computer bulletin board systems from the random sample just described were the basis for the following content analysis. Each system was entered, the public bulletin board or its equivalent on each of these systems was accessed, and whenever possible the last ten messages left on these boards were "downloaded" and later printed to provide a verbatim transcription of all messages extracted from these systems. This procedure generated 126 messages.

These messages were classified by their central topic, and listed in order of the most frequently discussed issues:

N	%	Topic*
24	18	Computer Information/Instructions Offered
18	14	Computer Equipment/Software for Sale
16	12	Self-Description Provided or Requested
14	11	Computer Information/Instructions Requested
10	8	Sexually-Oriented
10	8	Social Contact Sought
10	8	Request for other CBBSs Numbers
10	8	New CBBSs Numbers Announced
4	3	Computer Equipment/Software Needed
4	3	CBBS Operating Issues

2	2	Request to Contact a Specific User
2	2	Health/Fear Issues
2	2	Employment Listings
2	2	Employment Needed
2	2	Compliments to CBBS
2	2	Noncomputer Items for Sale

*Rounding off accounts for a total percentage of 103 percent. In addition, because a single message might contain more than one topic, the total number of topics equals 132.

Several observations regarding these topics are appropriate: First, these topics roughly parallel the results obtained by Danowski (1982) when he examined a single computer bulletin board in Boston; second, computer bulletin board systems, at this point in time, are predominantly a communication system for computer "hackers" and those seeking information regarding computer systems; third, some topics are apparently interpersonal in their central concern. Roughly 30 percent of the messages have such an interpersonal dimension. Messages offering or requesting self-descriptions, sexually-oriented notices, requests for social contacts, and requests for a specific user to contact another user all appear social rather than task related.

Yet, the interpersonal nature of these network exchanges remains an issue. Some researchers have concluded that computer-mediated communication reduces interpersonal satisfaction (Hiemstra 1982) and leads to "depersonalized, task-oriented discussions" (Fowler and Wackerbarth 1980), whereas others have argued that computer-mediated communication is "typically more personal than those in mass media" (Danowski 1982, 907) and richer than face-to-face conversations because participants drop the front that occurs during face-to-face interactions (Hiltz and Turoff 1978, 94). These conclusions are predominantly based on the reported responses of those using computer networks. Such methods of analysis have, thus far, produced only contradictory results.

Rather than rely on the responses of users to computer bulletin board systems, a system of analysis was sought which focused upon the content of the messages themselves. Most operational definitions of interpersonal communication presume that face-to-face interactions occur during an intimate exchange. Therefore, dimensions common to face-to-face interaction, such as incremental risk taking, progressive changes in the use of nonverbal cues, and so forth, were examined here.

After considering a number of options, Robert Shelby Frank's (1973) linguistic theory of verbal kinesics was ultimately employed as the basis for measuring the degree of interpersonal communication on these computerized networks. Because Frank has reported exceptional success with his mode of analysis in terms of both validity and intra- and intercoder reliability, the system seemed particularly appropriate for the ends governing this analysis.

Six conceptions embedded in Frank's mode of analysis are particularly relevant to this study. First, the method focuses on language choices or words as the basic unit of analysis. Second, the method presumes that language not only reflects but is used to create social and personal relationships. Third, the method assumes that all language constructions embody the use of varying but identifiable kinesic relationships to positively and negatively valued objects mentioned in any linguistic unit—specifically, forms of immediacy are linguistically employed to establish a closer relationship to positively valued objects of reference, while forms of nonimmediacy are linguistically employed to "distantiate" negatively valued objects of reference. Fourth, forms of immediacy, positively valued objects, and intimate relationships are characterized in discourses by the use of vertical symbols and metaphors. In contrast, forms of nonimmediacy, negatively valued objects, and stressful relationships are characterized in discourses by the use of horizontal symbols and metaphors. Fifth, particular coding rules and examples of each rule are provided for classifying specific sentences and words. Thus, for example, in more interpersonal relationships, a person is more likely to say, "We're moving up," while in a less involving or more stressful situation, a person is more likely to say, "Let's explore our options." Sixth, a mathematical ratio can be constructed in which the number of vertical symbols is divided by the number of horizontal symbols for a particular message, producing a V/H ratio. In practice, the higher the real number of the V/H ratio, the greater the use of vertical to horizontal symbols and therefore the higher the hypothesized interpersonal level of the symbol user. Conversely, the smaller the value of the real number, the greater the use of horizontal symbols to vertical symbols, and therefore the higher the hypothesized stress level of the symbol user. Frank notes that an equal number of interpersonal and stressful message units typically produces a norm of .589. Message units with a higher number of interpersonal or vertical symbols will have a higher real numeric value, while message units with a higher number of stressful or horizontal symbols will have a lower real numeric value, in terms of this .589 norm.

Applied to the messages derived from the computer bulletin board systems investigated in this study, three major results emerged: 1) When all of the messages are analyzed, these computerized networks are predominantly stressful in nature, with a V/H ratio of .312; 2) when only the topics expected to possess an interpersonal emphasis (sexually-oriented and requests for social contacts) are examined, the interpersonal measure increases dramatically to a V/H ratio of .875; 3) those topics most likely to be stressful were commercial in nature (offering computer equipment/software for sale), possessing a V/H ratio of .000.

Overall, it is concluded that computer bulletin board systems do possess a significant interpersonal dimension. In total, 31 percent of the codable message unit metaphors contained on these computer bulletin board systems constitute, by the measures employed here, interpersonal communication.

Conclusions

This examination of computer bulletin board systems generates several conclusions which are particularly noteworthy insofar as they may reveal broad social issues related to the computer and information revolutions.

First, computer-mediated communication does not function solely as the property of larger, complex societal organizations and institutions. While computer-mediated communication systems may foster task-orientations in institutional settings, there is nothing inherent within these systems which prevents them from functioning in the interpersonal realm to increase social presence and to establish and sustain more personal and social human linkages.

Second, the dramatic increase in the number and use of local computer bulletin board systems may signify an immediate fascination in the computer and information revolutions for a tremendous variety of individuals throughout the United States. Computer-mediated communication systems are becoming increasingly popular.

Third, while computer technology continues to be the dominant topic on computer bulletin board systems, some 30 percent of the messages exchanged on these networks are interpersonal in nature. Such a finding suggests that interpersonal communication may be the result of a merger of private and public, individual and mass, and personal and group technologies and contexts. Mediated interpersonal communication now appears to be a demonstrated entity.

Fourth, while computerized systems are now perceived as predominantly unfamiliar and impersonal, these systems can be employed in ways which reduce the indifference associated with such systems. How these systems are ultimately used appears to be as important as, if not more important than, how these systems are initially perceived.

Fifth, interpersonal communication need not be defined by an association with any particular medium. While face-to-face, telephone, and personalized letters may continue to provide channels for interpersonal communication, alternative technologies also appear to be emerging as vehicles for interpersonal relationships.

Sixth, the computer revolution appears capable of creating and sustaining interpersonal communication. Rather than separating people from others, computers may function as an outreach system in which new types of friendships are created and sustained through an electronic connection. Indeed, an "intimate" computer network may ultimately link people interpersonally throughout the United States. While face-to-face and physical intimacy will undoubtedly remain powerful motivating factors in human contacts, new modes of interactions may yet develop.

Although these explorations require additional research, they may subsequently alter traditional theories of communication in rather profound ways. Indeed, a speculative and creative perspective seems to be required if the full potential of these new media is to be fully understood. The media themselves are the determinants of the social uses and social consequences that emerge from their use.

References

Berlo, D. K. *The process of communication.* New York: Holt, Rinehart & Winston, 1960.

Cambron, J. Telecommunications glossary. *The On-Line Computer Telephone Directory,* 1983 (Summer), 12, 2-4.

Capron, H.L. and B.K. Williams. *Computers and data processing.* Menlo Park, Calif.: The Benjamin-Cummings Publishing Company, 1982.

Carpenter, T. Reach out and access. *The Village Voice,* September 6, 1983, pp. 8-12, 33.

Cathcart, R. and G. Gumpert. Mediated interpersonal communication: Toward a new typology. *Quarterly Journal of Speech,* 1983, 69, 267-277.

Danowski, J.A. Computer-mediated communication: A network-based content analysis using a cbbs conference. In M. Burgoon (ed.), *Communication yearbook* 6. Beverly Hills, Calif.: Sage, 19 82.

DeVito, J. *The interpersonal communication book,* 3rd ed. New York: Harper & Row, 1983.

Fowler, G. D. and M. E. Wackerbarth. Audio teleconferencing versus face-to-face conferencing: A synthesis of the literature. *Western Journal of Speech Communication,* 1980, 44, 236-252.

Frank. R. S. *Linguistic analysis of political elites: A theory of verbal kinesics.* Beverly Hills, Calif.: Sage, 1973

Glossbrenner, A. *The complete handbook of personal computer communications: Everything you need to go online with the world.* New York: St. Martin's Press, 1983.

Hiemstra, G. Teleconferencing, concern for face, and organizational culture. In M. Burgoon (ed.), *Communication yearbook* 6. Beverly Hills, Calif.: Sage, 1982.

Hiltz, S. R. and M. Turoff. *The network nation: Human communication via computer.* Reading, Mass.: Addison-Wesley, 1978.

Kerr, E. B. and S. R. Hiltz. *Computer-mediated communication systems: Status and evaluation.* New York: Academic Press, 1982.

Kerr, P. Now, computerized bulletin boards. *New York Times,* September 16, 1982, pp. C1, C7.

Knapp, M. L. *Nonverbal communication in human interaction,* 2d ed. New York: Holt, Rinehart & Winston, 1978.

McCroskey, J. C. and T. A. McCain. The measurement of interpersonal attraction. *Speech Monographs,* 1974, 41, 261-266.

Mehrabian, A. *Silent messages: Implicit communication of emotions and attitudes,* 2nd ed. Belmont, Calif.: Wadsworth, 1981.

On line computer telephone directory, the. Kansas City: OLCTD, Summer 1983.

Phillips, A. Attitude correlates of selected media technologies: A pilot study. Los Angeles: Annenberg School of Communication, 1982.

Phillips, A.R. Computer conferences: Success or failure? In R.N. Bostrom (ed.). *Communication Yearbook 7.* Beverly Hills, Calif.: Sage. 1983.

Pollack, A. Finding home computer uses. *New York Times,* May 11, 124, 1983a, pp. D1, D7.

_____. Moving data from a to b. *New York Times,* September 22, 1983b, p. D2.

User's reference guide. Dallas,: Texas Instruments, Inc., 1981.

Williams, F. and Rice, R.E. Communication research and the new media technologies. In R.N. Bostrom (ed.), *Communication Yearbook 7.* Beverly Hills, Calif.: Sage, 1983.

PART III

INFORMATION-PROCESSING AND INDIVIDUAL BEHAVIOR

13

Information-Processing Conceptualizations of Human Cognition: Past, Present, and Future

Elizabeth F. Loftus and Jonathan W. Schooler

Historically, scholars have used contemporary machines as models of the mind. Today, much of what we know about human cognition is guided by a three-stage computer model. Sensory memory is attributed with the qualities of a computer buffer store in which individual inputs are briefly maintained until a meaningful entry is recognized. Short-term memory is equivalent to the working memory of a computer, being highly flexible yet having only a limited capacity. Finally, long-term memory resembles the auxiliary store of a computer, with a virtually unlimited capacity for a variety of information. Although the computer analogy provides a useful framework for describing much of the present research on human cognition, it is insufficient in several respects. For example, it does not adequately represent the distinction between conscious and non-conscious thought. As an alternative, a corporate metaphor of the mind is suggested as a possible vehicle for guiding future research. Depicting the mind as a corporation accommodates many aspects of cognition including consciousness. In addition, by offering a more familiar framework, a corporate model is easily applied to subjective psychological experience as well as other real world phenomena.

Currently, the most influential approach in cognitive psychology is based on analogies derived from the digital computer. The information processing approach has been an important source of models and ideas but the fate of its predecessors should serve to keep us humble concerning its eventual success. In 30 years, the computer-based information processing approach that currently reigns may seem as invalid to the human mind as the wax-tablet or telephone switchboard models do today. (Roediger 1980, 248).

For thousands of years scholars have used contemporary technological instruments to model human cognition. Plato, in the third century B. C., described memory, a fundamental aspect of cognition, as a wax tablet that can take on endless impressions. In the seventeenth century, Descartes fashioned many mental processes after the complex clock robots that were popular at the time. During the first half of this century, human cognition was likened to a telephone switchboard, with incoming calls from the environment (stimuli) being connected to the appropriate telephone (response). By mid-century, human thought was suggested to resemble a symbol manipulation machine, similar to a mechanical calculator but considerably more powerful (Craik 1943). Other technological innovations that have been used to model the mind have included cameras, tape recorders, and even written text (Neisser 1982b; see also Roediger 1980).

Given the reliance on the technological advancements of a time by creators of contemporary models of cognition, it is not surprising that for the last several decades the popular model for understanding human thought has been the computer. One purpose of this article is to illustrate how many of the basic findings of cognitive psychology have been described within the computer model analogy. We then describe some current research that is less easily accommodated by the computer analogy. Finally, we suggest a new framework for discussing human cognition that may minimize some of the problems associated with machine models.

Generally, models that use computers as an analogy to human thinking are said to use an information processing approach to underscore the similarities between the functions of mental processes and those of a computer. Besides simply noting analogous functions of human thought and the computer, information processing models invariably suggest similar processes as well. For example, mental systems, like computers, have been argued to be divisible into two basic types of component processes: processes similar to computer software in the sense that they are learned (or programmed), and processes that are similar to computer hardware in the sense that they are wired into the system and cannot be changed by the environment.

Many other parallels between human thinking and computer processing have been suggested at one time or another (see Hintzman 1978; Loftus and Loftus 1976), however, there is one similarity that is central to almost all information processing models. Specifically, practically all computer models of mental processes include three basic stages of mental processing equivalent to the buffer, working memory, and auxiliary memory of computers (Hintzman 1978). In a computer, a buffer store, or register, holds each symbol of a given input string in memory while the remaining items are read. This stage allows the computer to read all of the symbols in a string together as a single entry. After an entry has been formed, it is combined and temporarily stored with other entries so that symbol manipulation can be performed. This process is often referred to as working memory. Working memory, though highly flexible, has only a limited capacity, hence most computers have an additional system often referred to as auxiliary memories that allow for the storage of information that is not currently in use. These three aspects of computers have almost identical analogies in the information processing framework and form the skeleton on to which much of cognitive research has been fit. In the next three sections we briefly review some of this research.

Sensory Memory

The first stage in the information processing framework is equivalent to the buffer of a computer and is often referred to as sensory memory. Like the buffer of a computer, sensory memory briefly stores information, providing sufficient time for meaning to be extracted. Actually, there are several sensory memories, each corresponding to a different sensory modality. With visual information, entrance is through the eyes and the information enters a visual sensory memory, often called iconic memory (Neisser 1967). With auditory information, entrance is through the ears, and the information enters into an auditory sensory memory, often called echoic memory. We can think of a sensory memory corresponding to each of the other modalities as well.

After a brief visual stimulus has been presented, a visual image or icon is held in the sensory memory while its features are being extracted. In a now classic series of experiments, Sperling (1960) demonstrated that the icon contains a substantial amount of information, but due to its brief life, most of the information is lost before it can be reported. Sperling's procedure involved briefly flashing rows of unrelated letters and then asking subjects to recall as many of the letters as possible. Generally, subjects could only recall a small percentage of the letters viewed. In a second condition, the partial report condition, subjects only had to recall a few letters. A signal indicated to subjects which portion of the letter array they should try to report. Since the signal varied from trial to trial, subjects had no way of knowing which letters they were going to have to recall until after the letter presentation had physically disappeared. Accuracy was very high. To perform so accurately subjects needed to retain at least briefly what letter occupied every position. One implication of this finding is that subjects can recall all of the briefly presented information, but only for a very limited duration; when subjects have to name everything they see, they simply forget much of it before they have a chance to say it. Apparently, the icon contains considerable information most of which is forgotten or becomes inaccessible if it is not immediately attended to.

While the discussion thus far has focused on the viewing of a single brief stimulus, natural viewing conditions are often much different. We generally look at enduring complex scenes, not individual tachistoscopic flashes. When we acquire information from our environment, the available evidence adds up to a picture of the eye sweeping over the scene to which it is directed with a series of eye fixations. The early portion of each fixation is spent extracting information from the stimulus while later portions appear to be spent making a decision about where to fixate next (Gould 1969; Loftus 1972). With each single fixation, the iconic image replaces the image from the previous fixation. Somehow, quite miraculously, each of the individual "looks" is then integrated into a smooth and complete representation.

Most of our conclusions rest, in large part, on experiments involving tachistoscopic stimuli (i.e., stimuli flashed for very brief durations). Recently, Haber (1983) has questioned the value of tachistoscopic experiments because the conditions they present are so dramatically different from any phenomenon observed in the real world. Specifically, a number of studies have demonstrated that the icon can be eradicated if a briefly flashed stimulus is followed by a flash of light or other type of visual stimulus (see Turvey 1973 for a review). Haber argues that with the exception of reading in a thunderstorm, there is no opportunity in real life situations to experience the icon, since under normal circumstances it will be constantly masked by subsequent visual stimuli. Although there is certainly some truth to Haber's observation that tachistiscopic conditions are rare, that does not necessarily prove his conclusion that it is worthless to study the icon. It is clear that some process is necessary in order to hold information long enough for it to be adequately sorted, evaluated, and acted upon. The icon appears to have all of the qualities of such a needed mechanism. It therefore seems likely that the icon, or some icon-like process, fulfills this buffer role.

It may be that in normal situations iconic images continue to exist after masking but are simply not consciously visible. In other words, the icon's function may be to provide information to mental systems that are operating at a less than fully conscious level. Indeed, there is now emerging evidence that complex processing occurs for stimuli that are never consciously perceived (e.g., Marcel 1983), a subject that we discuss later in this article.

With regard to other sensory modalities, memory for smell (Engen and Ross 1973), tactile information (Gilson and Baddeley 1969), and auditory information (Crowder and Morton 1969) have all been studied with the bulk of the research attention being devoted to audition. Auditory information, for example, is thought to enter an auditory sensory store called echoic memory, analogous to the one posited for visual information. Information from this store decays quite rapidly, although somewhat more slowly than the decay from the iconic store. Most estimates hover around two to four seconds (Crowder and Morton 1969; Darwin, Turvey, and Crowder 1972). It is typically the case that the last few items in a list are recalled at a higher level if they have been presented auditorily rather than visually, a finding that is usually attributed to the relatively long-lasting echoic memory traces.

As in a computer buffer that stores a series of individual entries together until they can be read as a control word, sensory memory requires a process whereby sensory information is converted into more meaningful information. This process is commonly known as pattern recognition and basically involves identifying sensory stimuli. Exactly how this process is accomplished is still not completely understood, but a common proposition is that it involves a matching of the current stimulus against a likely set of prototypes in long-term memory. The stimulus is then classified according to the name of the best matching prototype (Reed 1972).

In sum, sensory memory, as it has been studied within the information processing approach, closely resembles the buffer of a computer. It stores information briefly, with the precise duration depending on the modality. During this brief storage period information is sorted and some portion is interpreted and given meaning. According to the model, it is only this processed information that proceeds to the next stage in the system, generally known as short-term memory.

Short-Term Memory

Short-term memory, like the working memory of a computer, is viewed as the place or state where cognitive operations take place. Short-term memory has a number of general characteristics, many of which resemble the working memory of a computer. First, it has a limited capacity; that is, it cannot hold very much information. About four to six chunks of information is typical. Second, like a computer, access to information in short-term memory is generally believed to occur serially; that is, we can only access one item at a time. Third, information from this memory decays very rapidly—common estimates are between fifteen to twenty seconds—unless we engage in some activity, such as rehearsal, to prevent this decay.

Short-term memory is thought to have a limited capacity. Miller (1956) first argued that it holds seven + or – items or "chunks" of information, regardless of their size. The exact definition of a chunk is somewhat elusive, but it might be characterized as a stimulus that has some unitary representation in long-term memory. Thus, the letters "K C U D" constitute four chunks but in reverse order, the letter string "D U C K" is a single chunk.

Miller claimed that short-term memory could hold about seven chunks since this was about the number of items we could repeat back after a single hearing (the memory span). However, the memory span is in part aided by long-term memory and thus may not be a good estimate of the capacity of short-term memory. Waugh and Norman made this point in their 1965 article entitled "Primary Memory." The suggestion was that information could be copied into long-term memory quite quickly, while it was also maintained in short-term memory, a suggestion that implied that information about an item could reside in both systems simultaneously. This means that even a short time after presentation, the recall of some particular information will depend on the operation of both short- and long-term memory. This sort of thinking has led some investigators, including Mandler (1967), to argue that seven chunks is too large an estimate and the capacity of short-term memory is probably closer to four or five.

Access to information in short-term memory has been extensively investigated by Sternberg (1966). In his experiments, a subject is first read or shown a string of items called the memory set which the subject places in short-term memory. Next a test item is presented and the subject's task is to report whether the test item was or was not a member of the memory set. Thus, if the memory set consisted of the letters X P T V and the test item is T, the correct response would be "yes." The variable of interest is how fast the subject responds, and the typical finding is that the times increase with the numbers of symbols in the memory set. This finding indicated to Sternberg that access to items in short-term memory is serial, that is item by item. Although there is some dispute over this interpretation of Sternberg's results (see for example, Theios et al. 1973; Townsend 1972), the notion of serial processing has been one of the characteristics of short-term memory that has been commonly associated with the computer analogy to mental processing.

Estimates of the decay of short-term memory come from experiments using the Brown-Peterson paradigm (Brown 1958; Peterson and Peterson 1959). On each trial a subject is presented with a string of three consonants such as "BKG." Following the consonant trigram, a retention interval passes, ranging from 0 to 18 seconds. To prevent rehearsal of the trigram, the subject is required to perform a fairly demanding mental arithmetic task during the interval, namely counting backwards by threes from a three-digit number. At the end of the interval, the subject recalls the trigram. The typical result is that when the trigram is tested immediately—after a retention interval of 0—recall is nearly perfect. At longer intervals, however, memory performance drops steadily, reaching asymptote after an interval of about 15-20 seconds, at which point the trigram is recalled about 10 percent of the time. These results suggest that loss of information from short-term memory is complete by about 15-20 seconds.

In sum, short-term memory represents the processing stage in the computer model of memory. Like the working memory of a computer, short-term memory has a limited capacity and appears to process information in a serial manner. Information that is not processed or rehearsed decays rapidly and is not transferred to the next stage of the system, long-term memory.

Long-Term Memory

Long-term memory, like the auxiliary memory of a computer, is viewed as being an unlimited storehouse of information from which short-term memory can draw. Long-term memory includes information that we learned a few minutes ago as well as information learned years earlier.

Because of the proposed similarities between long-term memory and computers, many researchers have attempted to develop complex computer models to simulate these memory processes. These studies (e.g., Anderson 1976, 1983; Norman and Rumelhart 1975) have generated striking simulations of human performance. Nevertheless, in evaluating these simulations, it must be kept in mind that totally different processes can ultimately produce very similar final responses. Therefore, as effective as computer simulations may be, they may not necessarily describe what occurs during actual human information processing.

Propositional and Procedural Information

The information contained in long-term memory can be roughly broken down into two basic classes: propositional information and procedural information. Propositional information pertains to factual knowledge that can be explained. Procedural information pertains to skills that can be implemented. This distinction can be summarized as differentiating between "knowing that" and "knowing how." Possibly because of the computer analogy's emphasis on information processing, most studies of procedural information have focused on the procedures involved in processing propositional information. Thus, a discussion of long-term memory for propositional information incorporates much of what is known about procedural knowledge.

Tulving (1972, 1983) makes an important distinction between two basic types of long-term memory which he terms *semantic* and *episodic*. These

memories contain qualitatively different types of propositional information. Semantic memory contains general facts including words, historical
facts, rules, etc. For example, the statement "An apple is a fruit" is a
semantic fact. Episodic memory pertains to personal facts that can be
described in terms of the time and place of their occurrence. For example,
the statement "Yesterday I ate an apple" is an episodic fact. This distinction between semantic and episodic memory has been important both in
terms of elucidating the possible types of long-term memory stores as well
as providing a convenient way of dividing this memory into different areas
of study.

Some researchers maintain that semantic and episodic information are
actually stored separately. In his recent book, Tulving (1983) cites two basic
sources of evidence for the functional separation of these two types of
memory. First he notes the many studies that have demonstrated that
certain variables can affect subjects' performance on semantic memory
tasks without affecting their performance on episodic memory tasks and
vice versa. A second class of evidence to support this distinction is physiological. For example, one study observed differences in blood flow in the
brain depending on whether a task required semantic or episodic information (Wood et al. 1980). In addition many clinical reports describe patients
who experience deficiencies in one type of memory while the other type of
memory remains relatively intact (see Schacter and Tulving 1982). Although Tulving's gathering of evidence and his arguments are relatively
compelling, the distinction, at least at the functional level, has been rejected by some researchers (e.g., Anderson and Ross 1980). These critics
point to studies showing substantial interdependence between the two al-
legedly separate memory systems.

Although researchers disagree on whether this distinction represents two
functionally distinct systems, there is common agreement on the value of
the distinction as a heuristic device for classifying different types of long-
term memory information. Specifically, with the exception of a few global
lines of study (e.g., Anderson 1983) most investigations of long-term memory primarily apply to either semantic or episodic memory. It is therefore
convenient to discuss separately the basic principles that have been attributed to the processing of semantic and episodic memory.

Episodic Memory

In terms of episodic memory, two principles are particularly important
(Tulving 1983). The first principle involves the constructive nature of long-
term memory for episodes; that is, episodic memories are not only retrieved from the past, but they are also abstracted and distorted. What this
means is that in recalling episodic information from memory we often take

in an incomplete account and then use our general knowledge to construct a more complete description of what we experienced. For example, if we hear the sentence "The lawyer stirred her coffee," or actually see this happening, we might infer that she stirred it with a spoon, and we might add this inference to our memory of the incident. Our long-term memory for the episode will then contain not only the original information that we encoded from the initial event but bits of knowledge that we constructed and added to our memory.

This constructive process can help us accurately fill in forgotten details, but it can also produce errors. For example, when told "The floor was dirty because Sally used the mop," many people will infer that the mop must have been dirty and will later think that they had been presented with the statement "The floor was dirty because Sally used the dirty mop" (Bransford, Barclay, and Franks 1972). This occurs because the original information and the inferences are integrated into a single memory for the episode. One consequence is an inability, or at least a severe difficulty, in distinguishing between what was actually presented and what was only inferred afterward.

Similar effects are observed when people who have experienced complex events are exposed to subsequent misinformation about that event. It is common to find that the new information becomes incorporated into memory, supplementing or altering the original memory. For example, in one experiment college students were presented with a film of an automobile accident and then half of the subjects were asked "How fast was the white sports car going when it passed the barn while traveling along the country road?" whereas no barn existed. Later the subjects were asked whether they had seen the barn. The subjects who heard the statement suggesting the barn were considerably more likely than the control subjects to recall having witnessed the nonexistent barn (Loftus 1975). In this case the false information was integrated into the person's recollection of the event, thereby supplementing that memory.

In other studies, it has been shown that new information can do more than simply supplement a recollection; it can occasionally alter or transform a recollection. Subjects saw a series of slides depicting an auto-pedestrian accident; some saw the car come to an intersection with a stop sign and others saw a yield sign. Later subjects were exposed to misleading information about the sign, and finally they were asked to recognize which sign they had actually seen. In numerous studies using these materials, many subjects chose the sign that corresponded to the subsequent information, rejecting the sign that they had actually seen (Loftus, Miller, and Burns 1978). These results suggest that people will not only generate inferences, but they will use other information available to them to fill in the gaps in their memories. This process of rounding out fairly incomplete knowledge has been called refabrication.

Refabrications have been extensively investigated, yet, there remains considerable dispute over whether these new memories have replaced the original memories, or whether they are simply more accessible at the time of retrieval. A recent study by Bekerian and Bowers (1983) addresses the issue of whether suggested memories replace or coexist with original memories. They added an innovative condition to the stop sign/yield sign experiment conducted by Loftus, Miller, and Burns described above. Bekerian and Bowers presented the final recognition test slides in a sequential order that more closely approximated the order in which the slides were originally presented. In this condition, subjects were able to disregard the misinformation and recognize the sign that they had originally seen. Thus, the original information must have still existed. Bekerian and Bowers concluded that the sequential order of the test slides reinstated the context of the original presentation thereby allowing the original information to be retrieved. Unfortunately context reinstatement does not always induce the recall of original information (McSpadden and Loftus 1983). Thus, the issue of whether destructive updating ever occurs is still open.

The issue of context reinstatement leads naturally to a second important principle of episodic memory: encoding specificity (Tulving and Thomson 1973). According to this principle, the likelihood that an episodic bit of information is retrieved is a function of the similarity between the conditions under which it is encoded and the conditions under which it is retrieved. The encoding specificity principle is reminiscent of advice that is commonly offered when one can't find something: "If you can't find it, retrace your steps." By retracing your steps you effectively reinstate the context in which you originally encoded the placing of the object, thereby making the retrieval conditions similar to the encoding conditions.

The presentation of test slides in a sequence similar to that in which the slides were first viewed is another way of equating the conditions of encoding with the conditions of retrieval. This same basic principle has been used to explain why memorized words are often not recognized when they are presented in a context that suggests a meaning different from the meaning in which they were initially encoded (Tulving and Thomson 1971). It also accounts for why scuba divers who memorize words under water recognize more of the words when they are tested under water than when they are tested on dry land (Baddeley et al. 1975). Moreover, it explains why subjects' recognition of memorized words is facilitated by recreating the mood that they were in when they memorized the words (Bower 1981).

Thus, it appears that the recall of episodic information involves a combination of 1) the refabrication of new information, and 2) the retrieval of original information via successful recreation of the circumstances associated with the formation of a memory. Together the principles of constructive refabrication and encoding specificity powerfully determine what is recalled from episodic memory.

Semantic Memory

In some ways the processes involved in the retrieval of episodic and semantic long-term memories are different. Specifically, semantic information by definition does not include the context in which it was encoded, therefore the notion of equating context of encoding with context of retrieval is not really applicable. Moreover, for much of semantic memory—knowledge of words—the notion of refabrication does not apply. Thus, regardless of whether semantic and episodic memories are stored separately, they must certainly differ at least with regard to some of their respective retrieval processes.

In the case of the retrieval of semantic memory, considerable research has focused on the relationship between the semantic similarity of items and the speed of their retrieval. Generally, it has been observed that the time necessary to access memory is in part a function of the type of information that was previously accessed. For example, it has been shown (Meyer and Schvaneveldt 1971; Schvaneveldt and Meyer 1973) that the time to retrieve semantic information from memory is shorter if related information has been accessed a short time previously. In this research, subjects were required to classify letter strings as words or nonwords. In general the response time to classify a letter string as a word is shorter if the subject has just classified a semantically similar word, as opposed to a semantically dissimilar word. For example, subjects take less time to classify *butter* as a word if *butter* is preceded by *bread* than if it is preceeded by *nurse*. This result, along with others involving widely different paradigms (e.g., Loftus 1973), suggests that the retrieval of semantic information initiates a "spreading activation" of related semantic facts.

The spreading-activation theory (summarized by Collins and Loftus 1975) provides a vehicle for discussing these facilitation effects. The theory states that when an item in semantic memory is processed, other items are activated to the extent that they are closely related to the first item. Put another way, activation spreads through long-term memory from active portions, and along its pathway new portions of memory are temporarily more accessible. This basic notion of spreading activation has served as the springboard for many models that conceive of semantic memory as a network of propositions (Collins and Quillian 1972; Collins and Loftus 1975); accordingly, the position of semantic propositions within such networks can be examined by observing the spread of activation from one semantic fact to the next.

Recently, the simple notion of spreading activation of semantic memory has been complicated by the appearance of some experiments in which the retrieval of information was inhibited by the prior retrieval of related information (Brown 1979). This finding posed some difficulties for spreading-activation models that are based on the notion that the activation of semantic facts facilitates rather than inhibits the access of closely related facts. A recent study by Roediger, Neely, and Blaxton (1983), however, suggests that Brown's findings are due to particular response strategies associated with the type of material that he used in his research. Thus, the concept of spreading activation is alive and well (see also Anderson 1983).

Although the retrieval processes attributed to episodic memory, such as refabrication and encoding specificity, are not easily applied to semantic information, a number of spreading-activation models have attempted to include episodic information within their propositional network (e.g., Anderson 1976, 1983). In these cases the spread of episodic information is a function of 1) how closely associated the items were at the time of encoding, and 2) the number of different items that were associated with the activated information.

In sum, long-term memory represents the final stage in the computer analogy. It has a tremendous capacity and contains a great variety of information. Much of the information contained can be broadly classified as either episodic (personal) or semantic (general). The study of memory for episodic information has illuminated two important retrieval principles: refabrication, which involves the construction of new memories that fill in gaps, and encoding specificity, which suggests that the likelihood of recalling a memory is a function of the similarity between the conditions of its encoding and its retrieval. Considerable research in semantic memory has focused on the concept of spreading activation which helps delineate the structure of semantic networks.

Evidence for the Stage Model of Information Processing

In addition to providing a heuristically useful parallel between computers and the mind, the three-stage model of information processing is supported by both experimental work and clinical observation. Evidence for the sensory stage is primarily limited to research, described earlier, demonstrating the rapid decay of very detailed information that is apprehended in the same mode (e.g., acoustic, visual) in which it is experienced. The evidence for a short-term, as distinct from a long-term, memory is considerably more complicated and can be grouped into three relatively independent lines of research including: 1) free recall data, 2) observations in a clinical setting, and 3) memory consolidation studies.

Free Recall Studies

In numerous experiments, subjects have been presented with a list of words and then asked to recall them in any order. When the probability that a word is recalled is plotted against its location in the list (its serial input position), the typical serial-position curve results. The curve indicates that words at the beginning of the list (the primacy effect) and words at the end of the list (the recency effect) are recalled better than words in the middle. Why does the recency effect occur? Is it because the words at the end of the list were encountered most recently and so are less likely to be forgotten during the recall test? Those who believe in the short-term memory/long-term memory dichotomy have a different explanation. They argue that the words at the end of the list are very likely to be in short-term memory when the recall test begins, since no words followed that could displace them. Thus, they can be readily recalled.

A simple variation on the free-recall experiment provides support for this explanation and for the distinction between short-term and long-term memory (Postman and Phillips 1965; Glanzer and Cunitz 1966). In these experiments, subjects were presented with a list of words, followed by a distracting arithmetic task, followed by recall. Their recall was compared with that of control subjects who recalled the words immediately after they were presented. Control subjects showed the usual recency effect, whereas experimental subjects who had performed the distracting arithmetic task recalled the last few words poorly. In other words they showed no recency effect. Proponents of a dichotomous memory system explain this result by proposing that the last words on the list are still in short-term memory and can be readily recalled by control subjects. For experimental subjects, information from the arithmetic task enters short-term memory and interferes with recall of the words.

Clinical Evidence

The dichotomy between short-term and long-term memory has received further support by reports of clinical cases in which one memory system is apparently normal while the other is deficient or nonoperative (Milner 1970; Warrington and Shallice 1972.) One example is the patient H.M. who had temporal lobe surgery to treat a severe epileptic condition. Although his epilepsy was helped by the treatment, he developed a profound memory defect. His intelligence, as measured by standard tests, was even a bit higher than it had been before. His short-term memory was adequate, as was his ability to retrieve information from long-term memory that was acquired before the operation. However, he could not transfer information from short-term memory to long-term memory, a highly debilitating deficit. For example, ten months after H.M.'s operation his family moved to a new house situated only a few blocks away from the old one. A year after the move, H.M. had not yet learned the new address, nor could he be trusted to find his way home alone. The finding that one portion of the memory system can be grossly deficient while the other remains intact argues in favor of distinct systems.

Memory Consolidation Studies

Consolidation studies also provide evidence for the short-term/long-term distinction. In these studies electroconvulsive shock (ECS) is delivered to the brain soon after learning. This shock can prevent the new information from consolidating into long-term memory. The literature on memory consolidation is large and complex, but there is little doubt that when shock is given immediately after a learning experience it interferes with the normal link between short-term and long-term memory (McGaugh and Herz 1971).

Arguments against the Stage Model of Information Processing

Although the stage model is one of the most widely accepted concepts within the information processing field, it has not been without its critics. One of the most important criticisms of the stage model is that the stages are not always clearly defined. The duration of sensory memory can span from a fraction of a second to many seconds depending on the task (e.g., Phillips and Baddeley 1971). Differences between short-term and long-term memories are also often task dependent (Shulman 1971).

Levels of Processing Framework

Because of the fuzziness in distinguishing the stages, a number of researchers (e.g., Cermak 1972; Craik and Lockhart 1972) have suggested that information processing be viewed as a continuum rather than as discrete stages. This approach, often described as a "levels of processing framework" (Craik and Lockhart 1972) postulates that the strength of a memory trace is not a discrete matter of whether or not it is transferred to long-term memory, but is instead a continuous function of the "depth of processing" that that given piece of information receives. Thus, if rehearsal is of a shallow variety—like the type we perform to keep a new telephone number in mind long enough to dial it—little will be subsequently retrievable. If the rehearsal is of a deep, elaborative type, as when we try to set up meaningful connections among the items we are trying to remember, much more of the information will be available for later retrieval. Experiments by Craik and Watkins (1973) and Jacoby (1973) observed that when subjects viewed words in the context of tasks requiring meaningful processing, such as whether words would make sense in a particular sentence, they recalled considerably more words than subjects who encountered the same words in the context of a more superficial task, such as deciding whether a word was in upper or lower case. In "depth of processing" language these observations suggest that deeper processing results in stronger memory traces.

The depth of processing framework, while appealing in many ways, is not without its faults. Specifically, one corollary to the notion that depth of processing is responsible for the strength of a memory trace is the prediction that repeated rehearsal of an item at the same depth of processing should not result in any difference in the memory trace. At least one study, however, demonstrated that simply encountering the same word twice in a single superficial task increases the likelihood that that word will be subsequently recalled (Nelson 1978). The observation that increased rehearsal, even in the absence of deeper processing, can strengthen a memory trace, indicates that the concept of "depth" cannot exclusively account for the strength of a memory trace.

A second problem with the depth of processing approach is that it is not clear how depth is defined; for example, why is semantic processing inherently "deeper" than visual processing? Perhaps semantic processing is not deeper than other kinds, but is just different. Accordingly, differences between the retrieval of words processed visually, acoustically, or semantically is not a function of the depth of processing, but rather of the similarities between the context of encoding and the context of recall. Since free recall is in many ways a semantic task, it is not surprising that words learned in a semantic context would be better retrieved during semantic free-recall. If the retrieval situation asks subjects to recall words in an acoustic context such as by deciding what words rhymed with *train*, acoustically encoded words are better recalled (Morris, Bransford, and Franks 1977). The observation that retrieval is often a function of the type rather than the depth of encoding has led Baddeley (1982) to suggest that memory should be viewed as containing domains of processing rather than levels of processing.

Ultimately it seems likely that a compromise among the various approaches will become popular. Specifically, there is little doubt that items that are given greater attention—processed more deeply—are more readily retrieved (Anderson 1983). At the same time, it seems inappropriate to term one type of attention (e.g. visual, acoustic, semantic) as inherently more important than another. Moreover, neither the concept of levels of processing nor that of depth of processing necessarily excludes the existence of a central processor with a limited and temporary memory similar to that posited for short-term memory.

Thus, one compromise is the postulation of a central processor with limited capacity and temporary information retention that moves freely between different domains of memory. The strength of a memory trace would simply be a function of the amount of attention that the processor gives to a piece of information within a domain. The probability of retrieval would thus be a joint function of (1) the strength of the original trace and (2) whether the information is accessed in the same domain in which it was encoded. The above compromise is in many ways a simplified synthesis of some of the aspects of a number of recent memory models (e.g., Anderson 1983; Broadbent 1983; Tulving 1983). Its emphasis on a central processor that travels freely between domains is especially valuable because it addresses the common criticism of the computer-based stage model; namely that it does not adequately reflect the variety of processing strategies involved in human thinking.

The "Maltese Cross" Framework

Broadbent (1983), previously a major proponent of the stage model, recently outlined a series of criticisms of what he terms the "pipeline approach." He is troubled by the present model's inability to adequately identify the role of individual control processes. He argues that the tendency to reduce human processing to a strict flow chart serves to overly downplay the ways in which the individual can act on his or her environment.

He argues, for example, that present models portray the person as inappropriately passive, always the recipient of stimuli, and rarely initiating internal processes. Moreover, he argues that interpretation of stimuli is not just a matter of what stimuli are presented, but also reflects the personal cognitive operations of the individual. Finally, he notes that most models fail to adequately consider the multitude of different strategies that the individual may employ while engaging in mental tasks.

As an alternative, Broadbent proposes the "Maltese Cross," a theory which includes a central processor that is responsible for all of the flexible processing that is normally assumed to be under an individual's control. Broadbent suggests that the information processing system involves the interaction of a central processing system with four basic memory processes that he terms the sensory store, the motor store, abstract working memory, and the long-term store.

Broadbent uses the analogy of a man at his desk. On the desk are two baskets. One basket, equivalent to the sensory store, contains the incoming mail. A second basket contains all outgoing mail and is equivalent to the motor store which represents the executive's intended responses. The desk top on which he is working is analogous to working memory and includes all of the ideas that he is currently manipulating. Finally, the file behind him is equivalent to the long-term store and represents all of the information that is accessible to the man when he needs it. Using a set of flexible but preprogrammed rules, the man at his desk interacts with these four areas, processing new information as well as calling up old information that can be rearranged on his desk to produce seemingly spontaneous ideas.

Although Broadbent's model does not really posit any new components—ideas such as a motor memory (Sperling 1967), a central processor (Craik and Lockhart 1972), or flexible control processes (Atkinson and Shiffrin 1968)—it does help to highlight some of the difficulties with previous computer models. Specifically, by using a model that contains a conscious entity within it, it indirectly identifies a weakness of the computer model: its inability to naturally reflect the existence of conscious thought.

Conscious versus Nonconscious Processes

Previous attempts to include conscious and nonconscious processes in the computer model of human information processing have emphasized the distinction in computers between flexible and nonflexible processing strategies. For example, Atkinson and Shiffrin (1968) suggested an elaborate series of flexible control processes similar to computer software, which determined what stimuli were transferred to long-term memory. These variable control processes were contrasted with the invariable—(hardware)—structural processes. The difficulty with the hardware/software distinction is that it ignores the important distinction between what one is aware of and what one is in control of. For example, if control is defined as those responses whose outcome is dependent on a flexible set of prior experiences and resulting expectations, then it is possible that one could be "in control" of a perception, and at the same time be unaware of those control processes. For example, judging an ambiguous figure often involves the expectations of the perceiver, and yet a person may be unaware that he or she had control over what was perceived.

It could be argued that what an individual is aware of is ultimately inconsequential. All that is necessary to understand the mind is an appreciation of the internal processes that occur within an individual. Whether or not the individual is aware of those processes does not matter, since it is the processes themselves that are of interest. Indeed it is probably thinking of this sort that has enabled psychologists to so closely align the act of thinking with the processes of a nonconscious computer. Recent research conducted by Marcel (1980, 1983), however, indicates that whether or not individuals are aware of their thought processes has considerable bearing on the nature of those processes.

Marcel describes a series of subliminal perception experiments in which words are presented very briefly, followed by a flash of light that masks not only the meaning of the word but even the fact that it was presented. These unperceived words exert an effect on thought processes. In one paradigm, Marcel demonstrates that the presentation of a word, even subliminally, can affect subjects' reactions to later words. He observed, for example, that the letter string "Doctor" was recognized faster if its presentation was preceded by the subliminal presentation of the related letter string "Nurse." Apparently, the subliminal presentation of a word can activate related words. This finding indicates that complex semantic processing can occur without conscious awareness.

Marcel (1980) also demonstrated that nonconscious processing, though highly complex, differs in a fundamental way from conscious processing. He examined the priming effects of polysemous words (words that have more than one meaning, such as *palm*. When a polysemous word is presented long enough to be consciously perceived, only one of its meanings apparently exerts a priming effect. For example *palm* might prime *hand* or *tree,* but not both. When polysemous words are presented subliminally, however, both associated meanings can serve as primes. Marcel (1983) concludes: "Apparently more than one interpretation of an event in any one domain can be represented simultaneously nonconsciously but only one interpretation at a time can be represented consciously" (p. 252). Thus, it appears that nonconscious processing differs in a fundamental way from conscious processing in that at the nonconscious level various interpreting processes may occur simultaneously.

Marcel's research suggests that conscious attention itself is an important aspect of cognitive processing, an observation that is not naturally accommodated by the computer model. Specifically, computers cannot in any real sense be thought to be aware of some processes more than other processes. Thus, the distinction between conscious and nonconscious thought, unlike the hardware/software distinction, is not a natural one in the computer model.

Undoubtedly it is possible to design computer models that are able to simulate many of the properties of conscious and nonconscious thought, including the existence of parallel processing at the nonconscious level. Such pursuits make sense since the human mind, which operates in this way, is such an effective processor of information. Exclusive reliance on computers as a model of human thinking, on the other hand, makes little sense. By studying human thought in the context of the capabilities of present computers, we limit ourselves to considering only those aspects of human thought that are shared with the computer. The fact that, unlike computers, humans have both consciousness and an awareness of their own cognitive functioning, cries out for the development of new models that can more readily accommodate these neglected aspects of human thought.

An Alternative Model of Human Cognition

Broadbent's executive metaphor nicely accommodates conscious awareness and, therefore, represents a step in the right direction. Unfortunately, it fails to adequately represent parallel nonconscious processes. In Broadbent's view, if the executive does not process something, it simply does not get processed. Unless we are talking about a very busy person who is able to do many different things simultaneously, Broadbent's executive cannot readily handle parallel cognitive processes.

What executives need in order to process all of the information that is handled by the human mind is, quite literally, help. They need a secretary to handle files, various assistants to help sort the input and output piles, and, most importantly, helpers to do various different tasks simultaneously while the executives spend their time handling the important issues. In other words, the executive metaphor may have some applicability for the mind, if augmented by the rest of the corporation.

A model of the human mind as a well-run corporation accommodates many of the issues that are not well represented by the mechanistic computer model. It naturally provides a place for consciousness as the president of the organization. It captures the distinction between the serial processing at the conscious level—a leader can only think about one thing at a time—and the parallel processing at the nonconscious level: all of the subordinate members of the corporation can work simultaneously.

In addition, it offers a more natural way to envision the stages and levels of information processing; that is, by the status and departments of the various sensory and storage bureaucrats within the corporation. A bureaucracy contains members of differing status who receive differing amounts of the president's attention. Assume that what the president knows about any given member of the corporation is a function of the amount of attention that that member receives. In this sense, the corporate metaphor nicely exemplifies the relationship between the amount of attention that a stimulus receives and its subsequent retrievability. A bureaucracy may also contain various departments that are roughly equivalent in status. The president's ability to retrieve information would also be a function of whether or not he or she consulted the appropriate department, thereby capturing the notion that successful recall requires a correspondence between the memory domain involved in encoding and that involved in retrieval. Finally, different departments of a bureaucracy can work on similar issues—denoting the sometimes ambiguous distinctions between various aspects of memory.

By including a conscious entity, the corporate model is naturally susceptible to the "homunculus" criticism that including a conscious entity within a psychological model makes it circular. Yet, the president in the mental corporation is very different from a complete individual. Like all of the members of a corporation, this role is limited to specific functions; for example, the president does not have direct access to all of the information in the system but instead must rely on various sensory and storage assistants. Because the corporate president is not like a complete individual, the corporate metaphor is not a simple reintroduction of the concept being explained.

Although the corporate president does not have all of the qualities of a complete individual, he does have consciousness, so the corporate model can be viewed as circular to the degree that it attempts to define consciousness. However, the experience of consciousness is perhaps the one thing that we can take for granted. As Descartes understood long ago: "I think therefore I am." Thus, instead of ignoring consciousness, perhaps researchers should ask, "Given a consciousness, what does it experience?" Which processes are under its control, and which are not? How can it (I) have more control? Once the question becomes not what is consciousness but what is consciousness aware of, then the present model offers a way to conceptualize the conscious experience. For example, it hints at the many sides of our personalities, and suggests ways of understanding how individuals can think and feel dramatically different depending on the circumstances (or in the terminology of the present model, depending on whom in their mental corporation they are currently consulting).

Although the corporate model appears at first glance rather unusual, in many ways it resembles the once popular pandemonium model of pattern recognition (Selfridge 1959). In this model, perception is depicted as a series of demons who each shout when they see features that correspond to their particular bias. The closer the correspondence, the louder they shout. These lower demons are then heard by demons of a higher status, who begin shouting about feature combinations. This hierarchy of demons continues, as the message of the loudest shouting demons is passed from stage to stage, until finally the appropriate response is made by the individual at the top. Our corporate model resembles pandemonium in that it depicts the mind as a hierarchy of independent function-specific entities each simultaneously doing their own thing, but ultimately presenting a cohesive message to the consciousness at the top.

Besides accommodating a number of empirical findings that are not easily incorporated in the computer model, the corporate model of the mind also lends itself to more adequately responding to a number of other practical and philosophical issues that have been leveled against the mechanistic computer models. Cognitive psychology has often been criticized for failing to study what is meaningful or relevant to anyone outside of the field (Neisser 1982a). Indeed computer models are inherently alien to people outside psychology—it's just plain tough to honestly think about oneself as a computer. On the other hand, it is not difficult to imagine an organization, with a conscious leader who consults with various departments and individually handles those issues that require special attention. Moreover, the corporate model raises questions of a more socially relevant nature. Beyond asking how does information get from one stage to another, it would naturally emphasize the effective strategies for successful memory management. Research aimed at discovering strategies that individuals use to manage their "mental corporations" would not only facilitate our understanding of human cognition but might also indicate how information management might be improved.

Current machine models are also said to incorrectly depict human thought as a static, unchanging process (Samson 1981). By viewing the human condition as a static mechanism, cognitive psychology is making an ideological statement about the nature of people that may hinder our ability to observe how thinking can and does change. Comparing the individual to social organizations, on the other hand, focuses on the dynamic quality of human thought.

Depicting the mind as a social organization helps us to think about how society shapes the processes of human thought. Indeed, considerable evidence reveals that our cognitive operations reflect our social environment. For example, Vygotsky (1978) convincingly argues that thoughts are simply the internalization of conversations with others. Even in adulthood, the characteristics of one's internal thinking reflect the nature of the social organizations with which one is associated. For example, Kohn and Schooler (1983) observed a causal relationship between the characteristics of an individual's occupation and the characteristics of his or her internal thought, such that the more personal independence offered by one's occupation, the more apt one is to think independently.

Thinking about the relationship between the social environment and cognitive processes suggests the ways in which different types of mental corporations may be formed. For example, the amount of responsibility that individuals have within a social organization may affect the degree to which they actively supervise the processes occurring within their own mental corporations. Thus, social organizations that have decentralized control, may induce people within those organizations to develop more centralized mental corporations, and vice versa. Other models of cognition never made one think at all about the nature of society and its relationship to the individual.

As Neisser (1982b) aptly notes: "Models of the mind always follow the latest advances in gadgetry" (p.7). The computer model is no exception. As long as we are limited to mechanistic models, undoubtedly we will continue to change and update our models to keep up with modern technology, and undoubtedly people will continue to have difficulty applying these models to their own experience. Anderson (1977) has suggested that cognitive models are equally viable to the degree that they make similar predictions. Indeed, computer models have proven remarkably effective in providing accurate predictions of human performance. However, computer models, as predictively useful as they may be, do not provide a comfortable framework for describing human experience. There may be an important place for computer models in psychology, if only because of the level of precision that they permit. But the future demands that we also develop models that can describe the results of psychology in a framework that can be applied to our own experiences.

By including consciousness in its metaphors, cognitive psychology will be able to describe its findings in a more familiar context thereby allowing for increased communication between psychology and the rest of the world. Scholars previously estranged by psychological models may begin to see how psychological research fits into their own area of expertise, and additionally see how their own findings might apply to psychology. Moreover, nonscientists may finally be able to relate the findings of psychology to themselves. Since one of the goals of psychology is to illuminate the human experience, it seems only fitting that we describe thought processes in terms of humans. In the words of Neisser (1982b): "The dependence of popular conceptions of memory on current technology is obvious enough. Understanding that dependence may help us become free of it." (p.9)

Note

The writing of this article was supported by a grant from the National Science Foundation. The authors wish to thank Jim Jaynes and Mark Reinitz for their helpful comments. We are particularly grateful to Douglas Herrmann for his assistance in formulating the corporate metaphor.

References

Anderson, J.R. *Language, memory and thought.* Hillsdale, N.J.: Erlbaum, 1976.
_____. Arguments concerning representations for mental imagery. *Psychological Review*, 1978, 85, 249-277.
_____. A spreading activation theory of memory. *Journal of Verbal Learning and Verbal Behavior*, 1983, 22, 261-295.
_____, and B.H. Ross. Evidence against the semantic-episodic distinction. *Journal of Experimental Psychology: Human Learning & Memory*, 1980, 6, 441-446.
Atkinson, R.C. and R.M. Shiffrin. Human memory: A proposed system and its control processes. In K.W. Spence and J.T. Spence (eds.), *The psychology of learning and motivation: Advances in research and theory*, Vol. 2. New York: Academic Press, 1968.
Baddeley, A.D. Domains of recollection. *Psychological Review*, 1982, 84, 708-729.
_____, et al. Cognitive efficiency of divers working in cold water. *Human Factors*, 1975, 17, 446-454.
Bekerian, D.A. and J.M. Bowers. Eyewitness testimony: Were we misled? *Journal of Experimental Psychology: Learning, Memory & Cognition*, 1983, 9, 139-145.
Bower, G.H. Mood and memory. *American Psychologist.* 1981, 36, 129-148.
Bransford, J., J. Barclay and J. Franks. Sentence memory: A constructive versus interpretive approach. *Cognitive Psychology*, 1972, 2, 193-199.
Broadbent, D.E. The maltese cross: A new simplistic model for memory. *The Behavioral and Brain Sciences*, 1984, 7(1), 55-68.
Brown, J. Some tests of the decay theory of immediate memory. *Quarterly Journal of Experimental Psychology*, 1958, 10, 12-21.
Brown, A.S. Priming effects in semantic memory retrieval processes. *Journal of Experimental Psychology: Human Learning and Memory*, 1979, 5, 65-77.
Cermak, L.S. *Human memory: Research and theory.* New York: Ronald Press, 1972.
Collins, A.M. and E.F. Loftus. A spreading-activation theory of semantic processing. *Psychological Review*, 1975, 5, 85-88.
_____, and M.R. Quilian. How to make a language user. In E. Tulving and W. Donaldson (eds.), *Organization of memory*. New York: Academic Press, 1972.
Craik, F.I.M. and R.S. Lockhart. Levels of processing: A framework for memory research. *Journal of Verbal Learning and Verbal Behavior*, 1972, 11, 671-684.
_____, and M.J. Watkins. The role of rehearsal in short-term memory. *Journal of Verbal Learning and Verbal Behavior*, 1973, 2, 598-607.
Craik, K.J.W. *The nature of explanation.* Cambridge, England: Cambridge University Press, 1943.
Crowder, R.G. and J. Morton. Precategorical acoustic storage (PAS). *Percept & Psychophysics*, 1969, 5, 365-373.
Darwin, C.J., M.T. Turvey and R.G. Crowder. An auditory analogue of the Sperling partial-report procedure: Evidence for brief auditory storage. *Cognitive Psychology*, 1972, 3, 233-267.
Engen, T. and B.M. Ross. Long-term memory of odors with and without verbal descriptions. *Journal of Experimental Psychology*, 1973, 100, 221-227.
Gilson, E.Q. and A.D. Baddeley. Tactile short-term memory. *Quarterly Journal of Experimental Psychology*, 1969, 21, 180-184.

Glanzer, M. and A.R. Cunitz. Two storage mechanisms in free-recall. *Journal of Verbal Learning and Verbal Behavior*, 1966, 5, 351-360.

Gould, J.D. Eye movements during visual search. *Research Report* No. 2680. Yorktown Heights, N.Y.: IBM Thomas J. Watson Research Center, 1969.

Haber, R.N. The impending demise of the icon: A critique of the concept of iconic storage in visual information processing. *The Behavioral and Brain Sciences*, 1983, 6, 1-54.

Hintzman, D.J. *The psychology of learning and memory.* San Francisco: W.H. Freeman, 1978.

Jacoby, L.L. Encoding processes, rehearsal and recall requirements. *Journal of Verbal Learning and Verbal Behavior*, 1973, 12, 302-310.

Kohn, M.L. and C. Schooler. *Work and personality: An inquiry into the impact of social stratification.* Norwood, N.J.: Ablex, 1983.

Loftus, E.F. Activation of semantic memory. *American Journal of Psychology*, 1973, 86, 331-337.

———. Leading questions and the eyewitness report. *Cognitive Psychology*, 1975, 7, 560-572.

———, D.G. Miller, and H.J. Burns. Semantic integration of verbal information into a visual memory. *Journal of Experimental Psychology: Human Learning and Memory*, 1978, 4, 19-31.

Loftus, G.R. Eye fixations and recognition memory for pictures. *Cognitive Psychology*, 1972, 3, 525-551.

——— and E.F. Loftus. *Human Memory.* Hillsdale, N.J.: Erlbaum, 1976.

Mandler, G. Organization and memory. In K.W. Spence and J.T. Spence (eds.), *The psychology of learning and motivation*, Vol. 1. New York: Academic Press, 1967.

Marcel, A.J. Conscious and preconscious recognition or polysemous words: Locating the selective effects of prior verbal context. In R.S. Nickerson (ed.), *Attention and Performance 8*. Hillsdale, N.J., Erlbaum, 1980.

———. Conscious and unconscious perception: An approach to the relation between phenomenal experience and perceptual processes. *Cognitive Psychology*, 1983, 19, 238-300.

McGaugh, J.L. and M.J. Hertz. *Memory Consolidation.* New York: Albion, 1971.

McSpadden, M. and E.F. Loftus. Guided imagery and the retrieval of actual and suggested memories. Unpublished manuscript, University of Washington, 1983.

Meyer, D.E. and R.W. Schvaneveldt. Facilitation in recognizing pairs of words: Evidence of a dependence between retrieval operations. *Journal of Experimental Psychology*, 1971, 90, 227-234.

Miller, G.A. The magical number seven, plus or minus two: Some limits on our capacity to process information. *Psychological Review*, 1956, 63, 81-97.

Milner, B. Memory and the medial temporal regions of the brain. In K. Pribram and D. Broadbent (eds.), *Biology of Memory*. New York: Academic Press, 1970, pp. 29-50.

Morris, C.O., J.O. Bransford, and J.J. Franks. Levels of processing versus transfer appropriate processing. *Journal of Verbal Learning and Verbal Behavior*, 1977, 16, 519-533.

Neisser, U. *Cognitive Psychology.* New York: Appleton-Century-Crofts, 1967

———. Memory: What are the important questions? In U. Neisser (ed.), *Memory observed.* San Francisco: W.H. Freeman, 1982a.

———. On the trail of the tape-recorder fallacy. Paper presented at the symposium "Influence of hypnosis and related states on memory: Forensic implications,"

American Association for the Advancement of Science, Washington. D.C., January 1982b.

Nelson, T.O. Repetition and depth of processing. *Journal of Verbal Learning and Verbal Behavior*, 1977, 16(2), 151-171.

Norman, D.A. and D.E. Rumelhart. *Explorations in cognition.* San Francisco: W.H. Freeman, 1975.

Peterson, L.R. and M.J. Peterson. Short-term retention of individual verbal items. *Journal of Experimental Psychology*, 1959, 58, 193-198.

Phillips, W.A. and A.D. Baddeley. Reaction time and short-term visual memory. *Psychonomic Science*, 1971, 22, 73-74.

Postman, L. and L.W. Phillips. Short-term temporal changes in free-recall. *Quarterly Journal of Experimental Psychology*, 1965, 17, 132-138.

Reed, S.K. Pattern recognition and categorization. *Cognitive Psychology*, 1972, 3, 382-407.

Roediger, H.L. Memory metaphors in cognitive psychology. *Memory & Cognition*, 1980, 8, 231-246.

_____, J.H. Neely, and T.A. Blaxton. Inhibition from related primes in semantic memory retrieval: A reappraisal of Brown's (1979) Paradigm. *Journal of Experimental Psychology: Learning, Memory & Cognition*, 1983, 9, 478-485.

Samson, E.E. Cognitive psychology as ideology. *American Psychologist*, 1981, 36, 730-743.

Schacter, D.L. and E. Tulving. Memory, amnesia, and the episodic/semantic distinction. In R.L Issacson and N.E. Spear (eds.), *Expression of knowledge.* New York: Plenum Press, 1982.

Schvaneveldt, R.W. and D.E. Meyer. Retrieval and comparison processes in semantic memory. In S. Kornblum (ed.), *Attention and performance IV.* New York: Academic Press, 1973.

Selfridge., O.G. Pandemonium: A paradigm for learning. In D.V. Blake and A.M. Utreley (eds.), *Proceedings of the symposium on the mechanisation of thought processes.* London: H.M. Stationer Office, 1959, 511-529.

Shulman, H.G. Similarity effects in short-term memory. *Psychological Bulletin*, 1971, 75, 399-415.

Sperling, G. The information available in brief visual presentations. *Psychological Monographs*, 1960, 74, 1-29.

Sperling, G. Successive approximations to a model of short-term memory. *Acta Psychologica*, 1967, 27, 285-292.

Sternberg, S. High-speed scanning in human memory. *Science*, 1966, 153, 652-654.

Theios, J., et al. Memory scanning as a serial self-terminating process. *Journal of Experimental Psychology*, 1973, 97, 323-336.

Townsend, J.T. Some results concerning the identifiability of parallel and serial processes. *British Journal of Mathematical and Statistical Psychology*, 1972, 25, 168-199.

Tulving, E. Episodic and semantic memory. In E. Tulving and W. Donaldson (eds.), *Organization of memory.* New York: Academic Press, 1972.

Tulving, E. *Elements of episodic memory.* Oxford: Oxford University Press, 1983.

Tulving, E. and D.M. Thomson. Retrieval processes in recognition memory: Effect of associative context. *Journal of Experimental Psychology*, 1971, 87, 116-124.

Tulving, E. and D.M. Thomson. Encoding specificity and retrieval processes in episodic memory. *Psychological Review*, 1973, 80, 352-373.

Turvey, M.T. Peripheral and central processes in vision: Inferences from an information processing analysis of masking with pattern stimuli. *Psychological Review*, 1973, 80, 1-52.

Vygotsky, L.S. *Mind in society*, Cambridge: Harvard University Press, 1978.

Warrington, E.K. and T. Shallice. Neuropsychological evidence of visual storage in short-term memory tasks. *Quarterly Journal of Experimental Psychology*, 1972, 24, 30-40.

Waugh, N.C. and D.A. Norman. Primary memory. *Psychological Review*, 1965, 72, 89-104.

Wood, F., et al. Regional cerebral blood flow response to recognition memory versus semantic classification tasks. *Brain and Language*, 1980, 9, 113-122.

14

Information Exposure, Attention, and Reception

Lawrence R. Wheeless and John A. Cook

Selectivity processes of exposure, attention and reception have been of concern to social scientists and, specifically, those concerned with information and behavior. Human communication scholars, psychologists, and sociologists have accumulated a large body of theory and research literature about these processes. This article attempts to summarize and critique that literature. Hopefully, insights into warranted conclusions, problems, solutions, and directions for future thinking and research are provided. Based upon some explicit assumptions, the article examines first the psychological and sociological theory and research related to selective exposure to information. Then, the many and varied approaches to understanding attention and reception are explored. Both active and passive ways of dealing with information are applied to the literature on these selectivity processes.

Two assumptions which directed our survey on exposure, attention, and reception warrant consideration. First, we operated on the assumption that we live in an information-saturated environment. Stimuli capable of producing information-overload bombard us from all dimensions. Information selectivity processes function in this environment in ways that allow us to deal with these stimuli, often in a very passive manner. Second, we believe that much information selectivity is active and goal-oriented. The primary goal of most, if not all, of active information selection is decision making. Under this assumption concepts such as utility of information, stress, dissonance, and so on become relevant. Utilizing these two assumptions, we were better able to sort the literature and make sense of it.

Given the vast number of sources, however, this task was quite impossible in the space allowed. Therefore, citations and comments on all relevant theory and research are not given. Rather, we have been highly selective. The sources cited, however, provide a good contemporary bibliography and, in turn, cite other previous research.

The first half of the article deals with exposure to information and how selectivity operates within that process. Exposure has been previously defined as putting oneself in a position or situation physically to receive information (Wheeless 1974). Given electronic capabilities of television, computers, and microwaves, exposure may be thought of more completely as the initial accessing to gain physiological reception capability. From this perspective, accessing options include not only active processes of selecting and rejecting information but also more passive processes involving not-selecting or not-rejecting potential sources of information. In this sense, exposure, attention, and perception are highly interrelated processes. Often, selective exposure may be based upon previously formed or existing perceptions. Such perceptions, of course, are related to selective attention. However, theory and research on selective exposure usually treat it as the initial or prime act in information acquisition.

Research on selective exposure reached its apex in the mid and late 1960s. Operating from psychological and sociological perspectives, researchers attempted to ascertain if and why selective exposure occurs. Early studies found, for the most part, that audiences generally shared the viewpoints of the sources communicating with them (Sears and Freedman 1967). This early research was conducted after the fact (de facto selectivity) and provided no conclusions about causative processes on why audiences were composed in such a manner. However, this early research led to studies centering on an hypothesized preference for supportive versus nonsupportive information in selective exposure behavior. This psychologically-oriented research was rooted in concepts of reinforcement and cognitive dissonance (Festinger 1957). Sociological and applied research approaches have continued to focus upon demographic characteristics of group audiences and markets which enhance exposure to differing types of information and media. Here, our concern will be primarily with experimental and quasi experimental research into why selective exposure occurs and, secondarily, with social characteristics and processes.

The second half of the article deals with attention and reception as interrelated phenomena. Selective attention, perception, and retention are even more difficult to separate in time from these processes than exposure is. In a typical basic text discussion of communication acquisition processes, McCroskey and Wheeless (1976, 277) suggested that "attention may be defined as a selective physiological set to receive a stimulus (message)."

They said that tuning in channels to selectively sense some part of the environment constitutes the attentional process. This conceptualization is obviously oversimplified, given the extensive literature on the subject. (We will treat definitional issues from that literature in depth.) However, from this limited perspective, attention may be viewed as a receiver's state of receptivity to a chosen focal stimulus (or focal stimuli). This view forces us to combine attention and reception under one heading as inseparable, making definitions of the processes elusive.

Implicitly, Eysenck (1982) established the concept of attention as an intervening process between stimulus and response. Broadbent (1958) supported such a notion in suggesting that attention accounted in large measure for human functioning. Treisman (1969) wrote that attention was the selective aspect of perception and response. Vernon (1960) took the opposing view, stating that perception governs attention.

Another interesting perspective is that of Stankov (1983), who viewed attention as a mode of interpreting how one deals with increasing amounts of information. Stankov's correlations of certain types of attention with intelligence may be a precursor to equating intelligence and attention.

Implications for activation, control, switching, monitoring, and motivation of these receptive states are discussed under the topics of discussion that follow. Now, let us turn to more detailed discussions of exposure, attention, and reception. Exposure processes warrant first consideration.

Selective Exposure

Numerous studies of selective exposure centering on the hypothesized preference for supportive versus nonsupportive information produced mixed results. Some of the investigations found that receivers prefer or select messages that are perceived to be consistent with existing opinions, beliefs, or attitudes; others did not find a preference; and others found a preference for nonsupportive information (for citations and critiques of these studies, see Cotton and Hieser 1980; Donohew and Palmgreen 1971; Sears and Freedman 1967; Wheeless 1974.). Most of these studies were conducted in the 1960s. Since the late 1960s, research has focused on refining underlying constructs and testing the effects of related variables on selective exposure (e.g., Chaffee and Miyo 1983). This latter conceptualization and research has produced salient issues that must be considered when conducting or interpreting research on selective exposure. The following discussion addresses some of those issues.

Attitudes, Supportiveness, and Reinforcement

The central theme of selective exposure research has been the hypothesized preference for information consistent with or supportive of previous choices, commitments, attitudes, or opinions. Likewise, there often has been an assumed avoidance of nonsupportive information. In general, researchers thought agreeing information should be sought and disagreeing information avoided. As noted previously, most research has produced mixed and apparently inconsistent results.

One apparent problem is the definition of what constitutes supportive information and reinforcing information, as well as the question what does the information support. Wheeless (1974) argued with confirming research results that attitudes and beliefs about the source of the information affect its supportiveness in terms of its reinforcement value. The perceived competence of the source was the primary predictor of selective exposure. Perceived character, sociability, and composure were also significant discriminating variables between selected and rejected information. In realistic "real-world" communication, most communication comes from sources that are known or perceived to be known by the receiver. In most experimental studies the source of information probably has been perceived as the experimenter and constitutes a major, confounding variable.

Schramm noted earlier (Schramm and Roberts 1971, 32) that the likelihood of information selection is directly related to its reinforcement or reward strength:

$$\frac{(\text{perceived reward strength}) - (\text{perceived punishment strength})}{(\text{perceived expenditure of effort})}$$

In short, other attributes of information other than mere agreement or attitudinal consistency may make it "supportive" for the selector (see Chaffee and Miyo 1983).

What the selected information supports is also problematic. Cotton and Hieser (1980) proposed and confirmed that when people act on a real, dissonance-producing decision, they seek information supporting that decision and avoid dissonant information. In this paradigm, attitudes (ie., attitudinal agreement) are not the predictive agent. The researchers note: "It is important to realize that the information was relevant to the counterattitudinal decision. That is, dissonance theory would predict that subjects selectively exposed themselves to information not because they have changed their attitudes, but because of the decision" (p. 519).

Similarly, Aronson (1968) suggested that selected information may be supportive of less obvious underlying cognitions that neutralize apparently dissonant situations. Donohew and Palmgreen (1971, 414) argued, however, that "the non-obvious underlying cognition is the *most important* cognition of those being considered and thus plays an overriding rather than a neutralizing role." They noted, for example, that an underlying, overriding "open-mindedness" cognition would lead to exposure to discrepant as well as supportive information. Clearly, only two cognitions, postulated by dissonance theory (Festinger 1957), are inadequate to explain dissonance reduction through selective exposure when dissonant information is chosen. Selected information may, indeed, be supportive of salient underlying cognitions. Therefore, the need still exists today for the "multi-cognition approach" suggested by Donohew and Palmgreen (1971).

Other dimensions of cognitions and attitudes beyond agreement or consistency are operative in the selective exposure process. One attitudinal dimension, involvement, has been directly linked to information selectivity (Chaffee and Miyo 1983; Sherif and Hovland 1961; Sherif, Sherif, and Nebergall 1965; Schramm and Roberts 1971; Wheeless 1974). In replicated, step-wise analyses, for example, Wheeless (1974) found involvement to be one of three highly significant discriminators between selected and rejected information. Given freedom of choice, subjects selected more involving information and rejected less involving information. Apparently, levels of ego-involvement are more predictive of exposure than the degree of attitudinal agreement on specific issues, particularly in nondissonance paradigms. On the other hand, general attitude similarity—attitude homophily—with sources of information is predictive of exposure and rejection (Wheeless 1974). That is, attitude similarity across numerous issues or clusters, rather than similarity on a specific issue, appears to predict selection and rejection of information sources.

The Nature of Choice

The nature of choices presented to experimental subjects appears to mediate results and may account for inconsistent findings. In general, research appears to demonstrate exposure patterns in line with dissonance reduction when those involved are given a choice in regard to deciding to engage in dissonance-arousing activity (Cohen and Latane 1962; Frey and Irle 1972; Linder, Cooper, and Jones 1967; Thayer 1969). Frey and Wicklund (1978), for example, found that subjects who freely decided to enter dissonance-arousing activity selected supportive information and avoided the nonsupportive. They argued that using the choice variable minimizes the impact of extraneous influences such as curiosity, usefulness, and attractiveness of decision alternatives.

Likewise, Cotton and Hieser (1980) manipulated the freedom of choice decision to produce high and low dissonance conditions and confirmed cognitive dissonance predictions in regard to seeking consonant and avoiding dissonant information. Because most previous research failed to find an avoidance effect, McGuire (1969) was led to question that aspect of selective exposure. However, the later research cited appears to support active avoidance in the presence of free decision choice and the opportunity to reject actively some information.

In a different vein, Wheeless (1974, 330) considered the active and passive nature of selection and rejection choices: "In reality, individuals have the choices of selecting, not selecting or not rejecting information." Most earlier research provided highly restricted choice options and often assumed rejection of nonselected information. Clearly, allowing a full range of selection choices (Wheeless 1974) as well as allowing choice to participate or not in dissonance-arousing activity facilitates observation of more active information seeking and avoiding (Cotton and Hieser 1980; Frey and Wicklund 1978). At the same time, these procedures clarify at least one reason much early research failed to find avoidance effects.

Subsequent choices relevant to an individual's exposure remain largely unexplained. Sears and Freedman (1967) noted the importance of exposure history on subsequent information selection choices. Wheeless (1974) allowed subsequent choices but examined only first choices. Frey and Wicklund (1978) allowed additional information selection choices and analyzed all choices to support their findings. Research by Schwarz, Frey, and Kumpf (1980) provides some insight and Milburn's (1979) findings support the notion that selection choices vary across time, given previous choices. Sears and Freedman (1967) earlier summarized research findings indicating a general second choice preference for information, supporting opposing views to information selected on first choice. However, no firm findings for selective exposure patterns across significant time frames exist that provide a solid basis for prediction.

Dissonance Research Problems

Mixed and apparently conflicting research results based on cognitive dissonance theory are frequently attributed to theoretical inadequacy, method and design problems, or "intruding factors." Festinger (1964, 95-96) offered such an explanatory adjustment for the problems associated with predicted avoidance of dissonant information:

> When the person perceives that dissonance will be effectively reduced by exposing himself to and coping with the details of the dissonant information, one will certainly not observe avoidance of it. Avoidance of potentially dissonance-increasing information would be useful in the service of dissonance reduction only if the person feels unable to cope with the new information. . . . And, of course, such avoidance would be observed only under circumstances where other reasons for exposure, such as usefulness or curiosity, were absent.

So very early on, Festinger raised possible exceptions involving confidence, utility, and curiosity. Similarly, Frey and Wicklund (1978) summarized and documented the potentially disruptive effects of intruding factors of curiosity, intellectual open-mindedness, information utility, attractiveness of choice alternatives, and confidence in ability to refute counterarguments.

As noted earlier, Donohew and Palmgreen (1971) summarized the theoretical, "underlying cognition" problem associated with dissonance theory and its reliance on only two cognitions (e.g., "Considering these two alone, the obverse of one element would follow from the other," Festinger 1957, 13). These types of conceptual problems coupled with mixed research results have led many to question the theoretical adequacy of dissonance theory to explain selective exposure (e.g., Brown 1965; McGuire 1966).

Other approaches to this problem point to methodological and design deficiencies in research rather than to dissonance theory as the source of discrepant findings. In addition to failures to hold constant factors such as "utility, curiosity, relevance, familiarity, social class, education, etc.," Donohew and Palmgreen (1971, 413) note failure to measure actual exposure, inadequate distinctions and choice between seeking and avoiding, and the assumption rather than the validation of dissonance arousal.

Frey and Wicklund (1978), Wheeless (1974), and Cotton and Hieser (1980) argued for and utilized more adequate choice options with favorable results. By controlling for participation choice, for example, Frey and Wicklund (1978) and Cotton and Hieser (1980) believed that they were able to control for most intruding factors and also manipulated levels of dissonance arousal. They obtained results supporting dissonance theory.

Donohew and Palmgreen (1971) argued that stress was a necessary attribute of dissonance arousal. However, subsequent physiologic measurement of stress was found to correlate oppositely to exposure selections postulated in cognitive dissonance theory (Donohew, Parker, and McDermott 1972). Miller (1977, 126) found that the state of postdecisional regret, one index of dissonance arousal, "resulted in greater preferences for nonsupportive information than did the state of dissonance reduction," although both states produced choice of state-consistent information (ie, supportive of the state). Apparently, when there are successful manipulations of dissonance, adequate choices provided, actual exposure choices measured, and intruding variables dealt with, then results consistent with dissonance theory are more often obtained.

Intruding Factors

The major factors which may "intrude" in research and are inherent to selective exposure research from a cognitive dissonance perspective include confidence, refutability, and utility. Confidence and refutability may be considered together as related phenomena.

Festinger's (1964) notion that confidence in a decision could affect subsequent exposure to dissonant information resulted from and led to mixed research results. Broadbeck (1956) found persons preferred consonant information after a discrepant communication when their confidence had been lowered. Canon (1964) obtained an interaction between confidence in initial opinions and selection of subsequent information. Highly confident subjects approached supportive and nonsupportive information while those with low confidence sought only supportive and avoided nonsupportive information.

In a real attempt at replication, Freedman (1965) failed to obtain a selectivity effect for differential confidence levels. Likewise, Mills and Ross (1964) and Lowin (1967) failed to find any selectivity effect for confidence. Mills (1965) found lower-confidence subjects expressed more interest in consonant information, but confidence was not found to be related to selection of dissonant information. S. Thayer (1969) argued that, except for Freedman's (1965) and Mills' (1965) studies, efforts to experimentally manipulate confidence were either unattempted or unsuccessful. Even following sucessful validation of confidence manipulation, S. Thayer (1969, 119) found, "Most subjects read both consonant and dissonant information. No relationship was found between confidence and selective exposure either in actual reading or order of reading." More recently, Schultz (1974) also failed to detect any effects of confidence on selective exposure.

Apparently, research has been unable to detect an exposure effect for confidence such as that obtained by Canon (1964). Perhaps as Donohew and Palmgreen (1971) argued, confidence reflects self-esteem and, therefore, self-esteem would be a more reasonable predictor. Other personality variables such as dogmatism (open- and closed-mindedness) (Donohew, Parker, and McDermott 1972; Schultz 1974) and locus of control (Davis and Phares 1967; Williams and Stack 1972) have shown tendencies for predicting selective exposure behaviors.

Perhaps, confidence coupled with the refutability of counterinformation provides a better predictive conceptualization. Confidence in ability to refute dissonant information apparently enhances exposure to that information. Lowin (1967, 1969), Kleinhesselink and Edwards (1975), and Wellins (1977) detected preferences for nonsupportive, choice-inconsistent information when subjects felt able to counterargue. Lowin (1967, 1969) thought belief systems could be maintained through avoidance of nonsupportive information or through refutation of the nonsupportive information. Kleinhesselink and Edwards (1975, 788) argued, "Whether a person will approach or avoid depends upon the individual's subjective estimate of the vulnerability of the nonsupportive information to his own refutational efforts." Apparently, confidence in refutational ability as related to belief-discrepant information does impact information selection.

Likewise, there is strong, replicated support of Festinger's (1964) notion that usefulness or utility of information may override normal dissonance theory predictions, particularly those involving avoidance of nonsupportive, dissonant information. Sears and Freedman (1967) address some of this research. More specifically, research by Canon (1964), replicated by Freedman (1965), produced strong findings that perceived utility enhances exposure to dissonant as well as consonant information. Similar results were obtained by Lowe and Steiner (1968). While Brock and Balloun (1967) did not find utility to be an effective determinant of selectivity, subsequent research by Brock, Albert, and Becker (1970) clarified the issue by detecting selectivity efforts for utility when the information was new or unfamiliar. Not suprisingly, they detected no utility effect when "communications contained only old familiar statements" (p. 300). One could certainly question the utility of further exposure to old, familiar information. There is, of course, some evidence that familiarity, utility, and supportiveness of informatin function together in the production of selectivity effects (Brock, Albert, and Becker 1970; Hillis and Crano 1973).

The utility effect is further supported by experimental research by Hillis and Crano (1973) and R.L. Miller (1977), as well as by mass communication research focusing upon "uses and gratifications" (e.g., Swanson 1976). While attractiveness of decision alternatives appears to mediate subsequent selectivity (e.g., Mills 1965; Frey 1981), utility of information appears to be an overriding factor in selective exposure that may produce effects counter to original predictions of cognitive dissonance theory.

Social and Sociodemographic Factors

Social and sociodemographic factors may play an overriding role in the process of selective exposure, especially in regard to media and social channels of information. Outside of the controlled experimental laboratory, social rather than psychological factors may dominate how we access information. McGuire (1968), in his summary of literature on the selective exposure hypothesis, indicates that most studies supporting the selective exposure hypothesis may be spurious or actually support a sociological process rather than a psychological one. The social system and its underlying structure may produce biased selectivity in regard to exposure patterns. In their summary, Sears and Freedman (1967) noted that predispositional factors such as sex role, education, ethnic status, political affiliation, and so on, largely accounted for systematic bias in audience composition. Moreover, they concluded that "clearly the most powerful known predictor of voluntary exposure to mass communications of an informational or public affairs sort is the general factor of education and social class" (p. 209).

Examination of applied research in media and advertising journals illustrates the effective use of sociodemographics in predicting or determining exposure patterns of groups in regard to media, programming, coverage, and advertising. Likewise, audiences and markets for differing types of media institutions and formats are easily and frequently obtained through the use of sociodemographics. This approach is derived from what Defleur (1970) termed the "social categories theory." This theory "assumes that there are broad collectivities, aggregates, or social categories in urban-industrial societies whose behavior in the face of a given set of stimuli is more or less uniform" (pp. 122-123). Defleur further concluded that "knowledge of several very simple variables—age, sex, and educational attainment—provides a reasonably accurate guide to the type of communication content a given individual will or will not select from available media" (p. 123).

Related research of a more theoretical or scholarly nature utilizes such sociodemographics combined with assumptions of "social relationships theory" (Defleur 1970) which emphasizes the interrelatedness of media and interpersonal networks. For example, Atkin (1972, 1973) has emphasized the "communicatory utility" or interpersonal communication utility of selected media information for interpersonal communication by individuals within their social groups. Likewise, Chaffee (1972) has emphasized the complementary nature of the interpersonal context for understanding mass communication selection, effects, and interplay. Perhaps, the most extensive and researched sociological model is found in the literature on the diffusion or communication of innovations (Rogers and Adhikarya 1979; Rogers and Shoemaker 1971).

A few examples of recent research from this social perspective points to some possible trends. For example, Hirschman (1981) investigated parental socioeconomic status, sex, educational attainment, childhood use of print media, and occupational status/complexity in relation to exposure behavior of the individual—adult print media usage, television usage, and group membership/participation. While prediction of exposure patterns was good with the entire path model used in the study, educational attainment and occupational status/complexity surprisingly were not related in the manner expected especially in regard to group membership and newspaper/ magazine usage.

Some research on television use and information seeking has also focused on aging, "contextual age," viewing patterns, and motivations (e.g., Rubin and Rubin 1982). Other research has explored situational information seeking by children in relation to television and interpersonal sources of information (e.g., Atwood et al. 1982).

Research in the political arena has tended to focus on utility, certainty, and salience factors, as well as sociodemographics and the flow of campaign information (e.g. Becker and Krendl 1983; Swanson 1976; Ziemke 1980). While the concepts of information utility and salience apply to these research areas, some research is grounded in the functional "uses and gratifications" perspective (e.g., Blumler and Katz 1974) which views persons as active agents (rather than passive receivers), who seek media and use it in relation to their needs. Persons act on information by selecting and interpreting it in useful ways.

The social model involving the diffusion of innovations also speaks to the selective exposure process. Diffusion is a particular kind of societal communication which addresses how innovations spread to people in a social system over time (Rogers and Adhikarya 1979; Rogers and Shoemaker 1971). Based upon perceived and actual sociodemographic similarities plus some special information or knowledge possessed, opinion leaders (usually early adopters) are sought out for useful information regarding innovations and related decisions. They constitute dominant, interpersonal sources of selective exposure for others in the social system and can be readily identified. Moreover, opinion leaders and early knowers tend to have higher levels of exposure to mass media and to change agents, as well as more interpersonal exposure within the social system. Recently, Richmond (1977) reconfirmed that opinion leaders have more information on the topic, more media exposure, and acquire more information when exposure is controlled.

Certainly, much selective exposure in society relevant to decision making can be explained on the basis of such social models. Whether psychological models are needed from a practical point of view, can be readily questioned. However, while we can garner some explanation of how and why selective exposure occurs based on these social models, psychological models involving cognitive dissonance and reinforcement have the potential of providing more satisfying explanations of why.

Exposure Conclusions

Given these problems and limitations, is further theoretical research into selective exposure warranted? We believe that it is. Understanding the psychological as well as the sociological factors underlying selective exposure is extremely important if we are to explain fully how and why the process operates. Social models are, of course, predictive. But understanding and explanation warrant continued attention. And, given the overwhelming influence of social factors:

1. Social factors, structures, and sociodemographics must be controlled or assessed in the laboratory and must be assessed in the field.

Given this assumption, several additional conditions are necessary for prediction and explanation.

2. Underlying cognitions or overriding cognitions (e.g., open-mindedness) must be considered in determining what constitutes supportive and nonsupportive information.
3. Information relevant to the behavior associated with decisions, not necessarily attitudes, must be considered.
4. Freedom of choice and realistic choice options must be provided.
5. Initial choices must be distinguished from subsequent ones, over time.
6. Postdecisional dissonance must be operative in relation to previous decision alternatives. Validation is essential.
7. Attitudes and beliefs about the source of communication must be determined.
8. Confidence in ability to refute nonsupportive information must be assessed.
9. Utility and/or attractiveness of selectivity choice alternatives must be assessed.

The effects of the last three factors above are, we believe, sufficient to introduce enough error variance to produce confounded results even when controlled through randomization. Interactions with these factors should be expected when factors one through six are dealt with adequately. These specific conclusions, of course, need to be tested for their validity. We provide more general conclusions that may be useful in guiding theory and research following the next section on attention and reception.

Selective Attention and Reception

On first examining the literature on the selectivity processes of attention, two things are clear. There is a vast amount of experimental and theoretical discourse in this subject area from a variety of diverse academic disciplines, and the attention-reception processes are complex and difficult to explain.

It is probably true that the range of academic disciplines studying attention is at least one cause of such a volume of literature. These academic disciplines include but are not limited to human communication, mass communication, cognitive psychology, developmental psychology, neuro-psychology, education, special education, gerontology, and audiology.

The first and most obvious aspect of attention for the uninitiated is the concept of continually sustained directed energies. When one is asked to "pay attention," he or she is being required to focus upon a stimulus (stimuli) and discontinue reception from other sources. Sustained attention may be measured by cues as simple as eye gaze duration (Krull, Husson, and Paulson 1978) or other techniques requesting subjects to focus aurally or visually and subsequently report reception in some way (Meyer and Maguire 1981; Posner and Snyder 1975; Treisman 1977; Treisman and Gelade 1980).

The concentration aspect is but one element of operationalization, and each operational technique yields some insight into the attentional system complex. For this reason, it is appropriate that scholars utilize many different measures. Any operationalization may tap a different subunit or network of subunits within the processing complex. Thus, vigilance measures may be employed.

When vigilance measures are employed, attention is operationalized in terms of subject capacity to detect an infrequently occurring signal over a long period of time (Stankov and Horn 1980; H.J. Eysenck 1959, 1966; Mackworth 1969; Swets 1977; Nachreimer 1977; Davenport 1968, 1969; Sostek 1978; Warm et al. 1972; Schneider and Shiffrin 1977; Stajnberger 1972).

Researchers may use search measures of attention. Search is somewhat akin to vigilance. In this context, we are dealing with the detection of a signal from a set of signals (Moray 1969; Hartley 1981; Meyer and Maguire 1981; French 1951; Horn 1980). The search dimension has been labelled clerical-perceptual speed in the work of Horn (1980). Some scholars of attention have used various distractive artifices and competing tasks to examine processing capacities (Stroop 1935; Lowe and Mitterer 1982; Spelke, Hirst, and Neisser 1976; Cherry 1953; Moray 1969; Stacks and Sellers 1983; Hiscock and Kinsbourne 1980; ten Hoopen and Vos 1980).

Still other scholars examine the attentional processes based upon physiological and neurophysiological manifestations of an active attention state such as EEG, skin conductance, and pupil dilation (Corteen and Wood 1972; Shulman, Remington, and McLean 1979; Remington 1980).

One may derive consensus from these various perspectives on attentional processes regarding one issue: attention is not a single concept but a complex of subunits (Stankov 1983; Posner and Snyder 1975; Shiffrin and Schneider 1977; Johnston and Heinz 1978).

In the final analysis, the literature may best come to a meaningful synthesis by examining two areas of discussion. The first area of discussion would be stimuli characteristics, and the second would be the theories of explication of attention-reception.

Stimuli Characteristics

Basically we will combine two broad areas of concern regarding stimuli effects on attentional mechanisms. There should be a dichotomy in one sense because of the divergence of opinion on where selection takes place. Some say stimuli are selected at the level of sensory reception, others say stimuli are selected at the level of evaluation. Consequently, consideration will be given to both externally observable characteristics which generate the orienting reflex and attributional characteristics subsequent to cognitive processing.

The properties of stimuli which seem to increase the likelihood of attention are: (1) the setting in which the stimulus is located; (2) the intensity (volume, brightness, vividity, etc.) of the stimulus; (3) the extensity (size or quantity) of the stimulus; (4) the concreteness or complexity of the stimulus; (5) the contrast and velocity of the stimulus elements (novelty, variety, change, movement, animation, etc.); and (6) the impressivity of the stimulus (repetition, duration).

Such categories as these (McCroskey and Wheeless 1976; Woodworth 1934; Berlyne 1960; Vohs 1964) may vary from source to source but such variations are largely a matter of nomenclature. Minnick (1968) and O'Donnell and Kable (1982) described the external factors of attention stimuli in terms of that which is striking or conspicuous, contrasting, novel, in motion, or repetitive. Regardless of the labels, certain properties do engage sensory attention.

Some of the same perceived characteristics that engage selective exposure also engage selective attention. Familiarity, utility, and supportiveness of information affects attention just as it does exposure (Brock, Albert, and Becker 1970).

The mass communication research in television viewing also gives us some attention stimulus insight. Saliency of visual and auditory cues will enhance attention (Collins 1982). From the work of Alwitt et al. (1980) and Levin and Anderson (1976) we know that children are more attentive to women characters and women's and children's voices, auditory changes, peculiar voices, movement, camera cuts, sound effects, laughing, and applause. However, visual attention appears diverted when children hear male voices, view extended zooms and pans, and view animals and still pictures (Collins 1982). Program complexity was not an attention factor according to Krull, Husson, and Paulson (1978), yet Anderson et al. (1981) found that comprehensibility (which must be related in some way to complexity) did cause children to selectively attend.

Confirming the generalizations of basic communication textbooks, research in children's television viewing has shown that high action, fast pacing, cuts, fades, visual effects, sound effects, and music all hold visual attention (Collins 1982). Singer (1982) proposed that the attention-gaining quality of salience was a matter of being "affectively involving."

From the study of stimuli characteristics, we also know processing limits, based upon extensity (G.A. Miller 1956), frequency, duration, and difficulty of cues (Hansen and Hillyard 1983; Meyer and Maguire 1981; Flowers, Polansky, and Kerl 1981). We also know that early theorists saw the likelihood that physical characteristics might impact attention, for example Treisman's (1964a) hierarchy of stimuli characteristics (physical cues, syllabic patterns, specific sounds, words, grammatical structure, and meaning). In addition, the study of stimuli characteristics has demonstrated that that which is familiar, linguistically or otherwise, governs patterns of selective response (Lane 1980; Cherry 1981; Rice, Huston, and Wright 1982; Brock, Albert, and Becker 1970).

Theories of Attention

The basis of description of the receiver's selectivity processes of attention and reception comes primarily from experimental data and extrapolations of experimental data. The theories are somewhat difficult to classify, but fundamentally the explanations will be grouped according to perception, information processing, and developmental literature.

Receiver as Perceiver. In the case of perception and attention, it is not altogether clear which is the cart and which is the horse. Treisman (1969, 283) viewed attention as "the selective aspect of perception and response," thus, attention is viewed as the exploratory phase of perception. However, it may also be argued that perception governs attention (Woodworth 1921). Actually, both views are correct.

Sokolov (1960) described an orienting reflex: an alignment of the central attention to the source of stimulation. This alignment reflex is engaged because of attribution of perceived source and message characteristics, the psycho-biological state of the receiver, and (as seen in the discussion above) stimuli characteristics.

Paschal (1941), drawing on the work of Woodworth (1921, 1934, 1940), explicated the concept of a mental set, a drive which exerts an influence upon solving a certain problem or performing a certain operation, thus impacting selection. The mind set seems to be the amalgam of motivational and perceptual predispositions to respond (receive). Vernon's (1960) essay examined perception studies concluding that there was some process of filtering which was a learned part of consciousness governing judgment as to significance and choice of stimuli, and that this perceptual filter determined direction and magnitude of judgments.

Massad, Hubbard, and Newtson (1979) developed a perceptual selectivity model. The authors tell us that "perceivers do not comprehend all possible actions and then select the relevant ones to remember, rather perceivers select an initial framework of interpretation and use that as a guide for comprehension of an event" (p. 527). Thus, a body of experts claim that the cognitive frame of reference predisposes a receiver to attend. This is consistent with interpersonal perception research which indicates that positive perceptions increase the likelihood of interaction (McCroskey and Wheeless 1976; Schneider, et al. 1979). Logically, if the perceived credibility and homophily of a source tends to increase exposure, as Wheeless (1974) showed, it might also be true that attention mechanisms will be more likely engaged due to positive person perceptions.

This body of research suggests that perception governs attention. However, attention could logically precede the formation of perceptions. For example, Berger and Calabrese's (1975) model of proactive and retroactive attribution can be taken as a basis for this, since clearly one must attend at some level in order to formulate retroactive attributional perceptions.

Nielsen and Sarason's (1981) research supports the structure represented in Figure 14.1. They believe that there is a direct relationship between selective attention and the emergence of cognitions in awareness. Citing some shadowing experiments, Neilsen and Sarason suggested that the contents of the ignored channel can influence cognitive processing without entering awareness, and that preattentive processing includes a level of semantic analysis (Corteen and Dunn 1974; Corteen and Wood 1972). The authors stated that signal detection theory (Green and Swets 1974) is correct in postulating that subjects first either receive or do not receive, then decide whether their perceptual state allows them to report the presence of a signal. Ultimately, attending is linked by Nielsen and Sarason to emotion and personality. So, it would seem, there is a complex interaction of personality, emotions, semantic meaning and awareness. These interactions, contend the authors, comprise processing.

The introversion-extraversion continuum has produced much literature on attention. H.J. Eysenck (1966) suggested that introverts were more cortically aroused and therefore more subject to fears, anxiety, and other sources of fear conditioning. The potentially high arousal of introversion means that parallel processing (dual task) is more difficult for the introvert (M.W. Eysenck and M.C. Eysenck 1979a); thus there is some implicit evidence that introverts perform with more attentional selectivity.

Certain applied studies further the selectivity-introversion relationship. Revelle, Humphreys, and Gilland (1980) examined the impulsivity dimension of extraversion and noted that impulsivity was associated with attention deployment. Davies and Hockey (1966) noted that increased noise level and signal frequency improved extravert task performance but caused a decrement in introvert task performance. Other studies support the tendency of extraverts to prefer higher levels of sound stimulation (Elliot 1970). In Nideffer's (1976) work with a self-report measure, it was found that extraverts scan a broad environment while introverts focus on small segments.

The concept of such states as anxiety and hostility interacting with cognitive processing brings us back to M.W. Eysenck (1982), who believes that evidence from a variety of psychological disciplines provides support for an arousal state which determines the activation of attentional mechanisms.

M.W. Eysenck (1982) began with the work of Yerkes and Dodson (1980) that suggested that the relationship between the level of tension, motivation, or arousal and performance efficiency is an inverse parabolic function. That is to say, performance efficiency is maximized at moderate levels of arousal. Easterbrook (1959) noted that focus narrows (attention gets more selective) in arousal states so that difficult tasks, which involve the need for a greater number of cues, are impaired by maximal arousal.

Broadbent's (1971) landmark work, *Decision and Stress*, described cortical states of arousal which operate at two levels: (1) to execute decision processes; and (2) to monitor cognitive processes in order to maintain a given standard of performance. Perhaps the most compelling support for Eysenck's arousal perspective emerges from three works, all of which provided evidence that controlled attention takes place at a moderate state of psycho-biological arousal (Hasher and Zacks 1979; Kahneman 1973; R.E. Thayer 1970); thus an individual's active involvement is a determinant of selectivity. While too high an activation state (anxiety) may well impede performance, a moderately high level of arousal that increases the selectivity of attentional mechanisms to optimum levels will not.

This arousal state is described through a variety of predictors ranging from emotional involvement to caffeine. Many of these arousing agents appear to be causal factors of selective attention.

Incentive, the interaction of internal and external motivation factors, produces a goal-oriented state of arousal (Bindra 1969). Atkinson and Wickens (1971) suggested that attention was activated by reinforcement. Simon (1967) defined motivation as that which controls attention. However, there has been some study which indicates that selective focus is not always an asset in accomplishing task demands set forth by researchers.

For example, the work of Condry (1977) and Glucksberg (1964) reminds us that flexibility (required attention switching) may be reduced by incentive. Thus, experimenter reinforcement increases the tendency to persist with established models of thinking (such as patterns offered by the experimenter as to how to proceed in solving tasks involving memory tests).

McGraw and McCullers (1979) in their problem solving research noted that incentives may reduce cognitive flexibility such that a learned skill does not transfer to a different-solution similar task. Willett (1964) and Elliot (1970) both provided evidence that errors and mean reaction time are negatively affected by motivation.

These studies have implications for the relationship between an aroused (motivated) state and an anxious one. Anxiety produces exceedingly high levels of tissue activation, as does arousal. Exceedingly high arousal is an anxiety state which impedes performance. Moderate levels of motivation appear to facilitate active, flexible attention states.

Sleep deprivation has also been studied in its relationship to arousal (Bohlin 1973; Glaubman et al. 1978; Kjellberg 1977; Malmo and Surwillo 1960). Most of the results indicate that sleep deprivation increases the level of activation. Glaubman et al. (1978) believed the distinction between REM and non-REM sleep loss was important. REM sleep loss seems to have a greater deprecating effect upon flexibility. The deprivation of sleep is known to harm performance on sustained attention (Bergstrom, Gillberg, and Arnberg 1973), on vigilance (Wilkinson 1961, 1963, 1964), and on development of habituation (Kjellberg 1977).

Numerous factors such as those discussed above are related to the aroused (selective) state. What is clear from the attention-arousal links is that one tends to lose speed in order to maintain accuracy in a conscious attention task (M.W. Eysenck 1982). So, while an optimal state of arousal permits focus, it impairs flexibility, and such trade-offs should be borne in mind in future work on attention. The various emotional states, perceptual fields, and activation states described in the literature under this section lend us some insight as to that which affects attention. However, there is also some evidence which indicates that selective attention can be better understood in terms of information processing and the cortical central control mechanism.

Receiver as Processor-Central Control Mechanism. Any human environment may be thought of as stimulus-saturated. That is, regardless of where we are, the communication transactions in which we find ourselves permit virtually an infinite number of stimuli to attend to. This stimulus inundation is complicated by humans' finite capacity to process. Further complicating the stimulus-saturated milieu in which humans find themselves is the fact that we constantly expend a certain amount of attentional capacity monitoring the environment. One signal detected in monitoring may mandate a switch in focus. Ultimately, then, our finite information processing resources force the organism into a constantly selective mode. This view of receiver-as-information-processor centers upon capacity limitations and what we do with them (Atkinson and Shiffrin 1968; Baddeley and Hitch 1974; Baddeley, Thomson, and Buchanan 1975; Hitch 1980; Johnston and Heinz 1978; Navon and Gopher 1979; Norman and Bobrow 1975).

Our understanding of processing resource limits comes largely from competing tasks research and pathological data which indicate there is some means within humans which allows us to manage the flow through attentional selection. An important part of the explanation of input management is the dichotomy between that which is automatic (demanding little or no active control) and that which mandates control (focused attention effort). Automatic and controlled attention will be viewed as a pervasive tandem throughout this discussion.

Engaging automatic functions seems to allow tasks to get easier and make subjects less error prone as those subjects receive more practice time (Flowers, Polansky, and Kerl 1981; Posner 1978; Posner and Snyder 1975; Shiffrin and Schneider 1977; Spelke et al. 1976). On the other hand, controlled attention, activated by a complex of flexible resources, requires energy, awareness, and assets such as knowledge of language and effective use of the central control mechanism. Documentation for an active (controlled) element of attention stems from the "working memory" concept (Norman 1968; Baddeley and Hitch 1974; Reitman 1974; Hitch 1978), a limited capacity resource for storage and interpretation. This limited resource is the impetus for much of the theoretical discussion of attentional mechanisms. The conceptualization is one of a cognitive network of connected subunits, each of which may be engaged in some way in order to process information.

After a certain amount of time, learned activities take on the quality of automaticity in the attentional scheme, freeing actively controlled resources (Shiffrin and Schneider 1977). Thus a receiver as an information processor relegates some of the stimuli to the effortless (automatic) attention "file," while dealing with the effortful side of attention (controlled attention) as load capacity permits. The active side of attention is explained in terms of selective receptivity.

Active attention processes information in three stages: (1) construction of sensory inputs; (2) translation to semantic representations from sensory representation; and (3) admission of output from stages one and two into consciousness (M.W. Eysenck 1982). A bone of contention among theorists is where precisely that selection takes place. This point of theoretical debate parallels the explanations provided by input overload theories of information processing (J. G. Miller 1962). Broadbent (1958) argued that sensory inputs which are simultaneously transmitted are sequenced for processing based upon physical characteristics of the signal input. Thus Broadbent argued that selection takes place at stage one (sensory input). Deutsch and Deutsch (1963) held that selection occurs at stage three, at admission to consciousness. According to Deutsch and Deutsch, all inputs are perceived by the senses, the attention limits are upon selected conscious responses, or feedback responses to the sources of stimuli. Norman (1968) suggested that the primary or working memory selects such that interpreted input which is activated at stage two (semantic representation) and controlled at stage three provides the selective activation of controlled attention.

In his later work Broadbent (1971) argued that we tune in either at the level of sensory input interpretation (stage one) or at the semantic level (stage two), depending on conditions. Treisman (1964b) agreed. Treisman's work suggested that an attentional system analyzes and rarefies message inputs based on a hierarchy of stimuli characteristics.

Many current theorists lean toward a neuropsychological approach to the study of attention. The approach is not entirely novel. Four decades ago some of the psychopathology researchers viewed attention as a significant correlate of neural disorder (Paschal 1941). Neurophysiology has "gained ground" in recent time in part because of evidence which suggests that substantial amounts of processing transpire below the level of awareness (M.W. Eysenck 1982). If these researchers are correct, then we must examine the neural manifestations and physical symptoms of attention or else we may bypass some part of the attention-reception system. The theoretical basis for describing the attention-reception complex is rooted in the study of the central control mechanism.

The central control mechanism perspective is one in which attentional mechanisms manage the input flow by channeling inputs for interpretation based on the makeup of the receiver. Posner (1978, 1982) described a system of automatic activation to allow for unattended inputs. From Posner's perspective, conscious attention is involved when there is performance impairment in responding to unexpected events (see also Posner and Snyder 1975). When the automatic portion of attention is engaged, it is in some measure due to expected conditions, for example, driving a familiar route. A great deal of applied research showed that familiarity with signals or redundancy of signals or practice will lead to the low activation level of attention monitoring (Flowers et al. 1981; Hasher and Zacks 1979; Logan, 1979; Shiffrin and Schneider 1977). On the other hand, controlled processing takes much more time per unit of information (Hitch and Baddeley 1976; Posner 1978; Shiffrin and Schneider 1977; Treisman and

Gelade 1980), indicating that a complex of interpretations and representations are a part of that which is channeled through the conscious load. These interpretations and representations are, quite obviously, part of cortical functioning.

The evidence of brain asymetries is pertinent to the link between controlled attention and time lag required to process information. Since the sensory and semantic interpretations and the hemispheric transfers which take place account for some of the processing time in normal controlled processing, brain asymmetry is one point of focus (Andersen, Garrison, and Anderson 1979; Hiscock and Kinsbourne 1980). Posner (1982) believes that mental events are handled successively and that Pavlov's (1960) cortical processes of inhibition and facilitation exercise control over selective attention. The parcelling out of resources and the conveyance of inputs to the appropriate subunits of the attention system are believed to be governed by the central control mechanism, which may be thought of as a circuit switching system, a filing system, and a learning system.

Tyson (1982) explained the phenomena of attentional resources in terms of general systems theory. His attention metaphor is a link to open and closed biological systems. Tyson's explanation is that we have certain capabilities for attentional flexibility to overcome biases, thus allowing maximal communication input. Such a receptive state is described as the openness of meditation, which Tyson holds that we can maintain while carrying on the automated activities of day to day life. Tyson's analogic view is not entirely untenable, but there have been many in-depth descriptions of attentional mechanisms is terms of the processes of acquisition, storage, and retrieval of information. These more thorough explanations are experimentally based or are extrapolations of experimentation.

Norman (1968) posited a number of interactive processes linking operations of primary and secondary storage. Primary storage is the temporary activation of the storage system, or the so-called working memory. Secondary storage is long term. Norman (1968, 535) concluded:

> Complete interpretation of inputs and efficient selection among them require a continual reassessment of the permanently stored information associated with each input. This can be performed by a recursive process aided by temporary memory traces. The simple analysis of each input signal is not sufficient to resolve ambiguities and establish interpretation. This comes only when a pertinence input complements the sensory input. Thus, attention and selection use the primary storage to enable efficient selection among and analysis of sensory inputs.

The view of short-term storage being a proportion of total storage was expanded by Baddeley and Hitch (1974) and their working memory-articulatory loop mechanism. Working memory, according to this view, is primarily a limited capacity system for processing all kinds of input, while the articulatory loop is a proportion devoted exclusively to verbal short-term memory. Together, a general information processor and a specialized verbal processor comprise the attentional mechanism of working memory (Baddeley and Hitch 1974; Reitman, 1974). The articulatory loop is seen in required time-to-process words. There is as yet no empirical support for capacity shifts based on load (M.W. Eysenck 1982). There is, however, a theoretical base, which suggests a hierarchy of process capacities with lower levels being specific and automatic, and higher levels being general and consciously controlled. At the top of the hierarchy rests a central processor which flexibly assists the lower levels and transforms information and makes decisions. The effective functioning of that central processor is determined by the demands placed upon it (M.W. Eysenck 1982).

From this perspective, the central control mechanism is inclined to be at the core of both controlled and automatic reception of selected inputs. Figure 14.1 is a summary of the theoretical extrapolations discussed. Fundamentally, Figure 14.1 represents the signal pathways through working memory. Signal inputs will be interpreted first by sensory mechanisms, attended or not, then, if necessary, semantic representations will be constructed, and either attended or not.

The comparison and evaluation stage, surrounded by a perceptual filter, is where most attention determinations are made based upon past experience. Inputs are then confronted either by an activated state of arousal or by automatic attention, or judged unworthy of attention. Finally, that which is attended is stored either in habituation memory or in long-term (active) memory (Cook 1984). The alternative pathways are determined by the central control mechanism. The perceptual filter and activated arousal state are in the discussion above on the receiver as perceiver. These two aspects (perception and arousal) are believed to be products of the interaction of the attention cortical functions and the psychology of perception. *Developmental Research: Receiver Growing Selective.* We would be remiss in our duties to this state of the art if we did not mention the research which indicates that attention is learned. Lane and Pearson (1982) reviewed a large body of developmental research which indicated that the limited capacity of attention must be allocated by the flexible central control mechanism. As childhood growth and development proceeds, humans allocate progressively less capacity to irrelevant stimuli (Kahneman 1973; Pick, Christy, and Frankel 1972; Smith, Kemler, and Aronfreed 1975; Day and Stone 1980; Strutt, Anderson and Well 1975; Lane 1980). The preponderance of evidence in developmental psychology clearly shows an increase in selectivity over the years of childhood growth. Further support for

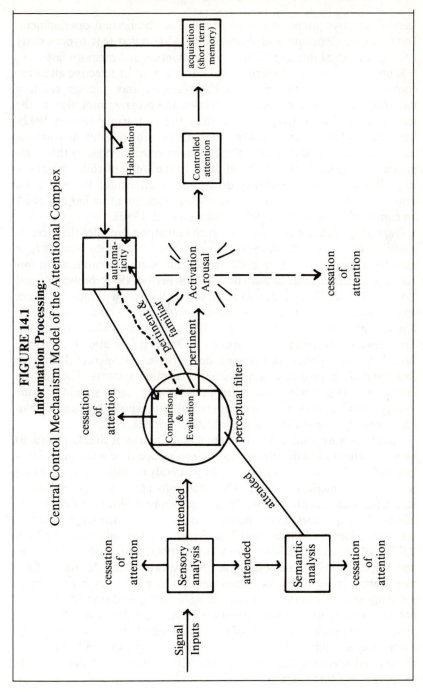

FIGURE 14.1
Information Processing:
Central Control Mechanism Model of the Attentional Complex

learned selective attention is seen in research on aberrant development. For example, Copeland and Wisniewski (1981) noted that hyperactivity and learning disability negatively impact attention and memory tasks.

Much of the recent research in the development of selective attention comes from mass communication. For example, since younger children have not learned to tune out irrelevancies, the programming flow of the television medium may be counterproductive to learning (Singer 1982). Likewise, children may possibly overfocus on violent content which is easier to comprehend (Singer 1982). Moreover, research suggests that comprehensibility should be a point of focus in the study of children's attention. While comprehensibility does impact attention, there is little evidence from mass communication research that attention has any effect on comprehension (Collins 1982; Anderson et al. 1980).

Ultimately research on children's media attention provides the receiver qualities that govern attention-reception quite clearly. Those factors which have been found to mediate children's attention are communication format and linguistic acquisition (Berndt and Berndt 1975; Newcomb and Collins 1979) and knowledge of media conventions such as music, camera angles, and program-commercial separators (Baggett 1979; Salamon 1979; Krull, Watt, and Lichty 1977; Collins 1978; Palmer and McDowell 1979). The term *media literacy,* or the capacity to extract information from television (Salamon 1979), is in essence a developmental construct. It may be that meaning is predicated on attention-reception systems. The authors believe that despite television's tremendous impact, more should be done with children's attention in the interpersonal vein, especially since much of classroom learning is through a face-to-face channel.

Commensurate with other developmental findings is Stankov's (1983) view that attention is directly related to intelligence. The intellectual functioning-attention link leads us logically to a body of cognitive research on the subject of intellectual aging. There has been a great deal of research in recent years on geriatric factors impacting attention. With the debilitating effects of aging, older adults perform less well on processes of selective attention (Hoyer and Plude 1980; Rabbitt 1977), attention switching (Craik 1977) and vigilance (Talland 1968). Horn (1982) and Horn, Donaldson, and Engstrom (1981) suggested that there was an aging decline in fluid intelligence, consequently the elderly tend to have difficulty in organizing, receiving new information, concentrating, avoiding irrelevant distraction, and maintaining and dividing attention. However, Willis et al. (1983) contend that attention is a matter of perceptual speed, not intelligence. The contention of Willis et al. seems to contradict the vast body of literature on the aroused attention state, but after all, Willis and company deal in training programs for the elderly.

The developmental research tells us that we acquire through the years the capacity to focus and screen out irrelevancies. Hence a certain amount of selective attention is learned. However, as we move into old age, the capacity to control (to select and concentrate) declines. Thus, the age and selective attention relationship is a nonlinear, inverted "U" function.

We noted in the introduction that attention may be that which mediates between stimulus and response. Scholars have studied response and inferred to the organismic processes. General patterns of selective attention and reception have been noted and published. It is now our job to delineate some patterns or trends based upon the literature.

Attention-Reception Conclusions

Despite the extent of the attention literature, the conclusions may be delineated briefly. These conclusions are considered to be indicative of that which is known or theorized in the significant work on attention. Future research should spring from these "truisms" enumerated below.

1. The capacity of a stimulus to engage attention results from an interaction of stimulus characteristics and
   * a receiver's form and symbol decoding capacity;
   * a receiver's perceptual frame of reference;
   * a receiver's prior experience;
   * a receiver's stage of cognitive development;
   * a receiver's psycho-biological state.
2. The attention mechanism is an organized system of subunits whose sum total processing capacity is finite.
3. Within the attentional mechanism, there is a strain toward less expended effort.
4. Over time, normally developing humans acquire more capacity to ignore signal inputs deemed irrelevant.
5. The growth and development of attention control peaks, and then declines in the latter years of life.

Concluding Remarks about Selectivity Processes

The question of the occurrence of selective exposure and attention is a moot one. Certainly, people engage in selective exposure to information and sources of information in interpersonal, public, and mediated contexts. Aside from confirming psychological and social research, one need only observe a teenager's social world or a parent in front of a television set. The question is, can we adequately predict and explain selective exposure and on what basis or bases? As noted immediately above, social roles and sociological models coupled with sociodemographics are quite able to tell how exposure occurs and predict selective exposure patterns, even over time. However, the psychological basis for the hypothesis that we select supportive and reject nonsupportive information, even in the short term, is less clear. Applying reinforcement theory to predecisional exposure and dissonance theory to postdecisional exposure is obviously helpful. Also, distinguishing between passive and active selection of information is useful. In the real world, we would obviously expect that there would be both active and passive predecisional, selective exposure. Perhaps, reinforcement theory is operative in passive, predecisional exposure patterns if organized into a reward/cost ratio of the type presented earlier by Schramm and Roberts (1971). We submit that underlying theories of reinforcement and cognitive dissonance are operative in a predictive manner in active, goal- or decision-oriented information seeking behavior. In this case, a certain level of involvement is a necessary psychological component.

Explanations of the causes and dynamics in the selectivity processes of attention and reception are a bit less clear than exposure issues. One of the cloudier issues is whether selective attention is actually an advantageous cognitive state in accomplishing tasks. On the one hand, selectivity permits a receiver to screen out irrelevancies (i.e., concentrate). On the other hand, some evidence points to the need for a wider scope of receptivity to stimuli in the case of complex tasks. At this juncture, we will simply say that the need to engage selective attention appears to be context specific. Of course, there should be further research in this area. Naturally, research in selective attention will continue, given its theoretical and practical importance to communication and information studies. The rise in concern is seen in the field of human communication in the printing of numerous texts on listening as well as in continued empirical explorations.

We believe that one of the weaknesses in human communication's early work was the focus on the sender of information. Clearly, communication skills training will perpetuate a sender-oriented perspective to some degree. Nevertheless, since attention is a receiver phenomenon, the emphasis in attention research must be on the receptivity state and how the receiver responds to stimuli. Given the focus just described, eventually the research scope should widen to the entire transaction process, incorporating the complex interactions of medium, context variables, and effects over time, as well as the receiver's receptivity and response. Finally, the causes and impact of automatic (passive) reception should be compared directly with controlled (active) reception.

References

Alwitt, L.F., et al. Preschool children's visual attention to attributes of television. *Human Communication Research*, 1980, 7, 52-67

Andersen, P.A., J.P. Garrison, and J.F. Andersen. Implications of a neu-rophysiological approach for the study of nonverbal communication. *Human Communication Research,* 1979, 6, 74-89.

Anderson, D.R., et al. Watching children watch television. In G. Hale and M. Levis (eds.), *Attention and the development of cognitive skills.* New York: Plenum, 1979.

_____, et al. The effects of TV program comprehensibility on preschool children's visual attention to television. *Child Development,* 1981, 52, 151-157.

Aronson, E. Dissonance theory: Progress and problems. In R.P. Abelson, et al. (eds.), *Theories of cognitive consistency: A sourcebook.* Chicago: Rand McNally, 1968.

Atkin, C.K. Anticipated communication and mass media information seeking. *Public Opinion Quarterly,* 1972, 36, 188-199.

_____. Instrumental utilities and information seeking. In P. Clark (ed.), *New models for mass communication research.* Beverly Hills, Calif.: Sage, 1973.

Atkinson, R.C. and R.M. Shiffrin. Human memory: A proposed system in its control processes. In K.W. Spence and J.T. Spence (eds.). *The psychology of learning and motivation: Advances in research and theory,* vol. 2. London: Academic Press, 1968.

_____ and T.D. Wickens. Human memory and the concept of reinforcement. In R. Glaser (ed.), *The nature of reinforcement.* London: Academic Press, 1971.

Atwood, A.R., et al. Children's realities in television viewing: Exploring situational information seeking. In M. Burgoon (ed.), *Communication yearbook 6.* Beverly Hills: Sage, 1982.

Baddeley, A.D. and G. Hitch. Working memory. In G.H. Bower (ed.), *The psychology of learning and motivation,* Vol. 8. London: Academic Press, 1974.

_____, N. Thomson and M. Buchanan. Word length and the structure of short term memory. *Journal of Verbal Learning and Verbal Behavior,* 1975, 14, 575-589.

Baggett, P. Structurally equivalent stories in movie and text and the effect on medium recall. *Journal of Verbal Learning and Verbal Behavior,* 1979, 18, 333-356.

Becker. L.B. and K.A. Krendl. Local voting decisions and the flow of campaign information. In R.N. Bostrom (ed.), *Communication yearbook 7.* Beverly Hills, Calif.: Sage, 1983.

Berger, C.B. and R.J. Calabrese. Some explorations in initial interaction and beyond: Toward a developmental theory of interpersonal communication. *Human Communication Research,* 1975, 2, 99-111.

Bergstrom, B., M. Gillberg, and P. Arnberg. Effects of sleep loss and stress upon radar watching. *Journal of Applied Psychology,* 1973, 58, 158-162.

Berlyne, D.E. *Conflict, arousal, and curiosity.* London: McGraw-Hill, 1960.

Berndt, T.J. and E.G. Berndt. Children's use of motives and intentionality in person perception and moral judgment. *Child Development,* 1975, 46, 904-912.

Bindra, D. The interrelated mechanisms of reinforcement and motivation, and the nature of their influence on response. In W.J. Arnold and D. Levine (eds.), *Nebraska symposium on motivation,* Vol. 17. Lincoln: University of Nebraska Press, 1969.

Blake, M.J.F. Temperament and time of day. In W. Colquhoun (ed.), *Biological rhythms and human behavior.* London: Academic Press, 1971.

Blechman, E.A. and E.A. Dannemiller. Effects on performance of perceived control over noxious noise. *Journal of Consulting and Clinical Psychology,* 1976, 44, 601-607.

Blumler, J.G. and E. Katz, (eds.) *The uses of mass communications: Current perspectives on gratifications research.* Beverly Hills: Sage, 1974.

Bohlin, G. The relationship between arousal level and habituation of the orienting reaction. *Physiological Psychology,* 1973, 1, 308-312.

Broadbeck, M. The role of small groups in mediating the effects of propaganda. *Journal of Abnormal and Social Psychology,* 1956, 52, 166-170.

Broadbent, D.E. Listening between and during practised auditory distractions. *British Journal of Psychology,* 1956, 47, 51-60.

_____. *Perception and communication.* London: Pergamon, 1958.

_____. *Decision and stress.* London: Academic Press, 1971.

Brock, T.C. and J.L. Balloon. Behavioral receptivity to dissonant information. *Journal of Personality and Social Psychology,* 1967, 6, 413-428.

_____, S.M. Albert, and L.A. Becker. Familiarity, utility, and supportiveness as determinants of information receptivity. *Journal of Personality and Social Psychology,* 1970, 4, 292-301.

Brown, R. *Social psychology.* Glencoe, Ill.: Free Press, 1965.

Canon, L.K. Self confidence and selective exposure to information. In L. Festinger (ed.), *Conflict, decision, and dissonance.* Stanford: Stanford University Press, 1964.

Chaffee, S.H. The interpersonal context of mass communication. In F.G. Kline and P.J. Tichenor (eds.), *Current perspectives in mass communication research.* Beverly Hills, Calif.: Sage, 1972.

_____ and Y. Miyo. Selective exposure and the reinforcement hypothesis. *Communication Research,* 1983, 10, 3-36.

Cherry, E. C. Some experiments on the recognition of speech with one and two ears. *Journal of the Acoustical Society of America.* 1953,25, 975-979.

Cherry, R.S. Development of selective auditory attention skills in children. *Perceptual and Motor Skills,* 1981, 52, 379-385.

Cohen, A.R. and B. Latane. An experiment on choice in commitment to counterattitudinal behavior. In J.W. Brehm and A.R. Cohen (eds.), *Explorations in cognitive dissonance.* New York: Wiley, 1962.

Collins, W.A. Children's comprehension of television content. In E. Wartella (ed.), *Development of children's communicative behavior.* Beverly Hills, Calif.: Sage, 1978.

_____. Cognitive processing in television viewing. In P.D. Bouthilet and J. Lazar (eds.), *Television and behavior: Ten years of scientific progress and implications for the eighties.* Washington: U.S. Government Printing Office, 1982.

Condry, J. Enemies of exploration: Self initiated versus other initiated learning. *Journal of Personality and Social Psychology,* 1977, 35, 459-477.

Cook, J.A. The attentional mechanism's processing of signal inputs: A descriptive model. Unpublished manuscript, Department of Speech Communication, Texas Tech University, 1984.

Copeland, A.P. and N.M. Wisniewski. Learning disability and hyperactivity: Deficits in selective attention. *Journal of Experimental Child Psychology,* 1981, 32, 88-101.

Corteen, R.S. and D. Dunn. Shock-associated words in a nonattended message: A test for momentary awareness. *Journal of Experimental Psychology,* 1974, 102, 1143-1144.

_____ and B. Wood. Autonomic responses to shock-associated words in an unattended channel. *Journal of Experimental Psychology,* 1972, 94, 308-313.

Cotton, J.L. and R.A. Hieser. Selective exposure and cognitive dissonance. *Journal of Research in Personality,* 1980, 14, 518-527.

Craik, F. Age differences in memory. In J.E. Birrien and K.W. Schaie (eds.), *Handbook of the psychology of aging*. New York: Van Nostrand Reinhold, 1977.

Davenport, W.G. Auditory vigilance: The effects of costs and values on signals. *Australian Journal of Psychology*, 1968, 20, 213-218.

_____. Vibrotactile vigilance: The effects of costs and values on signals. *Perceptual Psychophysiology*, 1969, 5, 25-28.

Davies, D.R. and G.R. Hockey. The effects of noise and doubling the signal frequency on individual differences in visual vigilance performance. *British Journal of Psychology*, 1966, 57, 381-389.

Davis, W.L. and E.J. Phares. Internal-external control as a determinant of information seeking in a social influence situation. *Journal of Personality*, 1967, 35, 547-561.

Day, M.C. and C.A. Stone. Children's use of perceptual set. *Journal of Experimental Child Psychology*, 1980, 29, 428-445.

Defleur, M.L. *Theories of mass communication*. 2d ed. New York: David McKay, 1970.

Deutsch, J.A. and D. Deutsch. Attention: Some theoretical considerations. *Psychological Review*, 1963, 70, 80-90.

Donohew, L. and P. Palmgreen. A reappraisal of dissonance and the selective exposure hypothesis. *Journalism Quarterly*, 1971, 48, 412-420.

_____, J.M. Parker and V. McDermott. Psychophysiological measurement of information: Two studies. Journal of Communication, 1972, 22, 54-63.

Duffy, E. *Activation and behavior*. London: Wiley, 1962.

Duncan, J. Divided attention: The whole is more than the sum of its parts. *Journal of Experimental Psychology* (HP&P), 1979, 5, 216-228.

Easterbrook, J.A. The effect of emotion on cue utilization and the organization of behaviour. *Psychological Review*, 1959, 66, 183-201.

Elliot, R. Simple reaction time: Effects associated with age, preparatory interval incentive shift, and mode of presentation. *Journal of Experimental Child Psychology*, 1970, 9, 86-107.

Eysneck, H.J. *The maudsley personality inventory*. San Diego: Educational and Industrial Testing Service, 1959.

_____. On the dual function of consolidating. *Perceptual and Motor Skills*, 1966, 22, 237-274.

_____. The biological basis of personality. Springfield, Ill.: Thomas, 1969.

Eysenck, M.W. *Attention and arousal*. New York: Springer-Verlag, 1982.

_____ and M.C. Eysenck. Memory scanning, introversion-extraversion, and levels of processing. *Journal of Research in Personality*, 1979a, 13, 305-335.

_____ and M.C. Eysenck. Processing depth, elaboration of encoding, memory stores, and expended processing capacity. *Journal of Experimental Psychology* (HL), 1979b, 5, 472-484.

Festinger, L. *A theory of cognitive dissonance*. Stanford: Stanford University Press, 1957.

_____, ed. *Conflict, decision, and dissonance*. Stanford: Stanford University Press, 1964.

Flowers, J.H., M.L. Polansky, and S. Kerl. Familiarity, redundancy, and the spatial control of attention. *Journal of Experimental Psychology* (HP&P), 1981, 7, 157-166.

Freedman, J.L. Confidence, utility, and selective exposure to information: A partial replication. *Journal of Personality and Social Psychology*, 1965, 2, 778-780.

French, T.W. *The description of aptitude and achievement tests in terms of rotated factors* (Psychometric Monograph No. 5). Chicago: University of Chicago Press, 1951.

Frey, D. Reversible and irreversible decisions: Preference for consonant information as a function of attractiveness of decision alternatives. *Personality and Social Psychology Bulletin,* 1981, 7, 621-626.

_____ and M. Irle. Some conditions to produce a dissonance and an incentive effect in a "forced compliance" situation. *European Journal of Social Psychology,* 1972, 2, 45-54.

_____ and R.A. Wicklund. A clarification of selective exposure: The impact of choice. *Journal of Experimental Social Psychology,* 1978, 14, 132-139.

Glaubman, H., et al. REM deprivation and divergent thinking. *Psychophysiology,* 1978, 15, 75-79.

Glucksberg, S. Problem solving: Response competition and the influence of drive. *Psychological Reports,* 1964, 15, 939-942.

Green, D.M. and J.A. Swets. *Signal detection theory and psychophysics.* Huntington, N.Y.: Kruger, 1974.

Hansen, J.C. and S.A. Hillyard. Selective attention to multidimensional auditory stimuli. *Journal of Experimental Psychology* (HP&P), 1983, 9, 1-19.

Hartley, L.R. Noise, attentional selectivity, serial reactions, and the need for experimental power. *British Journal of Psychology,* 1981, 72, 101-107.

Hasher, L. and R.T. Zacks. Automatic and effortful processes in memory. *Journal of Experimental Psychology* (Gen), 1979, 108, 356-388.

Hillis, J.W. and W.D. Crano. Additive effects of utility and additional supportiveness in the selection of information. *Journal of Social Psychology,* 1973, 89, 257-269.

Hirschman, E.C. Social and cognitive influences on information exposure: A path analysis. *Journal of Communication,* 1981, 31, 76-87.

Hirst, W., et al. Dividing attention without alternation or automaticity. *Journal of Experimental Psychology* (Gen) 1980, 109, 98-117.

Hiscock, M. and M. Kinsbourne. Asymmetries of selective listening and attention switching in children. *Developmental Psychology,* 1980, 16, 70-82.

Hitch, G.J. The role of short-term working memory in mental arithmetic. *Cognitive Psychology,* 1978, 16, 302-323.

_____. Developing the concept of working memory. In G. Claxton (ed.), *Cognitive Psychology: New directions.* London: Routledge & Kegan Paul, 1980.

_____ and A.D. Baddeley. Verbal reasoning and working memory. *Quarterly Journal of Experimental Psychology,* 1976, 28, 608-621.

Horn, J.L. Concept of intellect in relation to learning and adult development. *Intelligence,* 1980, 4, 285-319.

_____. The theory of fluid and crystallized intelligence in relation to apprehension, memory, speediness, laterality, and physiological function through the "vital years" of adulthood. In F. Craik and S. Trehub (eds.), *Aging and cognitive processes.* New York: Plenum, 1982.

_____, G. Donaldson, and R. Engstrom. Apprehension, memory, and fluid decline in adulthood. *Research on Aging,* 1981, 3, 33-84.

Hoyer, W.J. and D.J. Plude. Attentional and perceptual processes in the study of cognitive aging. In L.W. Poon (ed.), *Aging in the 1980's: Psychological issues.* Washington, D.C.: American Psychological Association, 1980.

Johnston, W.A. and S.P. Heinz. Flexibility and capacity demands of attention. *Journal of Experimental Psychology* (Gen), 1978, 107, 420-435.

Kahneman, D. *Attention and effort.* Englewood Cliffs, N.J.: Prentice-Hall, 1973.

Keister, M.E. and R.J. McLaughlin. Vigilance performance related to extraversion-introversion and caffeine. *Journal of Experimental Research in Personality,* 1972, 6, 5-11.

Keith, R., ed., *Central and auditory and language disorders in children.* Houston: College-Hill, 1981.

Kjellberg, A. Sleep deprivation and some aspects of performance 2: Lapses and other attentional effects. *Waking-Sleeping,* 1977, 1, 145-148.

Kleinhesselink, R.R. and R.E. Edwards. Seeking and avoiding belief-discrepant information as a function of its perceived refutability. *Journal of Personality and Social Psychology,* 1975, 31, 787-790.

Krull, R., W.G. Husson, and A.S. Paulson. Cycles in children's attention to the television screen. In B. Ruben (ed.), *Communication yearbook 2.* New Brunswick, N.J.: Transaction, 1978.

_____, J. Watt, and L. Lichty. Entropy and structure: Two measures of complexity in television programs. Communication Research, 1977, 4, 61-85.

Lane, D.M. Incidental learning and the development of selective attention. *Psychological Review,* 1980, 87, 316-319.

_____ and D.A. Pearson. The development of selective attention. *Merrill-Palmer Quarterly,* 1982, 28, 317-337.

Levin, S.R., and D.R. Anderson. The development of attention. *Journal of Communication,* 1976, 26(2), 126-135.

Linder, D.E., J. Cooper, and E.E. Jones. Decision freedom as a determinant of the role of incentive magnitude in attitude change. *Journal of Personality and Social Psychology,* 1967, 6, 245-254.

Logan, G.D. On the use of a concurrent memory load to measure attention and automaticity. *Journal of Experimental Psychology* (HP), 1979, 5, 189-207.

Lowe, D.G. and J.Q. Mitterer. Selective and divided attention in a stroop task. *Canadian Journal of Psychology,* 1982, 36, 684-700.

Lowe, R.H. and I.D. Steiner. Some effects of the reversibility and consequences of decision on post-decision information preferences. *Journal of Personality and Social Psychology,* 1968, 8, 172-179.

Lowin, A. Approach and avoidance as alternative modes of selective exposure to information. *Journal of Personality and Social Psychology,* 1967, 6, 1-9.

_____. Further evidence for an approach-avoidance interpretation of selective exposure. *Journal of Experimental Social Psychology,* 1969, 5, 265-271.

Mackworth, T.F. *Vigilance and habituation.* Middlesex, England: Penguin, 1969.

Malmo, R.B. and W.W. Surwillo. Sleep deprivation: Changes in performance and physiological indicants of activation. *Psychological Monographs,* 1960, 47, 1-24.

Mandler, J. and N. Johnson. Remembering of things passed: Story structure and recall. *Cognitive Psychology,* 1977, 9, 111-151.

Massad, C.M., M. Hubbard, and D. Newtson. Selective perception of events. *Journal of Experimental Social Psychology,* 1979, 15, 513-532.

McCroskey, J.C. & L.R. Wheeless. *Introduction to human communication.* Boston: Allyn & Bacon, 1976.

McGraw, K.O. and J.C. McCullers. Evidence of a detrimental effect of extrinsic incentives on breaking a mental set. *Journal of Experimental Social Psychology,* 1979, 15, 285-294.

McGuire, W.J. The current status of cognitive consistency theories. In S. Feldman (ed.), *Cognitive consistency: Motivational antecedents and behavioral consequents.* New York: Academic Press, 1966.

_____. Selective exposure: A summing up. In R. Abelson, et al. (eds.), *Theories of cognitive consistency: A sourcebook*. Chicago: Rand McNally, 1968.

_____. The nature of attitude and attitude change. In G. Lindzey and E. Aronson (eds.), *Handbook of social psychology*. Reading, Mass.: Addison-Wesley, 1969.

Meyer, G.E., and W.M. Maguire. Effects of spatial-frequency specific adaptation and target duration on visual persistence. *Journal of Experimental Psychology* 1981, 7, 151-156.

Milburn, M.A. A longitudinal test of the selective exposure hypothesis. *Public Opinion Quarterly,* 1979, 43, 507-517.

Miller, G.A. The magical number seven, plus or minus two. *Psychological Review,* 1956, 63, 81-97.

Miller, J.G. Information input overload. In M.C. Yovitz, G.T. Jacobi, and G.D. Goldstein (eds.), *Self organizing systems*. Washington: Spartan Books, 1962.

Miller, R.L. The effects of postdecisional regret on selective exposure. *European Journal of Social Psychology,* 1977, 7, 121-127.

Mills, J. Effect of certainty about a decision upon postdecision exposure to consonant information. *Journal of Personality and Social Psychology,* 1965, 2, 749-752.

_____, and A. Ross. Effects of commitment and uncertainty upon interest in supporting information. *Journal of Abnormal and Social Psychology,* 1964, 68, 552-555.

Minnick, W.C. *The art of persuasion*. Boston: Houghton-Mifflin, 1968.

Moray, N. Attention in dichotic listening: Affective cues and the influence of instructions. *Quarterly Journal of Experimental Psychology,* 1959, 11, 56-60.

_____. *Attention: Selective processes in vision and hearing*. London: Hutchison Educational, 1969.

Nachreiner, F. Experiments on the validity of vigilance experiments. In R.R. Mackie (ed.), *Vigilance: Theory, operational performance, and physiological correlates*. London: Plenum, 1977.

Navon, D. and D. Gopher. On the economy of the human processing system. *Psychological Review,* 1979, 86, 214-255.

Neely, J.H. Semantic priming and retrieval from lexical memory: Roles of inhibitionless spreading activation and limited-capacity attention. *Journal of Experimental Psychology* (Gen), 1977, 106, 226-254.

Newcomb, A.F. and W.A. Collins. Children's comprehension of family dramas: Effects of socioeconomic status, ethnicity, and age. *Developmental Psychology,* 1979, 15, 417-423.

Nideffer, R.M. Test of attentional and interpersonal style. *Journal of Personality and Social Psychology,* 1976, 34, 394-404.

Nielsen, S.L. and I.G. Sarason. Emotion, personality, and selective attention. *Journal of Personality and Social Psychology,* 1981, 41, 945-960.

Norman, D.A. Toward a theory of memory and attention. *Psychological Review,* 1968, 75, 522-536.

_____ and D.G. Bobrow. On data-limited and resource-limited processes. *Cognitive Psychology,* 1975, 7, 44-64.

O'Donnell, V. and J. Kable. *Persuasion: An interactive dependency approach*. New York: Random House, 1982.

Palmer, E.L. and C.N. McDowell. Program/commercial separators in children's television programming. *Journal of Communication,* 1979, 29(3), 197-201.

Paschal, F.C. The trend in theories of attention. *Psychological Review,* 1941, 48, 383-403.

Pavlov, I.P. *Conditioned reflexes.* New York: Dover, 1960.

Philip, R. *The measurement of attention.* Published dissertation, The Catholic University of America, 1928.

Pick, A.D., M.D. Christy, and G.W. Frankel. A developmental study of visual selective attention. *Journal of Experimental Child Psychology,* 1972, 14, 165-176.

Posner, M.I. *Chronometric explorations of the mind.* Hillsdale, N.J.: Erlbaum, 1978.

_____. Cumulative development of attentional theory. *American Psychologist,* 1982 (February), 37, 168-179.

_____ and M. Rothbart. The development of attentional mechanisms. In J.H. Flowers (ed.), *Nebraska symposium on motivation,* Vol. 28. Lincoln: University of Nebraska Press, 1980.

_____ and C.R.R. Snyder. Facilitation and inhibition in processing of signals. In P.M.A. Rabbitt and S. Dornie (eds.), *Attention and performance,* Vol. 5. London: Academic Press, 1975.

Poulsen, D., et al. Children's comprehension and memory for studies. *Journal of Experimental Child Psychology,* 1979, 28, 379-403.

Rabbitt, P. Changes in problem solving in old age. In J.E. Birrien and K.W. Schaie (eds.), *Handbook of the psychology of aging.* New York: Van Nostrand Reinhold, 1977.

Reitman, J.S. Without surreptitious rehearsal, information in short term memory decays. *Journal of Verbal Learning and Verbal Behavior,* 1974, 13, 365-377.

Remington, R. Visual attention, detection, and the control of saccadic eye movement. *Journal of Experimental Psychology* (HP&P), 1980, 6, 726-744.

Revelle, M.S., L.S. Humphreys, and K. Gilland. The interactive effect of personality, time of day, and caffeine: A test of the arousal model. *Journal of Experimental Psychology* (Gen), 1980, 109, 1-31.

Rice, M.L., A.C. Huston, and J.C. Wright. The forms of television: Effects on children's attention, comprehension, and social behavior. In P.D. Bouthilet and J. Lazar (eds.), *Television and behavior: Ten years of scientific progress and implications for the eighties.* Washington: U.S. Government Printing Office, 1982.

Richmond, V.P. The relationship between opinion leadership and information acquisition. *Human Communication Research,* 1977, 4, 38-43.

Rogers, E.M. and R. Adhikarya. Diffusion of innovations: An up-to-date review and commentary. In D. Nimmo (ed.), *Communication yearbook 3.* New Brunswick, N.J.: Transaction/International Communication Association, 1979.

_____ and F.F. Shoemaker. *Communication of innovations.* 2nd ed. New York: Free Press, 1971.

Rubin, R.B. and A.M. Rubin. Contextual age and television use: Reexamining a life-position indicator. In M. Burgoon (ed.), *Communication yearbook 6.* Beverly Hills, Calif.: Sage, 1982.

Salomon, G. *Interaction of media, cognition, and learning.* San Francisco. Jossey-Bass, 1979.

Schneider, D.J., A.H. Hastorf, and P.C. Ellsworth. *Person perception.* 2nd ed. Reading, Mass.: Addison-Wesley, 1979.

Schneider, W. and R.M. Shiffrin. Controlled and automatic human information processing I: Detection, search, and attention. *Psychological Review,* 1977, 84, 1-66.

Schramm, W. and D.F. Roberts, eds. *The process and effects of mass communication.* rev. ed. Urbana: University of Illinois Press, 1971.

Schultz, C.B. The effect of confidence on selective exposure: An unresolved dilemma. *Journal of Social Psychology,* 1974, 94, 65-69.

Schwarz, N., D. Frey, and M. Kumpf. Interactive effects on writing and reading a persuasive essay on attitude change and selective exposure. *Journal of Experimental Social Psychology,* 1980, 16, 1-17.

Sears, D.O. and J.L. Freedman. Selective exposure to information: A critical review. *Public Opinion Quarterly,* 1967, 31, 194-213.

Sherif, M. and C.I. Hovland. *Social judgment.* New Haven: Yale University Press, 1961.

_____, C. Sherif, and R. Nebergall. *Attitude and attitude change.* Philadelphia: Saunders, 1965.

Shiffrin, R.M. and W. Schneider. Controlled and automatic human information processing II: Perceptual learning, automatic attending, and a general theory. *Psychological Review,* 1977, 84, 127-190.

Shulman, G.L., R.W. Remington and J.P. McLean. Moving attention through visual space. *Journal of Experimental Psychology* (HP&P), 1979, 5, 522-526.

Simon, H.A. Motivation and emotional controls of cognition. *Psychological Review,* 1967, 74, 29-39.

Singer, J.L. Introductory comments. In P.D. Bouthilet and J. Lazar (eds.), *Television and behavior: Ten years of scientific progress and implications for the eighties.* Washington: U.S. Government Printing Office, 1982.

Smith, L.B., D.G. Kemler, and J. Aronfreed. Developmental trends in voluntary selective attention: Differential effects of source distinctiveness. *Journal of Experimental Child Psychology,* 1975, 20, 352-365.

Sokolov, E.N. Neuronal models and the orienting reflex. In M.A.B. Frazier (ed.), *The central nervous system and behavior: Transactions of the third conference.* New York: Josiah Macy, Jr. Foundation, 1960.

Sostek, A.J. Effects of electrodermal lability and payoff instructions on vigilance performance. *Psychophysiology,* 1978, 15, 561-568.

Spelke, E.S., W.C. Hirst, and U. Neisser. Skills of divided attention. *Cognition,* 1976, 4, 215-230.

Stacks, D.W. and D.E. Sellers. The effects of "pure" hemispheric reception of message acceptance. Paper presented to the Speech Communication Association, Washington, 1983.

Stajnberger, I. Uticaj; perceptualnih faktora i osobina licnosti na detekciju retkih kritienih signala. Ph.D. thesis, University of Belgrade, 1972.

Stankov, L. Attention and intelligence. *Journal of Educational Psychology,* 1983, 75, 471-490.

_____ and J.L. Horn. Human abilities revealed through auditory tests. *Journal of Educational Psychology,* 1980, 72, 19-42.

Stein, N. and C. Glenn. An analysis of story completion in elementary school children. In R. Freedle (ed.), *Advances in discourse processes,* Vol. 2. Hillsdale, N.J.: Erlbaum, 1979.

Stroop, J.R. Studies of interference in serial verbal reactions. *Journal of Experimental Psychology,* 1935, 18, 643-661.

Strutt, G.F., D.R. Anderson, and A.D. Well. A developmental study of the effects of irrelevant information of speeded classification. *Journal of Experimental Child Psychology,* 1975, 20, 127-135.

Swanson, D.L. Information utility: An alternative in political communication. *Central States Speech Journal,* 1976, 27, 95-101.

Swets, J.A. Signal detection theory applied to vigilance. In R.R. Mackie (ed.), *Vigilance: Theory, operational performance, and physiological correlates.* London: Plenum, 1977.

Talland, G. *Human aging and behavior.* New York: Academic Press, 1968.

Ten Hoopen, G. and J. Vos. Attention switching is not a fatigable process: Methodological comments on Axelrod and Guzy (1972). *Journal of Experimental Psychology* (HP&P), 1980, 6, 180-183.

Thayer, R.E. Activation states as assessed by verbal report and four physiological variables. *Psychophysiology,* 1970, 7, 86-94.

Thayer, S. Confidence and postjudgment exposure to consonant and dissonant information in a free-choice situation. *Journal of Social Psychology,* 1969, 77, 113-120.

Treisman, A.M. Selective attention in man. *British Medical Bulletin,* 1964a, 20, 12-16.

_____. The effect of irrelevant material on the efficiency of selective listening. *American Journal of Psychology,* 1964b, 77, 533-546.

_____. Strategies and models of selective attention. *Psychological Review,* 1969, 76, 282-299.

_____. Focused attention in the perception and retrieval of multidimensional stimuli. *Perceptual Psychophysiology,* 1977, 22, 1-11.

_____ and G. Gelade. A feature-integration theory of attention. *Cognitive Psychology,* 1980, 12, 97-136.

_____ and J.G.A. Riley. Is selective attention selective perception or selective response? A further test. *Journal of Experimental Psychology,* 1969, 79, 27-34.

Tyson, P.D. A general systems theory approach to consciousness, attention, and meditation. *Psychological Record,* 1982, 32, 491-500.

Underwood, G. Moray vs. the rest: The effect of extended shadowing practice. *Quarterly Journal of Experimental Psychology,* 1974, 26, 368-372.

Vernon, M.D. Perception, attention, and consciousness. *Advancement of Science,* 1960, 111-123.

Vohs, J.L. An empirical approach to the concept of attention. *Speech Monographs,* 1964, 31, 355-360.

Warm, J.S., et al. Motivation in vigilance: Effects of self evaluation and experimenter controlled feedback. *Journal of Experimental Psychology,* 1972, 92, 123-127.

Wellins, R. Counterarguing and selective exposure to persuasion. *Journal of Social Psychology,* 1977, 103, 115-127.

Wheeless, L.R. The effects of attitude, credibility, and homophily on selective exposure to information. *Speech Monographs,* 1974, 41, 329-338.

Wilkinson, R.T. Interaction of lack of sleep with knowledge of results, repeated testing, and individual differences. *Journal of Experimental Psychology,* 1961, 62, 263-271.

_____. Interaction of noise with knowledge of results and sleep deprivation. *Journal of Experimental Psychology,* 1963, 66, 332-337.

_____. Effects of up to 60 hours' sleep deprivation on different types of work. *Ergonomics,* 1964, 7, 175-186.

Willett, R.A. Experimentally induced drive and performance on a five-choice serial reaction task. In H.J. Eysenck (ed.), *Experiments in motivation.* Oxford: Pergamon, 1964.

Williams, J.G. and J.J. Stack. Internal-external control as a situational variable in determining information seeking by negro students. *Journal of Consulting and Clinical Psychology,* 1972, 39, 187-193.

Willis, S.I., et al. Training research in aging: Attentional processes. *Journal of Educational Psychology,* 1983, 75, 257-270.

Woodworth, R.S. *Psychology.* New York: Henry Holt and Company, 1921; 1934; 1940.

Yerkes, R.M. and J.D. Dodson. The relation of strength of stimulus to rapidity of habit formation. *Journal of Comprehensive Neurological Psychology,* 1980, 18, 459-482.

Ziemke, D.A. Selective exposure in a presidential campaign contingent on certainty and salience. In D. Nimmo (ed.), *Communication yearbook 4.* New Brunswick, N.J.: Transaction/International Communication Association, 1980.

15

Indexing Systems: Extensions of the Mind's Organizing Power

James D. Anderson

Just as machines can be seen as extentions of the strength and adaptability of human muscles, information retrieval systems can be viewed as extensions of the mind's own information analysis and retrieval capabilities. This article examines the indexing systems which perform the organizing functions of information retrieval systems and compares them to the indexing system of the mind. Both types of system are analyzed in terms of fundamental attributes of all indexing systems: media, codes, and channels; scope and domain; indexing methods; documentary units; indexable matter; exhaustivity; conceptual organization and control of index terminology; surrogation of documentary units; and searching and retrieving processes. Drawing on hypotheses and suggestions of cognitive science, possible improvements in the design of information retrieval systems are discussed, based on attributes of the mind's indexing and retrieval system.

The process of humanity's social evolution can be traced in the development of technologies (artifacts and techniques) which extend the power of the human body. Machines and tools extend the strength and adaptability of human muscles. Writing, written and printed documents, and sound and visual recordings extend the human capacity for storing information. These methods and media, together with postal, radio, television, and telephonic distribution channels extend the scope and range of human communication. Indexing systems represent attempts to extend the organizing capabilities of the human mind to these artificial (humanly devised) information storage and communication systems. Indexing systems turn information storage and communication systems into information retrieval systems analogous, in widely varying degrees, to the human information retrieval system of the mind.

This chapter compares human-devised (artificial) indexing systems with the most complicated and sophisticated indexing system of all, which humans develop unconsciously in their own minds. Descriptions of this human mind indexing system are based on analyses and hypotheses offered by cognitive scientists, which Morton Hunt (1982) has summarized in an excellent, relatively nontechnical overview. Descriptions of human-devised artificial indexing systems are based on the author's experience in designing, producing, studying, and using these systems. Important related studies include Christine A. Montgomery's (1972) investigation of the relationship between linguistics and information science, emphasizing the "common interest in natural language," which forms the basis of most information retrieval systems, and Linda Cheryl Smith's (1979) survey and analysis of possible contributions of artificial intelligence research to research and development in information retrieval systems. Working definitions of some key indexing terms are provided in Table 15.1.

TABLE 15.1
Definitions of Some Key Indexing Terms

Catalog: An index restricted in domain to a particular collection of documents or a particular set of collections (a union catalog).

Classification: The process of creating and/or placing items, concepts, ideas, etc. into categories; a principal component in indexing, since the predominant method for organizing information (or, more properly, documents in which information is represented) is by indicating the categories in which they have been placed or in which they fall by virtue of their characteristics.

Classification scheme: A set of categories ordered according to mutual relationships designed for indexing documents. Examples include the Dewey Decimal and Library of Congress classifications. The order of categories is maintained by the use of notation which can be arranged alphanumerically.

Code: "Any system of symbols for meaningful communication" (*Webster's*).

Communication system: A system for transferring information representations from one place or person to others. Examples include the mass media (radio, television, newspapers, magazines, books), the telephone system, the postal system.

Document: A physical medium on which information is represented.

Document unit or size: The document unit referred to in an index, such as a single line (in a concordance), a single page in a back-of-a-book index, a periodical article, an entire monograph (book), or an entire run of a periodical.

Exhaustivity: The detail with which the content and characteristics of a text is described in an index; usually defined operationally in terms of the number of index terms assigned to a documentary unit.

Index: A tool or device that indicates. Within the context of information retrieval, an index points to particular documents or parts of documents (passages, tables, illustrations) by describing content and other document characteristics (authors, titles, publishers).

TABLE 15.1 (Continued)
Definitions of Some Key Indexing Terms

Indexable Matter: Portions of documents used for deriving or selecting index terms. Examples include titles, title-pages, abstracts, reference citations, internal indexes, or the entire text.

Indexing: The process of creating an index.

Indexing system: The combination of input (e.g., documents), personnel, processes, procedures, and products which results in an index.

Information retrieval system: A system that combines the functions of information storage and communication systems with an indexing system, so that the information represented is organized to permit selective retrieval of particular information.

Information storage system: A system for preserving the representations of information, for example, in books, machine-readable files, etc.

Media: Physical entities or devices on which or through which information may be represented (paper, film, electronic circuits, and, within the brain, neurons).

Record: The description of or reference to a single document unit in an index, including index terms and/or classification notation.

Subject headings: Authorized index terms that are part of a controlled indexing vocabulary in which the number of terms are limited, synonymous and unused subordinate terms are connected to authorized terms, qualifiers distinguish among homographs, and relations among terms are often indicated through references to terms representing subordinate and related concepts.

Surrogate: A substitute, usually a brief, formal record which represents a complete, original document not directly included in the indexing system.

Surrogation: The creation of surrogate records for an index.

Artificial indexing systems became essential as soon as humans began collecting documents to augment the information storage capacity of human minds. Histories of the earliest libraries describe systems for indexing documents, usually based on classification of the documents themselves (Clark 1909; Jackson 1974; Johnson and Harris 1976; Parsons 1952). After the invention of printing, the ever growing number of documents made additional indexing systems essential, in the form of bibliographies, catalogs, and reference indexes (Wellisch 1981). These early systems share with contemporary systems and the human mind itself a number of fundamental characteristics, around which the remainder of this chapter is organized. These are:

1. Media on which index information is recorded, codes (symbol systems) by which this information is represented, and channels through which these representations are sent and received.
2. Scope of the document or document collection being indexed; domain (sources) of the information or documents included.
3. Indexing method: human processing of the concepts represented in documents versus machine processing of symbols used to represent concepts, or combinations of human and mechanized indexing.

4. Size of the document units being indexed.
5. Indexable matter: portions of documents considered for indexing.
6. Exhaustivity: level of detail with which the content of document units is described for the purpose of indexing.
7. Conceptual organization and control of index terminology:
- Specificity and number of index terms or categories.
- Control of synonyms, homographs, and contextually equivalent terms.
- Display of relationships among concepts or categories represented by terms.
8. Surrogation or representation of documentary units:
- Structure of surrogates.
- Surrogate display.
- Arrangement of surrogates in files and the display of files.
9. Procedures and processes for searching and retrieving index information.

Media, Codes, and Channels

Indexing systems include media on which information is represented (encoded) and processes by which these representations are created and organized for access and retrieval. For centuries the most prevalent media in artificial systems have been cards or slips of paper and printed pages. Representations of information, most often in the form of words, numerals, or similar symbols (e.g., mathematical, musical, choreographic, and chemical notation) were typically recorded on cards, which were then organized into files with predictable arrangements. These files could then be copied onto pages, as in back-of-the-book indexes or printed bibliographies and catalogs, or could remain on card media, as in the ubiquitous, at least until recently, library card catalog. Since these systems use visual signals for information representation (words, numerals, and related symbols), visual channels are used to transmit information representations to end users. (Of course, a human intermediary can translate the visual symbols of the indexing system to speech symbols and transmit them to the end user via sound channels.)

These paper media, and their micrographic counterparts, are rapidly being replaced by electronic computer-readable media, allowing computer manipulation of information representations and relieving humans from such tedious and error-prone processes as selecting and transferring information representations from indexed documents to index media and sorting these representations. In many cases, the resulting indexes do not differ significantly from earlier versions except in medium, but the new computer-based media also permit much greater flexibility in searching indexes, since terms and records can readily be rearranged to suit particular needs. Such rearrangement of index files on card or paper media is, in theory, possible, but highly impractical and rarely done.

Symbols used to represent information continue, for the most part, to consist of alphabetic letters, numerals, and analogous symbol systems, and final transmission to end users continues to be via visual channels. However, whereas the unit symbols for paper and paper-like media consist of the alphabetic letters (or syllabic or morphemic symbols of non-alphabetic writing systems), numerals, and similar symbols, the unit symbol for electronic media is the presence or absence of an electrical signal, or its representation in magnetic media. This underlying binary digital code permits the easy intermediate transmission and duplication of information representations via telecommunication channels prior to final transmission to end users.

The codes, media, and channels of the mind's indexing system are more varied and complex. Representations of information (or potential information) which the mind seeks or chooses to notice are received (perceived) via signals, media, and channels appropriate to the body's five sense organs: visual representations in the form of light, auditory and tactile representations in the form of mechanical deformation (Hubel 1979, 50), and olfactory and gustatory representations in the form of chemical stimuli (Loftus and Loftus 1976, 11; Popper and Eccles 1978, 250-74). These initial perceptions are translated into symbolic signals consisting of electrical impulses for transmission to the brain via neurons connected by sequences of synaptic linkages (Popper and Eccles 1978, 252). Chemical substances transmit these impulses from neuron to neuron across connecting synapses. The code in this symbol system is based on the number of impulses per second (Hubel 1979, 48).

Similarities between codes and symbol systems of the brain and computers ("both work with signals that are roughly speaking electrical," Hubel 1979, 46) has encouraged the widespread adoption of a computer-model of human information processing by cognitive scientists (Estes 1978, 3). This model suggests that the codes by which information is represented in the brain are based on patterns of electrical and chemical activity in and among neurons (Glass, Holyoak, and Santa 1979, 6), analogous to the representation of information in patterns of electrical activity in and among circuits in a computer memory or the records of these patterns in computer-sensible magnetic media. Differences of degree may constitute differences of kind, however. Hubel (1979, 45-46) estimates the number of nerve cells (neurons) in the brain to be a hundred billion, "give or take a factor of ten," and perhaps more significant, he considers an estimate of 100 trillion connections among neurons, or synapses, to be plausible, although "no one would want to be held to a guess." Furthermore, "the brain is not dependent on anything like a linear sequential program," as is the case with present computers, although future developments may add this kind of flexibility to computers.

The computer analogy also suggests the respective and synergistic roles of mind and brain. Simply put, the brain is the hardware, the physical entity designed to accommodate information storage, organization, and retrieval. "The brain is a tissue. It is a complicated, intricately woven tissue, like nothing else we know of in the universe, but it is composed of cells, as any tissue is" (Hubel 1979, 45). The mind, on the other hand, can be viewed as the software and the processes programmed by that software. "The brain is what *is*, the mind is what the brain *does*" (Hunt 1982, 81; for much more detailed discussion of the relationship between the brain and the mind, see Popper and Eccles 1978; Fodor 1981).

Scope and Domain

The concepts of *scope* and *domain* as they relate to indexing systems are developed in considerable detail by Marcia Bates (1976). Scope and domain relate to the information retrieval system as a whole, rather than the indexing subsystem which is used to organize the information within the overall system. *Scope* refers to the subjects and aspects of subjects covered and to the types of documents included. A generic definition of *document* is assumed here: a physical medium on which information is recorded by means of a system of symbols (code), for example, by language representation through writing or recorded speech, by visual images using a wide variety of artistic "codes," or by sound recordings of speech, music, or other aural codes. This definition is close to one in *Webster's Third New International Dictionary of the English Language, Unabridged* (1976): "a material substance . . . having on it a representation of the thoughts of men by means of some conventional mark or symbol."

Every human mind also has, obviously, a particular scope in terms of the subjects and aspects of subjects about which an individual has acquired, stored, and organized information in long-term memory. Just as a well-selected library is very discriminating in choosing what to add to its permanent collection, the mind selects only a small portion of its perceptions for processing into long-term memory. At each stage of processing, much potential information is ignored, and therefore discarded. Stimuli enter the mind's sensory register as a virtual "literal record" of the sensed image. It will quickly decay and disappear unless one chooses to pay attention to it, thus transferring it to short-term memory. It will remain there as long as it is rehearsed. If one chooses to analyze the content of short-term memory, this processing will place the information into long-term memory (Loftus

1980, 13-33). Consequently, the scope of the mind's information retrieval system is a very selective subset of the potential information encountered through life experience and initially perceived. It is this selected and stored information which is indexed by the mind as part of its information retrieval system.

In artificial information retrieval systems, scope is also defined by document characteristics such a particular authors or publishers, periodicity (monographs, serials), medium (paper, microform, magnetic tape, disk), code (written language, speech or other sound systems, visual images, and even scents), format (encyclopedia, handbook, bibliography, catalog, index), genre (fiction, poetry, treatise, dissertation), language (English, French), size, place and date of publication or creation, level or audience (children, young adults), and so on. Many of these aspects can serve also as bases for indexing and organization.

Knowledge of scope is important for effective use of any information retrieval system, since it indicates whether particular information might be found in a given system. This is as true for the human information retrieval system as for artificial systems. Humans are frequently able to react immediately to a question with an "I don't know" response when the query lies outside their scope of knowledge (Hunt 1982, 87). Similarly, well designed artificial systems inform their users of scope in order to help them avoid useless searches, but even the best of attempts do not approach the specificity of the mind's own knowledge of its scope. One is never sure that an artificial information retrieval system does not contain the answer to a query until after an exhaustive search, and even then, certainty is still only approximate.

The scopes of both artificial and human information retrieval systems change over time. Just as a well-run library (an artificial information retrieval system) will discard documents no longer needed, the human mind will forget information which, although indexed initially into long-term memory, was of less importance when received and therefore indexed less thoroughly (Craik and Lockhart 1972). Similarly, information which is never or rarely recalled and associated with new information will, over time, lose its "memory trace" and be, in effect, forgotten or discarded (Craik and Lockhart 1972).

The mind also revises information over time as new information is received and/or created and new relationships among concepts are established. This "continuous revision" is relatively uncommon in artificial information retrieval systems, especially bibliographic retrieval systems, although many encyclopedia publishers claim it for their encyclopedias, which are examples of fairly comprehensive information retrieval systems.

The human mind also relies on documents, in the sense of physical media in, on, or through which information is recorded. These "brain documents" consist of interconnected neurons, which are often compared to the interconnected electronic circuits of a computer's memory. Brain documents can be further characterized in terms of the types of codes used for the representation of information. Although, as discussed previously, the underlying code is based on signals or symbols consisting of varying electrical impulses, these symbols are used to encode or represent higher level codes for semantic information (Loftus and Loftus 1976, 70-71), visual images (Kosslyn 1975, 1976; Shepard, 1978), and acoustic and olfactory information (Craik and Lockhart 1972, 672-75; Loftus and Loftus 1976, 71; Glass, Holyoak, and Santa 1979, 79). These encoding levels are analogous to the similar levels in computer systems: the underlying binary code which is used to represent language, speech, graphics, and other higher level codes.

Domain refers to the sources of information included in the information retrieval system. Sources can be restricted to the collection of particular libraries, in which case the indexing system would produce a catalog, or to documents found in particular places. Frequently, the domain of an artificial system is not defined and is similar to most human domains— whatever the compiler stumbles across or discovers in purposeful searches. Just as well-defined scope can obviate pointless search, so can well-defined domains, since information known to reside in documents falling outside the domain of an information system need not be sought.

Indexing Method

Two fundamentally different approaches to indexing are used in artifical indexing systems. The faster, cheaper, and increasingly common approach is to focus only on the symbols recorded in documents, rather than the concepts which the symbols represent. Indexing can be based, for example, on the words recorded, without regard to any meaning. Current automatic, or computer-based, indexing relies on this approach. Simple KWIC (Key Word in Context) and permuted key word indexes based on titles have become widely accepted as effective, inexpensive, and quick-to-produce finding aids. Examples include the Permuterm Indexes for science, the social sciences, and the arts and humanities published by the Institute for Scientific Information to complement its Citation Indexes for these disciplines. Designers of automatic indexing systems experiment with statistical methods to select and assign variable weights to the more important words in abstracts or whole texts, considering frequency of words in particular documents as well as their frequency across collections (Salton and McGill,

1983). Weights can be modified by recording successful or unsuccessful retrieval associated with particular terms. Symbol indexing has so far been used exclusively with written language documents, since the number of symbols is limited, and they are well-defined. Symbol indexing is not yet feasible for visual images. Automatic indexing of sound appears to be less complicated, but it too remains largely hypothetical.

The second and more traditional approach is to base indexing on concepts represented by the symbols recorded in documents. This type of indexing requires the mind of a human indexer, which receives the symbols via normal perception processes, matches them against those stored in the mind, determines what concepts are represented and which are important, then chooses symbols to represent these concepts in the index. This is a highly subjective, individualistic procedure, and the source of both inter- and intra-indexer inconsistency (Leonard 1977). It inserts the mind of a usually anonymous indexer between the mind of the original creator of a message and the mind of the ultimate recipient. The creator of a message chooses particular symbols (e.g., words, sounds or images) to represent the message. These symbols are interpreted by an indexer and described in a new set of symbols which may or may not, in varying degrees, accurately represent or summarize the original message. This new "index" message, or rather its representation, is perceived by users of the index—potential recipients of the original message.

An indexer is, in many ways, like an editor. A good indexer, having an accurate understanding of the needs of information seekers, can produce index descriptions of the original document which will make it even more "findable" by interested recipients than would the terms used in the original text. On the other hand, indexers are just as capable of hiding documents by assigning inappropriate index descriptions. Whether, on the whole, the terms chosen by the original author or by a subsequent indexer are the more effective index descriptions is a widely debated issue among indexers and in indexing system research.

The distinction between symbol, or automatic, and conceptual indexing becomes tenuous in the human mind because of the exhaustive level of conceptual organization and control inherent in the mind's indexing system. At bottom, symbols are represented physically in ways analogous to their representation in artificial documents, but the myriad links created among symbols is precisely how the mind creates concepts, each of which is defined by its relations to all other relevant concepts (Lachman, Lachman, and Butterfield 1979). Thus, while indexing in the human mind is based on a system of physical representations, the system of linkages among them have created a metaphysical (beyond the physical) realm of concepts. Artificial indexing systems attempt to imitate this conceptual realm through thesauri which organize and control concept representations. (These will be discussed below under "Conceptual Organization and Control.")

Document Size

The size of documents in an information retrieval system determines the extent to which particular information can be quickly and easily located. Documents can be divided into subdocuments, which, at each level, can be treated as independent documents. Thus, the *Encyclopedia Britannica* is a single document published in many volumes, but each volume can be treated as a separate document; within each volume, each article can be treated as a separate document; within each article, each paragraph and illustration can be treated as a separate document; within each paragraph, each statement can be treated separately; and each phrase or word can be also treated separately. Similarly, a typical book or monograph can be considered a single document, or a collection of chapters, sections, paragraphs, illustrations, statements, phrases and words, each of which can constitute a separate document for indexing purposes. Thus, the distinction between document (in the usual sense) retrieval systems and passage (or information) retrieval systems is really a matter of document size. If every passage or distinct statement is treated separately, it can be separately retrieved, but if only large-scale documents are considered, then only large documents can be retrieved.

The documents of the mind tend to be treated at their most atomistic level, resulting in effective retrieval of small units of particular information. A person knowledgeable about bicycles may be able to retrieve specific information about appropriate gear ratios for mountain cycling, for example. Typical library indexing, at the other extreme, tends to treat documents at their maximum size or extent. Whole monographs, including multi-volume sets, and even complete runs of periodicals are indexed as single documents, making the retrieval of particular items of information difficult. Instead of retrieving a particular statement, paragraph, illustration, table, or page, an entire book, sometimes of several volumes, is retrieved as a unit, so that, in many cases, the actual information seeking has only begun. Detailed information about mountain bicycling gear ratios may be available in one or more books or journals, but this gross level of indexing will only retrieve books or periodicals on "bicycles" or "bicycling."

Back-of-the-book indexes, on the other hand, typically treat each page as a separate document, so that locating particular information is much easier. Such indexes can lead one to the particular pages on which gear ratios are discussed. If full-text databases uniquely identify each statement of the text, then the ease of locating particular information could approach that of the mind, assuming that other factors, such as exhaustivity of indexing, specificity of indexing terms, and conceptual organization and control (all of which will be discussed below), also approach the detailed level of the mind.

Indexable Matter

Indexable matter refers to those elements of a document considered in the indexing process. Since the "documents" of the mind's long-term memory tend to be so small, representing particular units of information, the entire content of the document constitutes indexable (and indexed) matter.

In artificial indexing systems, indexable matter frequently consists of some subset of the content of documents. At the limited end of the scale, documents may be indexed only on the basis of titles, in which case, titles constitute the indexable matter. More generous systems may consider also information from document title-pages (and analogous elements in non-book-type media) such as authors, editors, publishers, or abstracts, tables of contents, and internal (e.g., back-of-the-book) indexes. Citation indexes consider only reference citations as indexable matter. At the other extreme are systems which index the entire document. Until recently, however, these systems usually indexed quite shallowly, using very large documents and indexing them at a low level of exhaustivity. Library cataloging, for example, is, at least in theory, based on indexing the entire document, but documents are large (typically one or more physical volumes), and the content is summarized (indicated, indexed) using very few terms or subject headings and typically a single classification category. Computer databases, on the other hand, have begun to include full-texts of documents, in which every content-bearing word (as opposed to function words such as prepositions, articles, and conjunctions) is indexed. This approaches the level of indexing provided by the mind, both in terms of indexable matter (the entire document) and exhaustivity (discussed next), since it appears that every piece of information which is processed for storage in long-term memory is, by that very process, indexed as it is placed into the relational structure by which the mind organizes its semantic memory. However, such thorough indexing in artificial systems is rarely, if ever, accompanied by comprehesive conceptual organization and control of terminology, which is inherent in the mind's indexing system (discussed below).

Exhaustivity of Indexing

Exhaustivity or detail of indexing is usually closely associated with document size, although they are not inherently connected. Small documents (e.g., pages, paragraphs, statements) tend to be indexed in detail (exhaustively) while large documents (e.g., whole monographs) tend only to be summarized.

Indexing, simply put, is the process of indicating the content and related features of a document. Both words share the same Latin ancestor, *indicare*, "to point out, indicate." Maximum exhaustivity would result from using the entire content of the document to indicate its content. This is the level of the mind. Once a human being has decided that an item of information is worth recording in long-term memory, every element or factor of the item can be indexed. Craik and Lockhart (1972) suggest, however, that the mind takes a flexible approach to the care with which a new item of information is integrated into the mind's indexing system, depending on estimates of the importance of or interest in the information, as well as its similarity to previously indexed (analyzed) information. The more attention (processing) given to a new piece of information, the stronger the "memory trace." "Trace persistence is a function of depth of analysis, the deeper levels of analysis associated with more elaborate, longer lasting, and stronger traces" (Craik and Lockhart, 675).

Most artificial indexing systems summarize rather than index exhaustively. Typical library indexing assigns one to three subject headings to an entire monograph, a very low level of exhaustivity. Major indexing and abstracting services (e.g., *Chemical Abstracts, Psychological Abstracts*) tend to be much more generous in the detail of content indication but still only summarize the original information, albeit in considerable more detail than library catalogs. Full-text databases, which index every significant word, approach the exhaustivity of the mind. Artificial systems, however, tend to be quite inflexible in adjusting exhaustivity to match importance or interest. The mind's level of analysis or processing appears to vary from none at all to extensive analysis and integration into the indexing system in accordance with perceived importance. Artificial systems, once a general level of exhaustivity has been established, tend to treat every document equally, regardless of importance.

Conceptual Organization and Control

The mind creates concepts by linking its representation of a particular object, abstract idea, material, property, process, place, time period, event, number, or other datum of information with related data which are stored in the mind's semantic memory. Cognitive scientists have illustrated the resulting network with diagrams like Figure 15.1.

Long before cognitive scientists began modeling the semantic memory, indexers were constructing artificial "semantic memories" or thesauri to facilitate consistent and effective indexing and retrieval by organizing concepts and controlling their number and the terms used to express them. Figure 15.2 illustrates a very simple example, based on a single entry ("Animal") and two levels of related superordinate and subordinate entries from *Sears List of Subject Headings* (Westby 1972). Sears is used by school and small public libraries and was chosen because a small segment of it could be illustrated diagrammatically without undue difficulty. Much more sophisticated thesauri are described in detail by Soergel (1974).

In both the mind and artificial indexing systems, the conceptual organization and control provided by the semantic memory or thesauri includes three functions: (1) controlling the number and specificity of concepts; (2)

FIGURE 15.1
Conceptual Organization and Control in the Mind

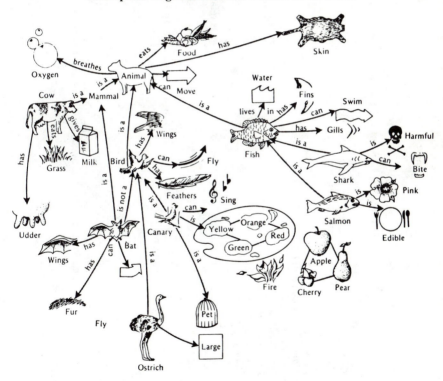

Source: From R. Lachman, J. L. Lachman, and E. C. Butterfield, *Cognitive Psychology and Information Processing* (Hillside, N.J.: Erlbaum, 1979), p. 325., as adapted by B.D. Ruben, *Communication and Human Behavior* (New York: Macmillan, 1984), p. 174.

Note: This adaption has not preserved an indication of "semantic distance," in which the length of connecting lines indicate the strength of association between concepts.

FIGURE 15.2

Artificial Conceptual Organization and Control

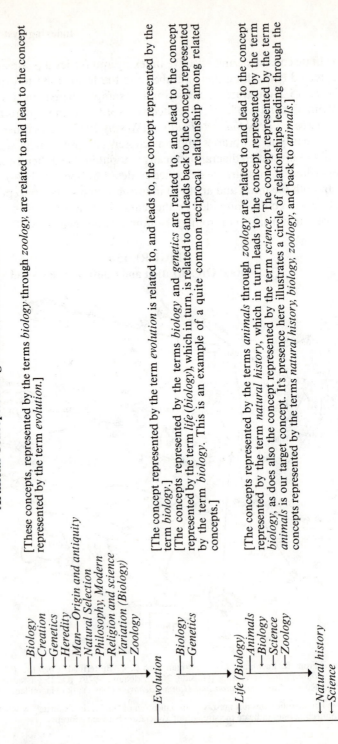

[These concepts, represented by the terms *biology* through *zoology*, are related to and lead to the concept represented by the term *evolution*.]

[The concept represented by the term *evolution* is related to, and leads to, the concept represented by the term *biology*.]

[The concepts represented by the terms *biology* and *genetics* are related to, and lead to the concept represented by the term *life (biology)*, which in turn, is related to and leads back to the concept represented by the term *biology*. This is an example of a quite common reciprocal relationship among related concepts.]

[The concepts represented by the terms *animals* through *zoology* are related to and lead to the concept represented by the term *natural history*, which in turn leads to the concept represented by the term *biology*, as does also the concept represented by the term *science*. The concept represented by the term *animals* is our target concept. It's presence here illustrates a circle of relationships leading through the concepts represented by the terms *natural history*, *biology*, *zoology*, and back to *animals*.]

┌─Biology
├─Creation
├─Genetics
├─Heredity
├─Man—Origin and antiquity
├─Natural Selection
├─Philosophy, Modern
├─Religion and science
├─Variation (Biology)
└─Zoology

└─Evolution

┌─Biology
└─Genetics

└─Life (Biology)

┌─Animals
├─Biology
├─Science
└─Zoology

┌─Natural history
└─Science

Biology

→ *Adaption (Biology)*
→ *Anatomy*
→ *Botany*
→ *Cells*
→ *Color of animals*
→ *Cryobiology*
→ *Death*
→ *Embryology*
→ *Evolution*
→ *Fresh-water biology*
→ *Genetics*
→ *Heredity*
→ *Life (Biology)*
→ *Marine biology*
→ *Microbiology*
→ *Natural history*
→ *Physiology*
→ *Protoplasm*
→ *Radiobiology*
→ *Reproduction*
→ *Sex*
→ *Space biology*
→ *Variation (Biology)*
→ *Zoology*

[The concept represented by the term *biology* is related to and leads to the concepts represented by the terms *adaption (biology)* through *zoology*. The concept represented by the term *zoology*, in turn, leads to our target concept, represented by the term *animals*.]

ANIMALS
→ *Color of animals*
→ *Desert animals*
└ *Camels*

[In general, the concepts represented by the preceding terms which have led to the concept represented by the terms *animals*, have been broader or more general than the concept represented by the term *animals*. In turn, the concept represented by the term *animals* is related to and leads to concepts which are, in general, narrower in scope, such as types of animals, activities related to particular types of animals, animal characteristics, animal environments, etc.]

FIGURE 15.2 (continued)
Artificial Conceptual Organization and Control

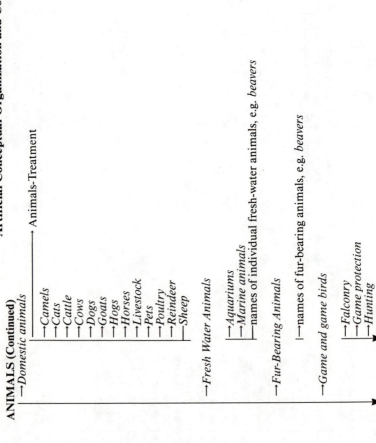

→Shooting
→Trapping
└─names of animals and birds, e.g. *deer; pheasants*

→*Geographical distribution of animals and plants*

→*Alpine plants*
→*Animals—Migration*
→*Birds—Migration*
→*Desert animals*
→*Desert plants*
→*Fresh-water animals*
→*Fresh-water plants*
→*Marine animals*
└─*Marine plants*

→ *Marine animals*

→*Corals*
→*Fishes*
└─*Fresh-water animals*

→*Natural history*   [But the concept represented by the term *animals* also leads to broad concepts, such as that represented by the term *natural history*, which in turn will lead to concepts such as *biology* and back to *animals*.]

→*Aquariums*
→*Biology*
→*Botany*
→*Fossils*
→*Fresh-water biology*
→*Geographical distribution of animals and plants*
→*Geology*
→*Marine biology*
→*Mineralogy*
→*Plant lore*
└─*Zoology*

FIGURE 15.2 (continued)
Artificial Conceptual Organization and Control

ANIMALS (Continued)
→Pets
    ———→*Domestic animals*
    ———→*names of animals, e.g. cats; dogs*
→*Zoological gardens*
    ———→*names of zoological gardens*
        [Here is another circular relationship, from *animals* to *zoology*; and back to *animals*.]
→*Zoology*
    →*Anatomy, Comparative*
    →*Animals*
    →*Embryology*
    →*Evolution*
    →*Fossils*
    →*Natural history*
    →*Physiology, Comparative*
    →*Psychology, Comparative*
    →*Variation (Biology)*
    →names of divisions, classes, etc. of the animal kingdom (e.g. *invertebrates; vertebrates; birds; mammals*)
    →*names of animals*
  →Names of orders and classes of the animal kingdom (e.g., *birds; insects*)
  →Names of animals, e.g. *dogs; bears*

Source: B. M. Westby, *Sears List of Subject Headings*, 10th ed. (New York: Wilson, 1972).

NOTE: This illustration is based on a single entry ("animal") and two levels of superordinate and subordinate related entries in the *Sears List of Subject Headings*, designed for school and small public libraries.

relating terms, including synonymous, homonymous, homographic, and essentially equivalent terms, to concepts; and (3) connecting associated concepts.

Conceptual Specificity and Repertoire

The mind's long-term memory "has an astronomical capacity" (Hunt 1982, 91). Loftus (1980, 15) suggests that long-term memory "records as many as one quadrillion separate bits of information." Nevertheless, no human mind includes all concepts. Particular concepts of no particular importance or interest to a person tend to be merged into broader concepts. For example, although the difference between an attorney and a lawyer (to say nothing of solicitors and barristers) has been explained to me, it neither interested me nor was important to me. These particular concepts are merged in my mind, although the terms "barrister" and "solicitor" are tagged as British. In the same way, artificial indexing systems merge less important concepts into broader concepts. *Sears*, for example, combines concepts for lawyers, attorneys, the bar, barristers, jurists, and the legal profession. Determining the optimum number (and therefore specificity) of concepts in indexing systems is an unresolved problem. Too few concepts do not sufficiently discriminate among documents, often resulting in the retrieval of too many irrelevant documents. Too many concepts scatter closely related documents, making it more difficult to retrieve the documents relevant to a particular query. The optimum conceptual repertoire will vary from person to person and from query to query, yet artificial systems generally provide a single set of concepts to all users. One approach is to opt for a large number of concepts with the option of conducting generic searches. To do this, genus-species relationships among concepts must be indicated, which is part of the third function of conceptual organization and control.

Terminological Control: Synonymy, Homography, Equivalence

Indexing system thesauri typically include large "lead-in" vocabularies with many more terms than the number of concepts included. In this way, users are afforded a variety of alternate terminological routes to particular concepts. Figure 15.3 shows the lead-in vocabulary provided for the concepts represented by headings in *Sears List of Subject Headings*. These lead-in terms, preceded by an equal sign (=), follow most "preferred terms" which are here printed in italic. Preferred terms are chosen to represent the concepts to be used for indexing. The lead-in vocabulary includes synonymous and essentially equivalent terms. In addition, when particular concepts are considered overly specific, they are merged into broader ones, and the terms indicating the particular concepts are also part of the lead-in vocabulary. For example, *Sears* subsumes "Christianity and Science" into the broader category of "Religion and Science." Finally, homographic terms representing different concepts are qualified to indicate context, e.g., Mercury (chemical element), Mercury (planet), Mercury (god).

Recent research by cognitive psychologists at Bell Laboratories has emphasized the diverse ways in which objects and concepts are named in natural language, suggesting that a weakness in many information retrieval systems is the limited accommodation of the many terms available and used for particular concepts (Furnas et al. 1983). In a domain of commonly known objects, such as household items listed for sale in want-ads, at most only 20 to 30 percent of persons studied agreed on terminology. If fifteen alternative terms were allowed in the system for particular items, 60 to 80 percent of terms actually used by searchers would be found in the system vocabulary. They conclude that "the lesson clearly is that systems must recognize many, many names" (p. 1797).

Terminological control appears to be organized in a similar way in the mind, with two separate subsystems, one for terms and one for concepts, with links between terms and concepts. The details of organization of the terminological, or "lexical," store are less understood than the organization of the semantic store, where conceptual associations are maintained (Lachman, Lachman, and Butterfield 1979, 334-44). Research suggests that the lexical store may be organized on the basis of word frequency, codability ("extent of agreement among people about the correct name for a thing," Lachman, Lachman, and Butterfield 1979, 338), and/or age of acquisition. Lachman, Shaffer and Hennrikus (1974) suggest that all three attributes are important: "the lexical store is structured around several principles" (Lachman, Lachman, and Butterfield 1979, 343).

Conceptual Associations

The semantic memory, as illustrated in Figure 15.1, connects a particular concept with all other concepts with which the mind has determined an association. Several types of association are illustrated in Figure 15.1: agent-process-object (animal eats food), species-genus (cow is a mammal), whole-part (fish has fins), environment (fish lives in water), property (shark is harmful, salmon is pink, is edible). Traditional indexing thesauri do not usually distinguish particular associations (e.g., Library of Congress, Subject Cat. Div. 1980; Westby 1972), but newer thesauri often specify genus-species and whole-part associations (broader-narrower concepts) on the one hand, and group all other associations as "related concepts" (Willets 1975).

Based on the proposition that "the mind has basically two main 'mechanisms' for interconnecting concepts: association and discrimination," each

FIGURE 15.3
Lead-in Vocabulary for Concepts from *Sears List of Subject Headings*

├──*Biology*=morphology
│ ←*Creation*=cosmogony, biblical; cosmology, biblical
│ ←*Genetics*
│ ←*Heredity*=ancestry; descent; genes; inheritance (biology)
│ ←*Man—origin and antiquity*=antiquities; man - antiquity; origin of
│   man
│ ←*Natural selection*=selection, natural; survival of the fittest
│ ←*Philosophy, Modern*=modern philosophy
│ ←*Religion and science*=Christianity and science; science and religion
│ ←*Variation (Biology)*=mutation (biology)
│ ←*Zoology*=animal kingdom; animal physiology; fauna

├──*Evolution*=Darwinism; development; mutation (biology); origin of species
│   ├──*Biology*=morphology
│   │ ←*Genetics*

←*Life (Biology)*
│   −*Animals*=beasts; fauna; wild animals
│   ←*Biology*=morphology
│   ←*Science*=discoveries (in science)
│   ←*Zoology*=animal kingdom; animal physiology; fauna

←*Natural History*=animal lore; history, natural
←*Science*=discoveries (in science)

Biology=morphology
│ →*Adaption (Biology)*=acclimatization; environment
│ →*Anatomy*=body, human; histology; human body; morphology
│ →*Botany*=flora; vegetable kingdom
│ →*Cells*=cytology; histology
│ →*Color of animals*=animal coloration; animals - color; pigmentation
│ →*Cryobiology*=freezing; low temperature biology
│ →*Death*
│ →*Embryology*=development
│ →*Evolution*=Darwinism; development; mutation (biology); origin of species
│ →*Fresh-water biology*
│ →*Genetics*
│ →*Heredity*=ancestry; descent; genes; inheritance (biology)
│ →*Life (Biology)*
│ →*Marine biology*=biology, marine; ocean life
│ →*Microbiology*
│ →*Natural history*=animal lore; history, natural
│ →*Physiology*=body, human; human body
│ →*Protoplasm*
│ →*Radiobiology*=radiation biology
│ →*Reproduction*=generation
│ →*Sex*
│ →*Space biology*=astrobiology; bioastronautics; cosmobiology; exobiology; ex-
│   traterrestrial life
│ →*Variation (Biology)*=mutation (biology)
├──*Zoology*=animal kingdom; animal physiology; fauna

FIGURE 15.3 (Continued)
Lead-in Vocabulary for Concepts from *Sears List of Subject Headings*

ANIMALS = Beasts; Fauna; Wild animals
→*Color of animals* = animal coloration; animals—color; pigmentation
→*Desert animals*
⊢——*Camels* = dromedaries
→*Domestic animals* = animal industry; animals, domestic; farm animals
→*Animals—Treatment* = animals, cruelty to; animals—protection; cruelty to animals; kindness to animals; prevention of cruelty to animals; protection of animals
→*Camels* = dromedaries
→*Cats* = cat
→*Cattle* = beef cattle
→*Cows* = cow; dairy cattle
→*Dogs* = dog
→*Goats*
→*Hogs* = swine
→*Horses* = horse
→*Livestock* = animal husbandry; animal industry; breeding; farm animals; stock and stock breeding; stock raising
→*Pets*
→*Poultry*
→*Reindeer*
⊢——*Sheep*
→*Fresh-water animals* = animals, aquatic; fresh-water; aquatic animals; water animals
→*Aquariums*
→*Marine animals* = animals, aquatic; animals, marine; animals, sea; aquatic animals; marine fauna; marine zoology; sea animals; water animals
→names of individual fresh-water animals, e.g. *beavers*
→*Fur bearing animals*
⊢——names of fur-bearing animals, e.g. *beavers*
→*Game and game birds* = wild fowl
→*Falconry* = hawking
→*Game protection* = game wardens; protection of game
→*Hunting* = the chase; field sports; gunning
→*Shooting* = gunning; marksmanship
→*Trapping*
⊢——names of animals of animals and birds. e.g. *deer; pheasants*
→*Geographical distribution of animals and plants* = animals—geographical distribution; biogeography; botany—geographical distribution; distribution of animals and plants; fishes—geographical distribution; plants—geographical distribution; Zoology—geographical distribution
→*Alpine plants* = mountain plants
→*Animals—Migration* = animal migration; migration of animals
→*Birds—Migration* = migration of birds
→*Desert animals*
→*Desert plants*
→*Fresh-water animals* = animals, aquatic, animals, fresh-water; aquatic animals; water animals
→*Fresh-water plants* = aquatic plants; water plants
→*Marine animals* = animals, aquatic; animals, marine; animals, sea; aquatic animals; marine fauna; marine zoology; sea animals; water animals

FIGURE 15.3 (Continued)
Lead-in Vocabulary for Concepts from *Sears List of Subject Headings*

ANIMALS (Continued)
    ├──*Marine plants*=aquatic plants; marine flora; water plants
   →*Marine animals*=animals, aquatic; animals, marine; animals, sea;
        aquatic animals; marine fauna; marine zoology; sea animals;
        water animals
     →*Corals*
     →*Fishes*=fish; ichthyology
     ├──*Fresh-water animals*=animals, aquaic; animals, fresh-water;
        aquatic animals; water animals
   →*Natural history*=animal lore; history, natural
     →*Aquariums*
     →*Biology*=morphology
     →*botany*=flora; vegetable kingdom
     →*fossils*=animals, fossil; animals, prehistoric; paleontology; pre-
        historic animals
     ├──*Fresh-water biology*
   →*Geographical distribution of animals and plants*=animals—geo-
        graphical distribution; biogeography; botany—geographical
        distribution; distribution of animals and plants; fishes—geo-
        graphical distribution; plants—geographical distribution; zo-
        ology—geographical distribution
     →*Geology*=geoscience
     →*Marine biology*=biology, marine; ocean life
     →*Mineralogy*=minerals
     →*Plant lore*=folklore of plants; plants—folklore
     ├──*Zoology*=animal kingdom; animal physiology; fauna
   →*Pets*
     →*Domestic animals*
     ├──names of animals, e.g. *cats; dogs*
   →*Zoological gardens*=zoos
     ├──names of zoological gardens
   →*Zoology*=animal kingdom; animal physiology; fauna
     →*Anatomy, Comparative*=animals—anatomy; comparative anat-
        omy; histology; morphology; zoology—anatomy
     →*Animals*=beasts; fauna; wild animals
     →*Embryology*=development
     →*Evolution*=darwinism; development; mutation (biology); origin
        of species
     →*Fossils*=animals, fossil; animals, prehistoric; paleontology; pre-
        historic animals
     →*Natural history*=animal lore; history, natural
     →*Physiology, Comparative*=comparative physiology
     →*Psychology, Comparative*=animal psychology; comparative
        psychology
     →*Variation (Biology)*=mutation (biology)
     →names of divisions, classes, etc. of the animal kingdom, e.g. *invet-*
        erbrates; vertebrates; birds; mammals
     ├──names of animals
   →Names of orders and classes of the animal kingdom, e.g., *birds;*
     insects
   →Names of animals, e.g. *dogs; bears*

Source: See Figure 15.2.
NOTE: Terms to the right of the = sign are considered synonymous or equivalent, within the
context of this thesaurus, to the terms to the left of the = sign.

of which "develops into three fairly well-defined stages," Farradane (1963, 1980) suggests a matrix of nine basic relations representing increasing levels of association on the horizontal axis, and increasing levels of discrimination on the vertical axis (Figure 15.4).

Willets (1975) has recommended that some of the more sharply defined of Farradane's relational categories be incorporated into thesauri to reduce the large numbers of undifferentiated relations indicated.

Members of the Classification Research Group (United Kingdom) have suggested that entity concepts are structured as well in series of "integrative levels," ranging from simple to complex, as illustrated in Figure 15.5.

Cognitive scientists suggest that there are at least three ways of accessing the mind's semantic memory: via a visual analogue channel, activated by seeing an object or its picture; via a lexical channel, activated by reading or hearing a word or phrase which represents a concept; and via a semantic-descriptive channel, activated by seeing (reading) or hearing a description (Lachman, Lachman, and Butterfield 1979, 334-35). Lachman and associates use the example of the concept *Robin*, which may be located via reading or hearing the name *Robin*, seeing a robin or a picture of a robin, or hearing or reading a description of a robin — "a bird about the size of a bluejay with a chestnut-red breast and belly" (1979, 335).

Indexing thesauri depend on the lexical channel primarily. Initial access must be via a word or phrase, but once in the thesaurus, the network of connections among concepts can lead from one concept to a myriad of related concepts in ways analogous to the semantic-descriptive channel of the semantic memory. Many modern thesauri seek to facilitate the semantic-descriptive channel by providing separate hierarchies for each set of terms sharing genus-species or whole-part relationships (e.g., *Thesaurus of ERIC Descriptors* 1984), and some go further by providing, in addition, an overall classification of concepts based on their primary relationships, much like Roget's original thesaurus (e.g., Aitchison's *Unesco Thesaurus* 1977). Some thesauri attempt to tap the visual analogue channel by illustrating concepts and relationships diagrammatically.

The indexing thesauri discussed thus far were created by humans drawing on their own mind's conceptual organization and that of experts in relevant disciplines. There are increasing attempts to generate effective thesauri automatically based on patterns of word or phrase use in texts. Automatic thesauri ignore concepts; instead, they group and relate terms on the basis of cooccurrence in documents. For example, after an initial search for documents in a collection, the frequency of terms occurring in the retrieved documents can be compared to their frequency in the entire collection. Terms are then ranked, with terms with the greatest difference in frequency (higher in the retrieved documents, lower in the collection as a

FIGURE 15.4
Farradane's Indexing Relations[1]

	Awareness	Increasing association → Temporary association	Fixed association
Concurrent conceptualization	1. Concurrence	4. Self-activity	7. Association
Not-distinct conceptualization	2. Equivalence	5. Dimensional	8. Appurtenance
Distinct conceptualization	3. Distinctness	6. Action	9. Functional dependence (causation)

Increasing discrimination ↓

Definitions[2]

1. Concurrence: the simple association of two concepts; the weakest of the relations.
2. Equivalence: the range of relations from partial similarity or quasi-synonymy to complete identity.
3. Distinctness: the relation of imitation or substitution of one thing by another.
4. Self-activity: the relation between an entity and its own activities or processes; in grammatical terms, the relation between a subject and its intransitive verb.
5. Dimensional: the relations of time and space, temporary states, and temporary properties.
6. Action: the relation of an entity and the activities or processes acting upon it.
7. Association: the relations of agent and action and of entity and certain abstract or calculated properties (as opposed to intrinsic properties).
8. Appurtenance: the relations of whole to part, genus to species, and of an entity to its intrinsic ingredients or properties.
9. Functional dependence: the relation of causation.

[1] J. Farradane, "Relational Indexing, part 1," *Journal of Information Science* 1(1980): 270.
[2] Taken from R. A. V. Diener, "A Longitudinal Study of the Informational Dynamics of Journal Article Titles." Ph. D. diss. Rutgers University, 1984.

whole) ranked first. Although the particular conceptual relationship between these high ranked terms and the search terms used to retrieve the initial set of documents is not determined, there is some kind of pragmatic association evidenced by the tendency of authors to use these terms together in the same text more frequently than their usual cooccurrence. The same algorithm can be applied to index terms assigned by indexers, displaying pragmatic relationships based on cooccurring index terms. The high ranked cooccurring terms are displayed in rank order to the searcher, perhaps accompanied with the number of additional documents which would be retrieved if each term were added to the original search argument, leaving it to the searcher to decide if terms are relevant to the query and should be added to the search statement (Doszkocs 1978; Salton and McGill 1983, p. 75-89).

Not all artificial indexing systems provide the kind of conceptual organization and control described here, but since any complete information retrieval system, which includes the sender and receiver of messages as well as the indexing subsystem, inevitably includes conceptual organization and control, the effect of omitting these operations from the indexing subsystem is to shift them to the end user. Searchers of simple keyword indexes, for example, must themselves think of relevant synonymous or equivalent terms and deal with homographic ones as well. If they wish to broaden, narrow, or shift the focus of their search, they must think of possible broader, narrower or related concepts and terms. Omitting these elements from the indexing system makes it much cheaper, of course, to create and operate, but it leaves open the question whether it is cost-effective to require every user to recreate the conceptual organization and control subsystem.

Most often, conceptual organization and control have been associated with concept-based, as opposed to automatic symbol-based, indexing, but

FIGURE 15.5
Integrative Levels

1. Fundamental particles
2. Nuclei
3. Atoms
4. Molecules
5. Molecular assemblages (natural objects and artifacts)
6. Cells
7. Organisms
8. Human beings
9. Human societies

Source: D. J. Foskett, "The Theory of Integrative Levels and Its Relevance to the Design of Information Systems," *Aslib Proceedings* 30 (1978).

this is not necessary, and index system designers are experimenting with combining automatic indexing with conceptual organization and control systems. An operational system developed for the Defense Documentation Center is described by Klingbiel and Rinker (1976). There is no reason that terms extracted according to some algorithm based on, for example, document and collection term frequencies could not be mapped against the lead-in vocabulary of a thesaurus, especially if the lead-in vocabulary is quite large. In fact, some index system designers believe that this pattern will become more and more common: potentially significant terms will be extracted from text through statistical frequency and cooccurrence algorithms. These terms will then be mapped against thesauri to link the extracted terms to the conceptual organization of a particular discipline or field. Searchers will then have the advantages of conceptual organization and control, yet the original indexing of documents will be based on the terms chosen by the original author; they will not pass through the private conceptual organization of an anonymous indexer.

Proponents of this view believe that it will be more cost-effective for human indexers to index fields of knowledge, rather than individual documents, creating thesauri which record, organize and display the conceptual relationships important to particular fields, together with the terms used for these concepts.

Surrogation

Most artificial indexing systems do not encompass whole texts, but rather rely on representations, or surrogates, in place of full documents and their texts. Typical document surrogates range from nothing more than the name of an author and the title of a text, to full transcriptions of bibliographic characteristics: authors, titles (main titles, subtitles, translated titles, series titles), edition statements, places of publication, publishers, dates of publication, size.

Surrogation of the subject matter and form characteristics of a text can range from a single term or heading through comprehensive lists of terms and/or detailed abstracts, to the full text itself. Subject surrogates based on subject terms range from lists of unconnected, uncoordinated terms, through precoordinated subject headings each consisting of a limited number of terms, to long strings of terms arranged in regulated order, often connected with various types of links and role indicators.

In relational indexing, for example, terms are connected by role operators based on the relations identified by Farradane (Farradane, 1963, 1980; see figure 15.4). The Preserved Context Indexing System (PRECIS) puts terms in a prescribed order, using computer-readable codes to indicate particular roles and relationships (Austin 1974). The Contextual Indexing and Faceted Taxonomic Access System (CIFT) orders terms within contextual strings in accordance with a predetermined sequence of subject facets and function categories (Anderson 1979). It is generally accepted that procedures for arranging terms into consistent strings or headings are essential for printed indexes designed to be searched manually. The contribution of prescribed term order and role indicators in computer-searchable indexes remains a controversial issue, however, since sets of terms unconnected by any syntax are known to permit successful retrieval.

The classified collections of typical libraries constitute the most common indexing systems in which entire documents are included. In effect, the document itself serves as its own surrogate. Electronic databases which include full-texts of documents are becoming more common.

The human mind is like a library collection or a full-text database in that its documents are included in its indexing system in full, not represented by abbreviated surrogates. The analogy is closer to electronic full-text databases, since the linear arrangement of classified collections in libraries provides only one avenue of access, while the electronic database (and, even more, the human mind) can accommodate a myriad of access paths.

There is an important difference between the "full-text" surrogates of artificial indexing systems and the "full-text" surrogates of the mind, however. In artificial systems, the text included in the indexing system is essentially identical to the original text which the system processes. In contrast, the mind does not store the original texts (whether experiential "texts" or actual document texts) which it processes, except in the case of verbatim memorization. Instead, from the mass of information perceived, it selects the information (concepts and their relations) of interest and creates new "documents" which are integrated into its semantic memory. It is the full text of these new documents which the mind stores and indexes. Once these new documents are created, they are not represented by abbreviated surrogates, but by the full texts of these documents.

Since the mind's indexing system is based on "full-text" surrogates which tend to be indexed thoroughly and exhaustively, the "indexable matter," which was discussed previously, is identical to the full-text surrogate. This is usually not the case in artificial indexing systems, in which the full text of documents is represented by an abbreviated surrogate. Indexable matter may then be taken from the entire original document, a subset of the original document, a subset of the surrogate or the entire surrogate. In a typical indexing and abstracting service, the full document, a monograph or periodical article for example, is represented by a surrogate consisting of a citation and an abstract. The indexable matter, the source of the index terms, is often the surrogate citation and abstract, not the original document.

Full-text databases in which the entire text of documents is indexed exhaustively are recent phenomena. The documents in these databases differ from those of the mind, however, in that they are not newly created documents of highly distilled information, nor is their content fully integrated into anything like the relatively structured semantic memory.

In artificial indexing systems, surrogation of documents involves the creation of records (or individual surrogates) to represent documents, the display of these records, and the arrangement of these records in files to accomodate one or more means of access.

Record Structure

Index designers have found it useful to provide a consistent structure for surrogate records, so that searchers can more easily interpret the content of the record or find particular pieces of data. Individual records in library catalogs, for example, follow a consistent pattern of elements (author, title, edition, place of publication, publisher, date, pagination, illustrations, size, series, notes, subject headings, etc.), so that persons seeking, for example, only the publisher will not have to read the entire record. Electronic records have become more complicated in an effort to accomodate directed access to more varied types of information. The MARC (Machine Readable Cataloging) record, for example, provides hundreds of fields and subfields for particular elements of information (some of which are illustrated in Figure 15.6), so that particular elements can be easily located and indentified in searches (Library of Congress, Automated Systems Office 1980).

Record Display

Individual records can be displayed in the same form in which they are stored, or they can be reformatted for display. Records on paper media are usually not reformatted; thus the record on a library catalog card is displayed in its original form. Electronic records are frequently displayed in abbreviated form, extracting the more important elements and arranging them for easier comprehension by the searcher. Thus, the record displayed in Figure 15.7 can be extracted from the much more complicated MARC record.

The display and storage of visual images in the mind can be considered analogous. Visual images are stored in long-term memory in an abstract format and are recreated upon recall. Kosslyn (1975, 342) suggests a "computer graphics metaphor" for this process: a visual image bears "the same relationship to its underlying structure as a pictorial display on a cathode

FIGURE 15.6

Example of a MARC Record for a Monographic Book

> NO HOLDINGS IN RUG — FOR HOLDINGS ENTER dh DEPRESS DISPAY RECD SEND

> OCLC:	6708730	Rec stat: c	Entrd: 800821	Used: 840322
> Type:	a Bib lvl: m	Govt pub:	Lang: eng Source:	Illus: af
Repr:	Enc lvl:	Conf pub: 0	Ctry: nyu Dat tp: s	M/F/B: 10
Indx:	1 Mod rec:	Festschr: 0	Cont:
Desc:	i Int lvl:	Dates: 1981,

> 1 010   80-8198//r82
> 2 040   DLC #c DLC #d m.c. |
> 3 020   0061811637 : #c $17.50 |
> 4 043   n————————— a n-us————— |
> 5 050 0 QL151 #b C62 1981 |
> 6 082 0 591.974 2 19 |
> 7 090   #b |
> 8 049   RUGG |
> 9 100 10 Collins, Henry Hill, #d 1905-1961. #w cn |
> 10 245 10 Harper & Row's complete field guide to North American
>   wildlife, Eastern edition / #c assembled by Henry Hill Collins, Jr. ;
>   ill. by Paul Donahue . . . [et al.]. |
> 11 250   1st ed.
> 12 260 0 New York : #b Harper & Row, #c c1981. |
> 13 300   xi, 714 p., [55] leaves of plates : #b ill. ; #c 23 x 12 cm.
> 14 500   Includes index. |
> 15 650 0 Zoology #z North America. |
> 16 650 0 Zoology #z Atlantic States. |
> 17 650 0 Animals #x Identification. |
> 18 740 01 Complete field guide to North American wildlife, Eastern edition.

Source: OCLC, *On-line Systems—Cataloging: User Manual* (Columbus: Ohio, 1979). p. 5-36.

NOTE: The subject headings assigned to the document represented by this record are located in fields 650 on lines 15-17. These headings place this work into a network of concepts and terms similar to that illustrated in Figures 15.2 and 15.3. The classification notation in fields 050 on line 5 (Library of Congress Classification) and field 082 on line 6 (Dewey Decimal Classification) place the work in two additional conceptual networks. More informal networks are created by key words in the title (field 245 on line 10): the term *wildlife*, for example, not only provides an additional access term, but links this document with all other which use that term.

FIGURE 15.7
Examples of Alternative Record Displays

Standard Catalog Record Display

Collins, Henry Hill, 1905-1961.
  Harper & Row's complete field guide to
North American wildlife, Eastern edition /
assembled by Henry Hill Collins, Jr. ; ill. by
Paul Donahue . . . [et al.]. — 1st ed. — New
York : Harper & Row, c1981.
  xi, 714 p., [55] leaves of plates : ill.;
23 x 12 cm.

  Includes index.
  ISBN 0-061-811637 : $17.50.

  1. Zoology — North America. 2. Zoology —
Atlantic States. 3. Animals — Identification.
I. Complete field guide to North American wildlife,
Eastern edition.
QL151.C62 1981   591.974'2'19   80-8198/ /r82
                                        MARC

Alternative Record Display

Author: Collins, Henry Hill, 1905-1961.
Title: Harper & Row's complete field guide to North
  American wildlife, Eastern edition.
Illustrator: Donahue, Paul
Edition: 1st
Place published: New York.
Publisher: Harper & Row.
Date of publication: 1981.
Length: 714 pages.
Language of text: English
Notes: Includes index; 55 leaves of plates; illustrated
Price: $17.50
Subjects: 1. Zoology — North America. 2. Zoology —
  Atlantic States. 3. Animals — Identification.
Available in library under call number: QL151.C62 1981

Source: J. D. Anderson "Essential Decisions in Indexing Systems Design." In *Indexing Specialized Formats and Subjects,* ed. H. Feinberg (Metuchen, N.J.: Scarecrow, 1983), p. 16.

ray tube does to the computer program that generates it. The underlying 'deep' structure is abstract and not experienced directly."

Shepard (1978) suggests that the mind accommodates alternative displays of similar information, and that individuals differ with respect to their preferred displays. He cites Albert Einstein and Samuel Taylor Coleridge as examples of persons who preferred or at least made extensive use of visual or mental images: "As is well known, Einstein later stated quite explicitly that he 'very rarely' thought in words at all" (p. 126, citing Wertheimer 1945, 184). Coleridge claimed that he did not compose his poem "Kubla Khan" in the usual way, rather "all of the images rose up before him as things" (p. 127, citing Ghiselin 1952, 85).

File Structure and Display

Two or more index records constitute a file. Records in inflexible media such as print-on-paper must be arranged in an order known (or learnable) by the searcher so particular records can be located. Two principal choices are available—alphanumeric arrangement based on letters and/or numerals in headings chosen to represent the record, or classified arrangement based on relationships among concepts represented by the record. A classified arrangement is maintained by assigning alphanumeric notation which, when sorted according to the sequence values assigned to letters and/or numbers, will preserve the order specified by the classification scheme. Within these types of arrangement, a multitude of alternatives exist. Anderson (1982) has compared alternative codes currently in use for the arrangement of alphabetical catalogs. The alternative arrangement of classified files are as many as the vast number of classfication schemes, some of which attempt to cover all knowledge, most of which organize the concepts of single or related disciplines or fields.

The arrangement of files is less central to retrieval of records from electronic media. In fact, file structures are invisible to the searcher and are used only to facilitate economic storage and speedier response times. It is possible to arrange records in random order and examine all of them for every search, but this is usually an inefficient use of computer time.

The files of the mind differ most markedly from those of artificial indexing systems in that they are not linear, but multi-dimensional, so that searches can go in multiple directions more or less simultaneously. Computer scientists are working on nonlinear computer systems, but current systems rely on linear sequential programs (Hubel 1979, 46). In contrast, evidence suggests that "people probably do not use a successive-scanning process to retrieve information from semantic memory" (Loftus and Loftus 1976, 121), but rather access directly appropriate categories and sub-categories until a sought item is found.

Searching and Retrieving

At least two models of searching indexing systems exist. The most prevalent is based on Boolean algebra and the search for sets of records which meet criteria specified in a Boolean statement. In such searches, the Boolean search statement, consisting of terms connected by 'ORs', 'ANDs', and/or 'NOTs', is matched sequentially against the records file, or subsets of the records file. Records either match or do not match the search statement.

An alternative model, vector searching, is probably closer to the mind's search processes. Each record is described by a set of vectors, or dimensions, with the value of each dimension indicating the strength of association between the dimension and the document represented by the record. For example, this paper might have a high value on the "indexing systems" dimension, but a lower value on the "cataloging" dimension, and an even lower value on the "library" dimension. Similarly, queries are translated to sets of vectors or dimensions, also with the possibility of variable weights based on the importance of various elements. In matching query vectors with record vectors, success and failure are a matter of degree, rather than a dichotomous yes or no, and records can be ranked according to the degree to which they match the query. Searchers may then examine only the first few, or more, depending on how comprehensive a search they desire (Salton and McGill 1983, 201-4).

Conclusions

The comparison of the human mind's indexing system and the artificial indexing systems created to extend the range and power of the mind brings to light many significant similarities and differences between the two types of system. Although detailed knowledge and understanding of the human indexing system is still very skimpy, its attributes do suggest several areas which might improve the performance and usefulness of artificial systems.

A principal difference appears to be the evaluation of the potential importance or interest of information at several stages of processing. A large mass of information is received by the mind, but most of it is ignored or permitted to evaporate after quick, initial evaluation. A second stage of evaluation is performed on the information which is transferred to short-term memory, but only the most important or interesting of this information is permanently documented and indexed in long-term memory. Even here, however, the extent of this indexing varies with respect to importance and similarity to existing knowledge.

In contrast, artificial indexing systems rarely include evaluation of importance or interest after the initial definition of scope or domain. Once a certain collection of documents is included within the scope and domain of a system, every document tends to be treated more or less equally. An obvious problem is making evaluations of importance or interest on behalf of large numbers of unknown future users, but several researchers have suggested the inclusion of evaluative information in indexing systems as a way to make retrieved information of more immediate usefulness (e.g., Hall 1982).

The other major difference is the mind's integration of operations and elements which tend to be separate, although linked, in artificial systems. Artificial systems usually include a collection of documents, a separate set of surrogates representing these documents, in part through index terms, and a separate, but linked vocabulary control system. Current understanding of the mind's indexing system suggests that the original document (or experience) is quickly discarded after the information judged to be important or interesting is extracted. This information then becomes a part of the semantic memory, which maintains conceptual associations. Thus, the actual information stored is integrated into and modifies the counterpart of the artificial system's vocabulary control thesaurus.

Artificial systems might imitate this treatment of information by extracting the important/interesting information from texts and integrating it, through comprehensive conceptual relationship displays and scope notes, into the thesaurus, making the thesaurus a true treasury of the knowledge of a particular discipline or subject field. This type of thesaurus, in addition to recording concepts, terminology, and the relations among them, would make substantive statements about these concepts and their relations, which together constitute the fundamental elements of information and, in turn, knowledge.

Brookes (1980, 254-55) has proposed an exploratory project along these lines, using Farradane's relations to "graph" the concepts contained in the literature of a compact field of science. "When the cumulative graph becomes reasonably stable, it can be regarded as both a data base and also the current 'knowledge structure' of the subject field. . . . As a knowledge structure it offers the opportunity of observing growth by accretion of new ideas and also the internal restructuring that will occur from time to time."

The mind's indexing system also suggests a possible symbiotic relationship between automatic, mechanical indexing of symbols and the intellectual indexing of concepts. Just as all information is initially perceived by the mind through coded signals or symbols and is interpreted by matching these symbols against the "lexicon" of symbols stored in the mind, so the initial indexing of documents by artificial systems might best be done automatically and mechanically on the basis of the symbols chosen by authors to represent the messages they wish to transmit. These symbols, or a subset chosen according to one or more of the many algorithms under investigation, could be mapped against the symbols (i.e., terms) in a thesaurus of the concepts and terms of the subject field, so that information seekers could find the representations of concepts via alternative sets of symbols, representing synonymous, essentially equivalent, as well as related concepts. Once a document receives actual use, demonstrated, for example, by retrieval, circulation, or citation in a new document, its vocabulary could be analyzed for inclusion in the thesaurus. This approach would provide an operational way to select the more important documents for detailed indexing (through the integration of their vocabulary and concepts into a thesaurus), rather than investing expensive intellectual labor in the indexing of all documents as if they were equal.

Systematic research in artificial indexing has been pursued actively since the 1950s, yet here, too, as in the case of the mind's indexing system, knowledge is rudimentary. Understanding of the effect of the many, complex, and interrelated variables is based more on intuition and insight than verifiable data (Sparck Jones 1981). One hopes that the ongoing research in cognitive science and in information organization and retrieval systems can together contribute to a better understanding of both types of system.

Note

I thank the persons who have reviewed drafts of this chapter and made valuable suggestions for clarification and revision, especially Pauline Atherton Cochrane, Brent Ruben, and my doctoral students David Johnson, David Oettinger, and Barbara Kwasnik. In no case did I incorporate all of their suggestions; remaining problems are my own.

References

Aitchison, J. *UNESCO thesaurus: A structured list of descriptors for indexing and retrieving literature in the fields of education, science, social science, culture and communication.*Paris: United Nations Educational, Scientific, and Cultural Organization, 1977.

Anderson, J. D. Contextual indexing and faceted classification for databases in the humanities. *Proceedings of the Annual Meeting, American Society for Information Science*, 1979, 16, 194-201.

_____. Catalog file display: Principles and the new filing rules. *Cataloging and Classification Quarterly*, 1982, 1(4), 3-23.

_____. Essential decisions in indexing systems design. In H. Feinberg (ed.), *Indexing specialized formats and subjects*. Metuchen, N.J.: Scrarecrow Press, 1983, pp. 1-21.

Austin, D. *PRECIS: A manual of concept analysis and subject indexing.* London: Council of the British National Bibliography, 1974.

Bates, M. Rigorous systematic bibliography. *RQ,* 1976 (Fall), 16(1), 7-26.

Brooks, B. C. Measurement in information science: Objective and subjective metrical space. *Journal of the American Society for Information Science,* 1980, 31 (4), 248-255.

Clark, J. W. *The care of books: An essay on the development of libraries and their fittings, from earliest times to the end of the eighteenth century.* Cambridge, England: The University Press, 1909; reprinted: Folcroft, Pa.: Folcroft Library Editions, 1973.

Craik, F. I. M. and R. S. Lockhart. Levels of processing: A framework for memory research. *Journal of Verbal Learning and Verbal Behavior,* 1972, 2, 671-684.

Diener, R. A. V. A longitudinal study of the informational dynamics of journal article titles: A treatise in the science of information. Ph.D. diss., Rutgers University, 1984.

Doszkocs, T. D. An associative interactive dictionary (AID) for online bibliographic searching. *Proceedings of the Annual Meeting, American Society for Information Science,* 1978, 105-109.

Estes, W. K. The information-processing approach to cognition: A confluence of metaphors and methods. In W. K. Estes (ed.), *Handbook of learning and cognitive processes, volume 5: Human information processing.* Hillsdale, N.J.: Erlbaum, 1978, pp. 1-18.

Farradane, J. Relational indexing and classification in the light of recent experimental work in psychology. *Information storage and retrieval,* 1963, 1(1), 3-11.

———. Relational indexing, part 1. *Journal of Information Science,* 1980, 1(5), 267-276; part II. *Journal of Information Science,* 1980, 1(6), 313-324.

Fodor, J. A. The mind-body problem. *Scientific American,* 1981, 244(1), 114-123.

Foskett, D. J. The theory of integrative levels and its relevance to the design of information systems. *Aslib Proceedings,* 1978, 30(6), 202-208.

Furnas, G. S., et al. Statistical semantics: Analysis of the potential performance of key-word information systems. *The Bell System Technical Journal,* 1983, 62(6), 1753-1806.

Ghiselin, B. *The creative process.* New York: American Library, 1952.

Glass, A. L., K.J. Holyoak, and J. L. Santa. *Cognition.* Reading, Mass.: Addison-Wesley, 1979.

Hall, H. J. Services for the analysis and evaluation of information. Final report for the National Science Foundation, Division of Information Science and Technology, Research Grant 78-16628. New Brunswick, N.J.: School of Communication, Information, and Library Studies, Rutgers, the State University of New Jersey, 1982.

Hubel, D. The brain. *Scientific American,* 1979, 241(3), 45-50.

Hunt, M. *The universe within: A new science explores the human mind.* New York: Simon & Schuster, 1982.

Jackson, S. L. *Libraries and librarianship in the west: A brief history.* New York: McGraw-Hill, 1974.

Johnson, E. D. and M. H. Harris. *History of libraries in the western world.* 3rd ed. Metuchen, N.J.: Scarecrow Press, 1976.

Klingbiel, P. H. and C. C. Rinker. Evaluation of machine-aided indexing. *Information Processing and Management,* 1976, 12, 351-366.

Kosslyn, S. M. Information representation in visual images. *Cognitive Psychology,* 1975, 7(3), 341-370.

_____. Can imagery be distinguished from other forms of internal representation? Evidence from studies of information retrieval time. *Memory and Cognition*, 1976, 4(3), 291-297.

Lachman, R., J. L. Lachman, and E. C. Butterfield. *Cognitive psychology and information processing*. Hilldale, N.J.: Erlbaum, 1979.

_____, J. P. Shaffer, and D. Hennrikus. Effects of stimulus codability, name-word frequency, and age of acquisition on lexical reaction time. *Journal of Verbal Learning and Verbal Behavior*, 1974, 13, 613-625.

Leonard, L. E. *Inter-indexer consistency studies, 1954-1975: A review of the literature and summary of study results*. Urbana: University of Illinois, Graduate School of Library Science, 1977 (occasional papers, no. 131).

Library of Congress. Subject Cataloging Division. *Library of Congress subject headings*. 9th ed. Washington, D.C., 1980.

_____. Automated Systems Office. *MARC formats for bibliographic data*. Washington, D.C., 1980.

Loftus, E. *Memory: Surprising new insights into how we remember and why we forget*. Reading, Mass.: Addison-Wesley, 1980.

Loftus, G. R. and E. F. Loftus. *Human memory: The processing of information*. Hillsdale, N.J.: Erlbaum, 1976.

Montgomery, C. A. Linguistics and information science. *Journal of the American Society for Information Science*, 1972, 23(3), 195-219.

OCLC Inc. *On-line systems—cataloging: User manual*. Columbus, Ohio, 1979.

Parsons, E. A. The classification and cataloging of the material assembled. In *The Alexandrian Library*. Amsterdam: Elsevier, 1952, pp. 204-218.

Popper, K. R. and J.C. Eccles. *The self and its brain*. New York: Springer International (Springer-Verlag), 1978.

Ruben, B.D. *Communication and human behavior*. New York: Macmillan, 1984.

Salton, G. and M. J. McGill. *Introduction to modern information retrieval*. New York: McGraw-Hill, 1983.

Shepard, R. N. The mental image. *American Psychologist*, 1978, 33(2), 125-137.

Smith, L. C. Selected artificial intelligence techniques in information retrieval systems research. Ph.D. diss., Syracuse University, 1979.

Soergel, D. *Indexing languages and thesauri: Construction and maintenance*. Los Angeles: Melville, 1974.

Sparck Jones, K., ed. *Information retrieval experiment*. London and Boston: Butterworths, 1981.

Thesaurus of ERIC descriptors. 10th ed. Phoenix: Oryx, 1984.

Webster's third new international dictionary of the English language, unabridged. Springfield, Mass.: Merriam, c1976.

Westby, B. M., ed. *Sears list of subject headings*. 10th ed. New York: Wilson, 1972.

Wellisch, H. H. How to make an index—16th century style: Conrad Gessner on indexes and catalogs. *International Classification*, 1981, 8(1), 10-15; Discussion, 1981, 8(3), 185.

Wertheimer, M. *Productive thinking*. New York: Harper, 1945.

Willets, M. An investigation of the nature of the relation between terms in thesauri. *Journal of Documentation*, 1975, 31(1), 158-184.

16

Communication, Information, and Adaptation

Young Yun Kim

This chapter addresses the issue of pervasive changes in today's dynamic world and the corresponding need for individuals to increase their capacity to adapt to change. Based on the system-theoretic perspective, the human organism is viewed as an open communication system with its self-reflexive and self-organizing capacities. Through the exchange of information with its environment, an individual organism continually undergoes change while maintaining its overall integrity. Adaptation, in this sense, is an interactive process in which the human organism pursues its maximum "fit" between its internal information-processing system and the complexity of its informational environment. The author maintains that change (and the accompanying stress) needs not be viewed as essentially problematic, but as a vital condition for an organism's growth in its adaptability and thus, in its life activities. The article concludes with a number of basic conditions under which the internal adaptive capacity of an individual may be maximally cultivated.

Life is a process that inevitably brings change. No two events are quite alike; we continuously face the challenge of uncertainty. Particularly in today's fast-changing world, we must adapt to unprecedented change—in technology, social structure, values, politics, and in human relations. In many tribal and traditional societies, change has encroached upon nearly every stable pattern of life—cultural values, the structure and functions of the family, and the relations between generations.

In this global context, the concept of adaptation takes on a profound personal and social significance. People struggle to cope with, and to adapt to, the complexities and uncertainties of life. Some individuals feel lost and

324

try to withdraw. Some violently resist change and fight for the old ways. Others desperately try to keep up with the changes, often experiencing a sense of failure and despair.

To be effective in dealing with change in our environment, we need to clarify our understanding of the fundamental dynamics of human adaptation. How do we humans adapt, and adapt to, our sociocultural world? What are some of the crucial conditions that are likely to promote or deter effective adaptation to change? Answers to these questions may serve as a basis upon which we can design methods for dealing creatively with specific situations of the dynamic interface between ourselves and the environment.

Contexts of Change

There are numerous events and circumstances that present significant contexts for individuals to modify their customary life patterns. The degree to which such adaptive response is needed, of course, depends on the extent to which the new environment challenges us by confronting us with psycho-social demands that differ from those to which we are accustomed.

An extensive list of examples of contexts that involve significant change and adaptation is presented by Taft (1977) under five headings: sojourning, settling, subcultural mobility, segregation, and change in society at large. While the first two necessarily accompany geographic mobility from one sociocultural environment to another, the others do not. Examples of non-locomotive changes include such situations as adapting to a new profession, retirement, marriage, divorce, aging, death of loved ones, and transition from school to work. Also, there are situations in which one is incorporated into a new social system such as the armed services, religious orders, residential university, prison, concentration camp, rehabilitation center, or mental hospital.

Contexts of change that result from geographical relocation include those of foreign students, Peace Corps volunteers, business men and women working overseas, missionaries, diplomats, technical advisors and administrators of international organizations. Also, there are immigrants and refugees who have, voluntarily or involuntarily, moved from one society to another with the intention of becoming a full member of the new society. Similarly, individuals moving from a small rural town to a large metropolitan area within the same country are in situations which involve some degree of adaptation to the new environment.

There are also cultural and technological changes that occur in the larger, societal context of rapidly industrializing or postindustrial societies. A societal environment that becomes increasingly dynamic and complex presents individuals with an increasing level of unfamiliarity and uncertainty.

The mass media also acquaint the average individual with alternatives to their own country, job, wife, religion, political system, and life style. These media images often help the individual set standards and expectations in all these aspects that are often at variance with their usually more constricted reality. Phenomena such as information overload, over-choice, relative deprivation, and the revolution of rising expectations, stimulated by the penetration of the mass media images into everyday life, are pervasive in developing as well as in postindustrial societies. To a larger or smaller degree, each of us is an alien in the midst of the waves of change that surround us within our own society and around the world.

All changes are stress-producing. As Ruben (1983) notes, "There are always stressors to be dealt with. . . . In effect, living systems wage an uphill struggle to adjust" (p. 137). Many contexts of change can, indeed, present traumatic experiences in individuals who must deal with them. Accordingly, literature in mental-health related areas of study in the human and social sciences has amply documented the variety of stressful situations which can result in maladaptation. These studies have typically approached the phenomena of change and adaptation with an implicit or explicit assumption that stress-producing changes are problematic and undesirable for an individual's psychological well-being. This problem-oriented view is reflected in many widely popularized terms such as culture shock, future shock, stress management, alienation, nervous breakdown, and mental illness.

Yet, there are numerous cases which demonstrate the impressive human capacity to deal with changes in environment without damage to one's psychological integrity. Experiences of many immigrants, refugees, and sojourners who have successfully adapted in a new cultural environment are invaluable in providing us with insights into human resilience. Confronted with situations where assumptions and premises acquired in childhood are called into question, the individual must weather a conflict in which they abandon identification with cultural patterns that symbolize personal integration. Gradually, these individuals undergo the process of adaptive change in the new sociocultural environment, demonstrating the profound plasticity and adaptibility of humans.

Even more impressive tests of the limits of human adaptive capacities occur in concentration camps and prisons. Under these circumstances, individuals must live in tightly controlled environments which are contrary to their values and goals and in which they are victims of external power. We know from such situations that people often lose hope and become apathetic, accepting their fate without a fight. But we also know that some individuals succeed in developing competing forms of social organization that allow them to resist their oppressors and may allow survival under the most unusual circumstances. Kogon (1958), a long-term resident of Buchenwald, describes the underground organization of inmates which was developed under the most severe life conditions.

It is clear, then, that individual responses to stressful environments vary widely—from severe psychological disturbances and breakdowns to triumphant control over one's internal and external conditions. How can we explain the variations in adaptive responses? Why are some individuals better than others at coping?

Communication and Adaptation

Each individual can be viewed as an open system interacting with, and making sense of, his or her environment through the process of communication. Communication is to human feeling and intellect what physical metabolism is to the body's physiological processes. Thus, communication can be looked upon as one of the two basic life processes of all human systems (Thayer 1968; Ruben 1975). The environment which we take-into-account is an infinite array of ongoing event data. As our sense and minds impose structure, meaning, and utility upon these data, the product is information—functional units which our brains are uniquely equipped and "programmed" to process. Information is the raw material for thinking, decision-making, problem-solving, attitude development, learning, and all of the specifically human activities that concern us about our psychological functioning and behavior.

Humans' systems are characteristically homeostatic. Like most living biological organisms, we as open communication systems attempt to hold constant a variety of variables in our internal meaning structure and thus achieve an ordered whole. Homeostasis is achieved when an increase in some variable in the system is met by compensatory actions, so that the joint effect is "deviation reducing" (Krippendorff 1977, 159). The so-called defensive mechanism of individuals toward changes is considered to be the result of the homeostatic tendencies of human communication systems. When the individual receives messages which conflict with the existing meaning system, its first impulse is to reject them as in some sense untrue. Suppose, for instance, that someone tells us something which is inconsistent with our picture of a certain person. Our first impulse is to reject the proffered information as false.

The literature of psychiatry and closely related fields overwhelmingly gives the impression that we avoid the painful elements at all costs and reject them as part of ourselves—even if this requires extensive self-deception. We rely heavily on avoidance and reduction of information. Also, firmly rooted in psychiatric thinking is an appreciation of the importance of a stable, predictable environment which helps us to develop and preserve the skills and psychological organization needed for effective adaptation. Whether the emphasis is placed on the maturing child or on the grown-up individual, it appears that some measure of internal balance is required in order for effective performance to prevail.

On the other hand, we humans also manifest "morphogenetic" or "deviation amplifying" features, continually engaging in internal structural change and self-modification (Krippendorff 1977). With the "reflexive" and "self-reflexive" capacities, the human mind becomes a creative factor not only in image-formation but also in the active transformation of outer reality. In Jantsch's (1980) words, the human mind "mirrors an outer reality which it rebuilds in the inner world. This mirror image does not simply enter from the outside but emerges from exchange processes between a mosaic of sensory impressions and tentative models which the reflexive mind projects outward. The most significant characteristic of the reflexive mind is apperception, the capacity of forming alternative models of reality" (p. 163).

When this role of the mind is fully activated, the symmetry or equilibrium between the inner and the outer world is broken without losing the ecological relation of the organism with its environment. Our experiences become emancipated: not only can the experience of the past become effective in the present, as it does in biological evolution, but the new capability of anticipation makes the future also effective in the present.

In this process of perceptual transformation of the outer reality, the subjective aspect plays an equally important role with the objective one. The iterative feedback process between inner and outer world may lead to the creative evolution of the mental structure. We can learn to see a situation "with new eyes" and become aware of it through a change in our emotional attitude. In this light, our existing mental capacity is responsible for its subsequent development. Calling this uniquely human capacity to further one's own growth, "self-organizing," Jantsch (1980) writes, "We live, so to speak, in co-evolution with ourselves, with our own mental products" (p. 177).

This self-organizing characteristic of human systems enables human thought to generate more adaptive alternatives than lower animals are capable of generating. More meanings can be attributed to objects, and a greater number of connections between meanings arise. In this way, human thought is less stimulus-bound. Action can be delayed, and a given stimulus gives rise to a greater number of outcomes, creating more uncertainty and ambiguity. Taking an extreme case, the moth has no alternative when faced with a light and immediately flies toward it, whereas a human engaging in complex thought processes can perceive stimuli in many ways and can consider many ways of interrelating these perceptions for his or her adaptive purposes. In this sense, human thought has more degrees of freedom.

Thus, no complex system, such as the self-organizing human system, is ever truly stable. The two tendencies of humans described above—deviation-reducing (striving toward stability and symmetry) and deviation-amplifying (striving toward fluctuation and asymmetry), operate dynamically. As a result, humans can be characterized as "metastable" systems (Jantsch 1980, 255). From the moment of birth, humans are active agents engaged in the adaptive process of constructing reality. As we change, there is a necessary progression of stages which integrates previous behaviors, concepts, and actions into the present condition. We simultaneously go through qualitative change and yet maintain integrity throughout the lifespan.

At the heart of adaptation, then, lies the relationship between a person's internal experiences, conditioned by the existing adaptive system and the external reality. In other words, a person's communication system is an entity which forms a dynamic balance between internal, fixed constraints that impose its enduring structure, and external, unrestrained forces that mold and evolve the entity. Rappaport (1971) defines adaptation as "the process by which organisms or groups of organisms, through responsive changes in their states, structures, or compositions, maintain homeostasis in and among themselves in the face of both short-term environmental fluctuations and long-term change in the composition or structure of their environment" (p. 60).

In this definition, adaptation implies maximizing the social life chances. But maximization is almost always a compromise, a "vector" in the internal structure of a person and the external pressure of the environment. This dynamic interplay of internal forces and between internal and external conditions is described by Piaget (1977) as "equilibration."

Successful adaptation, then, can be conceived as successful equilibration, achieving "goodness-of-fit" between the characteristics of the person's internal world and the properties of his or her environment (Brody 1969, 14; French, Rogers, and Cobb 1974, 316). Successful adaptation further implies modifying oneself as well as modifying the environment. Piaget (1977) refers to self-modification as "accommodation" which is an adjustment on the part of the organism itself so it fits better with the existing environment. On the other hand, the action of the organism on surrounding objects is called "assimilation"[1] That is, the organism, instead of submitting passively to the surroundings, modifies them by imposing a structure of its own. To put it differently, the assimilating organism changes the substance of the environment into something compatible with its own structure. Behavior that is maladaptive may result when individuals are challenged with environmental events that exceed their ability to accommodate and assimilate, and thus implement their goals and intentions.

Change, therefore, should not be conceived of as a change in an immutable environmental constellation presenting itself to the individual. It always results from the cumulative interaction between the person and the environment. A situation of change is defined by the interaction of two groups of factors: (1) the nature of the environmental input and (2) the state of the person's communication system at the time of the input reception. In this framework, successful levels and forms of equilibrium are conceived as levels of adaptation, with change being caused by a mismatch or discrepancy between our internal world and the external reality.

Internal Complexity

Given the nature of human adaptive processes, how can we explain individual differences in their adaptive capacities? The crucial factor in addressing this question is their internal fixed constraints—internal cognitive structure. Due to different responses in their internal logic, they "select" different responses. These internal differences among individuals (and groups) depend on their built-in and learned differences.

In psychology, this intrapsychic characteristic of humans is called cognitive complexity. It refers to the individual's capacity to process environmental inputs with the aid of a differentiated set of categories that are integrated on an abstract level and thereby disengaged from concrete environmental situations. Making a sharp distinction between what a person thinks and how he or she thinks, Schroder, Driver, and Streufert (1967, 1975) identify two basic interdependent properties of information processing structures: the parts or dimensions and the integrating rules. Individuals who possess a complex cognitive structure tend to better differentiate the number of dimensions in their personal constructs, to better articulate and abstract the dimensions and integrate them into a meaningful whole.

In addition to these cognitive capacities of human systems—to differentiate and to integrate—there is yet another dimension which has not been fully explored in psychological studies. This dimension has been proposed by German sociologist and systems theorist, Luhmann (1978; in Geyer 1980), as "structural selectivity."[2] The selectivity of cognitive structure, according to Luhmann, refers to "the ability to selectively decide which of the latently possible interrelations between its elements should be 'actualized' not only as a necessity, but also as a chance for the system; a chance, because it enables the system to deal selectively with its own internal relationing potential according to need and circumstance" (Geyer 1980, 140). Such an increase of structural selectivity makes for an increase of both the contingency and the nonarbitrary character of choice of structure.

The point of departure for Luhmann's conceptualization of intrapsychic complexity is the number of elements of a system and the number and the nature of the possible interrelations among them. Regardless of how one defines the system's elements and the interrelations among them, it can easily be shown that an increase of the system's elements precludes any over-proportional increase of the possible interrelations. In somewhat larger systems, such a complete interdependence would soon become unworkable when actualized, and can serve only as an indeterminate background, which functions to increase the system's selectivity.

Cognitive complexity, in Luhmann's conceptualization, does not point simply to the quantity of structurally possible relations, but to their selectivity. It finds its unity in the form of a relation—the relation of mutual facilitation of series of elements and reductive ordering. The complexity of a system is higher when there is an increase in the selectivity of the interrelations that are possible given a certain system size and system structure. It is as though the individual with a high structural selectivity is equipped with fully functioning inbuilt feedback mechanisms. When these feedbacks start operating above a certain threshold value, they automatically help the internal system to regulate itself by helping it temporarily come to a standstill.

There comes a moment for everyone when internal complexity becomes a more immediate and pressing problem than reduction of environmental complexity. Suddenly we long for the simplicity of the good old days and for peace of mind. Where internal complexity has increased to the point where the system tends to lose sight of its own structure, selectivity should be turned inward rather than outward. It should serve to underline what really are the important modes and interconnections in the system's internal structure, in accord with its value hierarchy—the "need for the real self"—and irrespective of the pressure of the environmental demands. Individuals with high structural selectivity in their internal cognitive system are more capable of doing this reflective information processing.

Based on the conception of Schroder and his associates as well as of Luhmann, it would appear that individuals with a high degree of internal complexity are better able to differentiate and integrate environmental data in a way which maximizes the adaptation or fitness between the internal priorities and values and the external constraints. They have more alternative ways of reacting to a given environmental condition. They are less randomly or blindly influenced by the environmental pressures, and

thus less prone to experiencing psychological disturbances and social mal-
functions. In other words, the higher the degree of one's internal complexity, the greater his or her autonomy and freedom from the dictates of the
environment. This means that increased autonomy also increases one's
dependence on one's own history, hence the "burden of co-ordination"
(Geyer 1980, 140).

The complexity of an individual's internal system of information processing is considered developmental and cumulative (Schroder, Driver, and
Streufert 1967, 1975). During our lifetime, we remain subject to the
negentropic pressure towards internal information build-up. Maturity itself involves development of cognitive complexity, which enables us to see
a situation in broader, relativistic perspectives. As this relativistic orientation increases, we express our desire for an absolute core by developing
faith in an activity, a movement, or a religion. These we hope will compensate for the progressive disillusionment which we experience when we face
the relativistic nature of things (Ruesch 1968, 87). Also, a mature person's
image contains not only what is, but what might be. It is full of potentialities as yet unrealized. In rational behavior, we contemplate the world of
potentialities, evaluate ourselves according to our value system, and choose
the "best."

Thus, as we mature, we also acquire various aspects of personal and
social competence by which the activity inherent in our living organism is
given its clear representation—the power of initiative and exertion that we
experience as a sense of an agency in our lives. This experience can be
termed a "feeling of efficacy" (Moos and Tsu 1976, 6).

Maladaptation

Maladaptation, and other related terms—maladjustment, alienation,
mental illness, culture shock, and future shock—refer to those situations
which stem from the lack of an individual's internal capacities to compe-
tently deal with environmental fluctuations and complexities. Maladaptation, like adaptation, reflects a relative degree of such misfit or mismatch
between a person and the environment. In this sense, adaptation and maladaptation are viewed as two ends of the same continuum of fitness between an individual's internal complexity and environmental complexity.

Even the severest forms of maladaptation that are often considered in
the domain of psychopathology can be viewed as a state of misfit (Ruesch
1968). For example, the condition which the psychiatrist labels "psychosis"
is interpreted by Ruesch as the result of the patient's misinterpretation of
messages received due to a cognitive structure that lacks the ability to
differentiate and selectively integrate the information received from the

environment. Similarly, the condition which we commonly label "neurosis" is viewed as the result of the unfortunate attempts of a patient to manipulate social situations with the purpose of creating a stage to convey messages to others more effectively. The messages are usually not understood by others, and the result is frustration for the patient and for others. The patient is forced to develop ways of handling the frustration, often distorting the process of communication further.

Similarly, schizophrenic individuals can be viewed as those whose cognitive structure is incapable of discriminating the multiplicity of human affairs. Thus, they assign themselves a role and neglect the fact that roles are determined through the reciprocity of relationships. On the other hand, psycho-neurotic patients, according to Reusch, tend to flood others with messages in an attempt to coerce them into accepting roles they are not willing to assume. These compulsive attempts to shape situations and coerce people obviously reflect and result in maladaptive behavior.

One does not, however, need to be a psychiatric patient to experience the symptoms of maladaptation. Every so-called normal person is subject to occasional psychological disturbances under various forms of environmental complexities. For example, the widely acknowledged contemporary phenomenon of alienation presents patterns of maladaptation that are basically isomorphic to psychiatric symptoms. In both cases, problems can be attributed essentially to the lack of complexity in an individual's information processing system vis-à-vis its environment.

Based on this perspective, Geyer (1980) presents an in-depth analysis of various facets of alienation. Powerlessness, for example, is viewed as primarily a problem of the ineffective information output process of the individual. According to Geyer, the main characteristics of powerlessness is that the number and/or the effectiveness of output alternatives are diminished. Whatever behavior alternative the individual decides upon, it is ineffectual in bringing about the reinforcement he or she seeks.

Similarly, Geyer explains the experience of meaninglessness as primarily a problem of the input process of the individual. Human beings function optimally within a certain stimulus range, and the upper and lower limits are distinct in each individual case depending on the complexity of his or her cognitive structure and that of the environment. Given this perspective, individuals show temporary maladaptive or pathological reactions which indicate a breakdown of normal information-processing.

Such deviant reactions can result in either overstimulation or understimulation. When the environmental complexity overrides the internal complexity of the individual, overstimulation occurs. If the system is overstimulated, the system does not admit the majority of the inputs offered (the denial mechanism in psychiatry). When inputs are admitted, they are necessarily distorted, because they are forced in categories that are not sufficiently subtle and differentiated to contain them adequately. On the other hand, if the system is understimulated, it either tries to find a more stimulating environment or, when limited mobility makes this impossible, it starts manufacturing its own inputs such as daydreams, some kinds of fantasies, and even hallucinations.

Other commonly employed terms such as culture shock and future shock reference various patterns of maladaptation that result from the significant discrepancy between an individual's internal cognitive structure and the external environment. In the situation of culture shock, it is usually the immigrant or sojourner who has entered, and must adapt to, an unfamiliar sociocultural environment. During the initial phase of immigration, at least, the individual is unable to adequately differentiate and integrate the host milieu in a way that maximizes his or her survival and adaptation. As Schuetz ([1944] 1963) describes it: "The cultural patterns of the approached group is to the stranger not a shelter but a field of adventure, not a matter of course but a questionable topic of investigation, not an instrument for disentangling problematic situations but a problematic situation itself and one hard to master" (p. 108).

Individuals in a rapidly changing technological and cultural environment must deal with many aspects of their environment that are neither familiar nor stable. As the pervasiveness of unfamiliarity increases, more individuals find themselves in a situation quite similar to that of immigrants or sojourners. Their internal cognitive structure is no longer as capable of handling the varied situations in the environment as before. In both situations—situations of international migrants and of individuals in rapidly changing societies—individuals may be, temporarily at least, subject to severe or mild maladaptive experiences that are observed in psychosis, neurosis, schizophrenia, powerlessness, and/or meaninglessness.

Adaptation and Growth

The central remedy for the increased unpredictability of complex environments is increased internal complexity. As the environmental input increases in its complexity, a higher level of internal complexity is demanded. The more complex the sector of the social environment one is interacting with, the more the chains of consequences stretch out synchronously, and more persons, institutions, processes are involved simultaneously in complex decision-making procedures. At the same time, the decision chains become wider diachronically, as decisions made now have their consequences further in the future, and often in unforeseen ways.

Such environmental conditions naturally produce strain in individuals' internal systems as they strive to adapt to the new, changed external conditions. The strain, or stress, can be viewed as the internally-felt resistance of the human organism against its own evolution. It implies the psychological pain in the destruction of the old structure. The more rigid one's internal structure, the more one could be expected to change. Compared to young children, for example, the internal structure of fully grown adults is quite rigid, and thus, more resistant to change. This is clearly evidenced in studies of immigrants in which age has been observed to be inversely related to the rate of adaptation to the host environment (Kim 1977).

No human system can stabilize itself forever, even if it has to defend itself to its utmost and dampen the fluctuations. If it did not do this, nothing much would come of change. Human systems are characterized by a phenomenal capacity for adaptation, internal growth, and development quite independent of external forces. Paradoxically, the higher the resistance, the more the fluctuations. When these fluctuations ultimately break through, "the richer and more varied the unfolding of self-organization dynamics at the platform of a resilient structure . . . the more splendid the unfolding of mind" (Jantsch 1980, 251).

Such fluctuations, then, comprise the very fundamental basis for our psychic growth, of "becoming" more complex and adaptive internally. It is the process which Jourard (1974) describes as the process of integration-disintegration-reintegration: "Growth is the dis-integration of one way of experiencing the world, followed by a reorganization of this experience, a reorganization that includes the new disclosure of the world. The disorganization, or even shattering, of one way to experience the world, is brought on by new disclosure from the changing being of the world, disclosures that were always being transmitted, but were usually ignored" (p. 456). Similarly, Hall (1976), in proposing a psychic growth of individuals beyond their own cultural parameters, calls this fundamental nature of psychic growth processes "identity-separation-growth dynamism."

Empirical research has provided some indirect indication that the stresses in adapting to a different culture may indeed lay the groundwork for subsequent psychic growth. On the basis of their study among Canadians in Kenya, Ruben and Kealey (1979, 41) found tentative empirical data to suggest that, in some cases at least, the person who will ultimately be the most effective in adapting to a new culture can be expected to undergo the most intense culture shock during the initial transition period. Other acculturation studies of immigrants and foreign students in the United States have shown that, once the initial phase has been successfully managed, individuals go through a gradual process of adaptation and psychic growth in the new cultural environment (see, for example, Coelho 1958; Kim 1979).

What is significant about the individuals who have successfully adapted to a new environment, in spite of the pervasive and adversary nature of challenges, is that they may well have acquired capacities for understanding their old environment, which is still very much part of their psyche. Furthermore, they possess the necessary cognitive capacity to differentiate the new and the old environments, which in turn helps them to develop an objective eye in dealing with the new environment. In other words, their cognitive structures have been reorganized on a higher level of complexity and awareness.

In the case of long-term sojourners or immigrants, for instance, the process of psychic growth is manifested in their increased capacity to overcome cultural parochialism. It is a process of growth beyond the psychological parameters of any one culture and of becoming a truly "intercultural person" (Kim 1982). Upon successfully overcoming the multitude of stressful life conditions that inevitably accompany the process of cultural adaptation, one develops cognitive capacities that are more adaptable, flexible, and resilient than those who have limited exposures to the challenges of continuous intercultural encounters.

This level of growth in internal complexity corresponds to what Piaget, Dewey, and Kohlberg label as the postconventional or autonomous stage of moral development. Individuals who have developed this level of moral character (which reflects a refined internal complexity) tend to define right action by a decision in accord with self-chosen ethical principles appealing to internal logical comprehensiveness, universality, and consistency. These contain the dual criteria of universality and respect for the human personality. Such modes of judgment reflect the comprehensiveness of the individual information processing structure, which includes fine differentiation of individuals, circumstances, and contexts. At the same time it enables one to comprehensively integrate information concerning a particular event or act, and, most of all, to develop a clear internal value structure which discriminates the irrelevant from the relevant and the insignificant from the significant.

Indeed, the process of internal growth can reach an optimal level of adaptability, in which one is capable of creatively resolving and integrating seemingly contradictory characteristics of peoples and events, as well as inconsistencies between internal experiences and external circumstances of change and uncertainty. Furthermore, one is able to transform them into complementary, interacting parts of an integral whole. This state of development is what Harris (1979) refers to as the optimal level of communication competence. As one achieves the highest capacity to interact with one's environment, one may be able to make deliberate choices for actions in specific situations rather than simply being bound by the culturally normative courses of action. These optimal, comprehensive, inclusive, selective, and critical information-processing capacities can be considered the full blossoming of the uniquely human symbolizing capacities variously termed reflexiveness, self-reflexiveness, and self-organization.

Conditions for Internal Growth

Given the critical relationship between a person's internal complexity and adaptability, individuals must make a continual updating of the existing information, and thus, develop a highly complex internal cognitive structure in order to adapt to the ever-changing, unfamiliar, and complex environment. How, then, can we maximize our potential for internal growth? What are some of the basic conditions which facilitate the process of developing our inner capacities to adapt and grow?

First, we must realize that the problem is no longer to acquire a stable personality-concept in the traditional sense, but to improve the ability to cope with the changing structures of modern societies, and to make the best possible use of the opportunities it affords. Identity—in the sense of having a world view that may show its very coherence in its consequent attribution of incoherence to the world around—is a thing of the past. This is the case at least for the increasing number of people who are sensitive to extreme environmental complexity, and who try to maintain an open interaction with their environment, in spite of it. In this context of change, Zurcher (1977) presents a model of self-concept for social change which is aptly labeled as the "mutable self."

Second, to be adaptive and to be able to maximize internal complexity of information processing, the openness of our inner world is of critical importance. This openness, like children's innocence, enables us to minimize our resistance to change and to maximize our capabilities to embrace changes as they occur. As Jantsch (1980, 255-56) states:

> To live in an evolutionary spirit means to engage with full ambition and without any reserve in the structure of the present, and yet to let go and flow into a new structure when the right time has come. Such an attitude is meant by the Buddhist virtue of "non-attachment" which is so frequently confused with non-engagement. Once we open our mind and fully respond to environmental changes, the profound plasticity of our human systems begins to work.

Third, we must recognize that our "either-or" way of categorization, derived from simpler times and civilizations, is utterly inadequate to contain the increased complexity of internal codes necessitated by growing environmental complexity. A more flexible "both-and" way of categorizing would increase tolerance of ambiguity to the higher levels needed to maintain ourselves in a complex and ambiguous environment, and would move in the direction of an increasingly complex internal state of being. As Rogers (1980, 104) states:

> Can we today afford the luxury of having "a" reality? Can we still preserve the belief that there is a "real world" upon whose definition we all agree? I am convinced that this is a luxury we *cannot* afford, a myth we dare not maintain. . . . It appears to me that the way of the future must be to base our lives and our education on the assumption that there are as many realities as there are persons, and that our highest priority is to accept that hypothesis and proceed from there.

At this point, the question of the limits to human complexities may be posed. In principle, these limits define the parameters of stability. Stability, in turn, is limited by the degree of coupling with the environment. In a static view, higher complexity implies a loss of stability. Self-organizing nonequilibrium systems, however, may be unstable and yet remain functional through evolution. As long as the bonds among subsystems are strong enough to dampen smaller and medium fluctuations, the system can maintain a state of metastability. Thereby, the shift to a new structure is delayed during a finite period which is sufficient for the adaptive unfolding of life processes.

Humans, like all living systems, eventually decay—reaching the state of entropy. But life itself is continually reorganized, giving rise to higher and higher, more and more complex organization. With greater acceptance of change, openness, and flexibility, it is possible for individuals to participate in a more open-ended communication process and to increase their adaptability in this dynamically changing world.

Notes

1. The usage of the term *assimilation* by Piaget is distinguished from the conventional views taken in studies of immigrants, such as the "relinquishing of cultural identity by immigrants" (Padilla 1980) or the "highest degree of acculturation theoretically possible" (Kim 1979).
2. Luhmann's term *structural selectivity* should not be confused with the selectivity in perception as commonly used in cognitive psychology. While the former refers to the cognitive capacity to reductively order and integrate input data, the latter refers to the selectivity in receiving of input data.

References

Brody, E. B. Migration and adaptation: The nature of the problem. In E.B. Brody (ed.), *Behavior in new environments: Adaptation to migrant population.* Beverly Hills, Calif.: Sage, 1969.

Coelho, G. *Changing images of America.* Glencoe, Ill.: Free Press, 1958.

French, J., W. Rodgers, and S. Cobb. Adjustment as person-environment fit. in G. Coelho, D. Hamburg, and J. Adams (eds.), *Coping and adaptation.* New York: Basic Books, 1974.

Geyer, R. F. *Alienation theories: A general systems approach.* New York: Pergamon, 1980.

Hall, E. T. *Beyond culture.* New York: Doubleday, 1976.

Harris, L. Communication competence. Paper presented at the annual conference of the International Communication Association, Philadelphia, May 1979.

Jantsch, E. *The self-organizing universe: Scientific and human implications of the emerging paradigm of evolution.* New York: Pergamon, 1980.

Jourard, S. Growing awareness and the awareness of growth. In B. Patton and K. Giffin (eds.), *Interpersonal communication.* New York: Harper & Row, 1974.

Kim, Y. Y. Communication patterns of foreign immigrants in the process of acculturation. *Human Communication Research,* 1977, 4(1), 66-77.

———. Toward an interactive theory of communication-acculturation. In D. Nimmo (ed.), *Communication yearbook 3.* New Brunswick, N.J.: Transaction-International Communication Association, 1979.

———. Becoming intercultural. Paper presented at the annual conference of the Society for Intercultural Education, Training, and Research. Long Beach, Calif., March 1982.

Kogon, E. *The theory and practice of hell.* New York: Berkeley Medallion Books, 1958.

Krippendorff, K. Information systems theory and research: An overview. In B.D. Ruben (ed.), *Communication Yearbook 1.* New Brunswick, N.J.: Transaction-International Communication Association, 1977.

Moos, R. H. and V. D. Tsu. Human competence and coping: An overview. In R. H. Moos (ed.), *Human adaptation: Coping with life crises.* Lexington, Mass.: D. C. Heath, 1976.

Padilla, A. *Acculturation: Theory, models, and some new findings.* Boulder, Colo.: Westview-American Association for the Advancement of Science, 1980.

Piaget, J. Problems of equilibration. In M. Appel and L. Goldberg (eds.), *Topics in cognitive development.* New York: Plenum, 1977.

Rappaport, R. A. Ritual, sanctity, and cybernetics. *American Anthropologist,* 1971, 73(1), 59-76.

Rogers, C. *A way of being.* Boston, Mass.: Houghton Mifflin, 1980.

Ruben, B. D. Intrapersonal, interpersonal, and mass communication process in individual and multi-person systems. In B. D. Ruben and J. Y. Kim (eds.), *General systems theory and human communication.* Rochelle Park, N.J.: Hayden, 1975.

———. A system-theoretic view. In W. B. Gudykunst (ed.), *Intercultural communication theory.* A publication of the Speech Communication Association's Commission on International and Intercultural Communication. Beverly Hills, Calif.: Sage, 1983.

———, and D. J. Kealey. Behavioral assessment of communication competency and the prediction of cross-cultural adaptation. *International Journal of Intercultural Relations,* 1979, 3(1), 15-48.

Ruesch, J. Communication and mental illness: A psychiatric approach. In J. Ruesch and G. Bateson (eds.), *Communication: The social matrix of psychiatry.* New York: Norton, 1968.

Schroder, H., M. Driver, and S. Streufert. *Human information processing: Individuals and groups functioning in complex social situations.* New York: Holt, Rinehart & Winston, 1967.

———. Intrapersonal organization. In B. D. Ruben and J. Y Kim (eds.), *General systems theory and human communication.* Rochelle Park, N.J.: Hayden, 1975.

Schuetz, A. The stranger. *American Journal of Sociology,* 1944, 49, 499-507. Reprinted in M. Stein and A. Vidich (eds.), *Identity and anxiety.* Glencoe, Ill.: Free Press, 1963.

Taft, R. Coping with unfamiliar cultures. In N. Warren (ed.), *Studies in cross-cultural psychology, Vol. 1.* New York: Academic Press, 1977.

Thayer, L. *Communication and communication systems.* Homewood, Ill.: Irwin, 1968.

Zurcher, L. *The mutable self: A self concept for social change.* Beverly Hills, Calif.: Sage, 1977.

PART IV
SOCIAL AND POLITICAL PROCESSES

17

Information and Political Behavior

Dan Nimmo

The role played by information in influencing political behavior has long been a concern of political theory, particularly democratic political theory. However, empirical research has demonstrated that the requisites of ideal citizenry posited by democratic theory—popular interest, discussion, motivation, knowledge, principle, and rationality—are not broadly distributed among mass electorates. Discovery of popular apathy and ignorance has led to a search for their sources and to a reassessment of the information requisites of democracy. In the process more systematic attention has been given to uncovering the nature of the information polity, how political information is diffused, and the character of information networks. More recent research interests have shifted from levels of popular information to the examination of how citizens process information/misinformation, both with respect to voting and to other forms of political behavior. Remaining open is a question of whether developing communication technology will enhance citizens' information processes or produce an information overload with the consequent incapacity of citizens to participate in meaningful political behavior.

Politicians posture about accountability to an informed electorate. Journalists justify their craft in the name of serving the people's right to know. Political scientists measure levels of public knowledge on controversial issues, then ponder the consequences of rising levels of public apathy and political ignorance. Philosophers debate whether an informed citizenry is a requisite of democratic politics. Civics teachers lecture their students on the nuts-and-bolts of governing systems. And lay persons, in a common sense way, take for granted the importance of being informed—although not always practicing the virtues of political education. In short, it seems

343

that everyone assumes a link between information and politics. But how close is that tie?

It is the purpose of this review to examine the role information plays in political behavior. Obviously the expanse of that role is almost boundless. It extends, for example, to such areas as the character of political information; its sources, generation, control, and flow; the processing of information, for both governmental and individual political decision making; concepts of political systems based upon cybernetics and/or information theory; the relationship between evolving technologies of information generation and delivery and developing governing arrangements; vested interests in the management of political information; futurist conceptions of the information polity; and many, many more.

Given such a broad expanse, it is beyond the capacity of this review to be comprehensive. It focuses instead upon selected areas, namely, the role generally assigned information in democratic politics and approximations of whether that role is fulfilled, analyses of the information polity, the processing of political information at the citizen's level, and the problems of information overload in a democratic society. Although the linkages between information and political behavior traced in this review may apply in a variety of political systems, the assessment here is confined to democratic politics and primarily based upon studies derived from research in the United States.

The Democratic Ideal of an Informed Citizenry

Although philosophers have debated for centuries the nature of democratic government and which members of the populace should and should not have their opinions counted in making decisions, there has been a general understanding that those citizens who do take part should be informed and enlightened. Although not always explicit in the writings of all democratic theorists, there does exist in democratic theory at least the ideal of what Downs (1957) has labeled "rational action." Drawing upon economic theory, Downs spoke for a key tradition in democratic thinking when he characterized the essentials of rational behavior. The rational person (1) "can always make a decision when confronted with a range of alternatives," (2) "ranks all the alternatives . . . in order of preference . . . in such a way that each is either preferred to, indifferent to, or inferior to each other," (3) constructs a transitive preference ranking, (4) always chooses from among the possible alternatives that rank highest in preference ordering, and (5) "always makes the same decision each time" when "confronted with the same alternatives" (p. 6).

Downs recognizes that such an economic notion of rational behavior may be more an ideal than practiced act in political reality, yet he is speaking for a major tradition in classical democratic theory that says rational behavior consists of calculating appropriate means to achieve desired ends. Such a conception establishes stringent requirements upon those seeking to qualify as rational political actors.

In what has become a seminal statement on the matter, Berelson, Lazarsfeld, and McPhee (1954) spelled out the requirements of democracy's textbook citizen prior to Down's idealized depiction. First, the democratic citizen was expected to be interested in and to participate in politics. This should extend beyond voting to political discussion. Second, Berelson and his colleagues set a requirement of knowledge: "The democratic citizen is expected to be well informed about political affairs" (p. 308). This, they said, includes knowing what the issues are, their history, relevant facts pertaining thereto, and likely consequences—a list very much in keeping with the ideal of rational behavior. Third, the democratic citizen is principled, supposed to cast a vote "not fortuitously or impulsively or habitually, but with relevance to standards not only of his own interest but of the common good as well" (p. 310). Finally, Berelson, Lazarsfeld, and McPhee spoke directly to the requirement of rationality noting that democracy's "average" citizen, in theory, should arrive at principles by reason and rationally calculate implications and consequences of policy alternatives.

Berelson and his colleagues inventoried these requirements of the democratic citizen—i.e., interest, discussion, motivation, knowledge, principle, and rationality—in a concluding chapter of their classic study, *Voting*, an examination of how voters behaved in Elmira, New York, in the 1948 presidential election. It was because they found that their subjects fell short of the ideals of democratic citizenry that they addressed the discrepancy between the textbook and practicing democratic citizen. The latter they found to be voting without substantial involvement in politics, uninformed about the details of political campaigns, responding to traditional social allegiance rather than principle, and far less likely to be rational in voting than in the purchase of a car or home. This study represented the first systematic effort to contrast the ideal link between information and political theory posited by democratic theory with the relationship found in practice. The result was threefold: (1) Later studies continued to document relatively low levels of political information among the citizenry, (2) there was a search for reasons why information levels were so low, and (3) there were revisions in the requirements posed by democratic theory itself.

Levels of Public Information

As political researchers explored the role played by information in political behavior it became apparent that, as conventionally defined, levels of political knowledge were so low in the citizenry as to question whether there was a linkage except among relatively few members of the populace. Consider information about current events, for example. Pollster Louis Harris (1974) reported typical findings: fewer than 40 percent of Americans considered themselves up-to-date on public affairs; less than 60 percent could name one U.S. Senator from their state, less than 40 percent both; less than one-half could name their congressman; less than one-half could name a member of the U.S. Supreme Court. Levels of information about textbook matters appeared no higher: Harris found that 38 percent of respondents did not know that Congress consists of both the House of Representatives and the Senate; 20 percent thought the Supreme Court part of Congress; and 8 percent had no idea of what Congress consists.

With increases in the potential exposure of citizens to public affairs information via news media and higher proportions of citizens moving through public schools, there is reason to speculate that levels of public knowledge might be on the rise. Recent studies, conducted among young Americans, do not so indicate. In 1979, for example, George Gallup Sr. reported on the results of a nationwide "citizenship test" conducted under the sponsorship of the National Municipal League (Pierce 1979). Again the results indicated that whether the information is of direct relevance—for example, about a specific public issue—or contextual—historical, institutional, or background information—vast areas of political ignorance persist. For instance, a scant 4 percent of 17- and 18-year olds could name the three men to serve as president prior to Gerald Ford; 38 percent knew that voters can split party choices between president and other offices when voting; only 29 percent knew that national party conventions make the final choice of presidential nominees; and one-third did not know which party had control of Congress.

What nationwide surveys such as those conducted by pollsters Harris and Gallup document has been echoed in more intensive soundings. Stein (1983) reported results of his focus group studies of information levels among college students. Repeatedly he found that his subjects did not know when World War II was fought, which nations were U.S. allies and enemies, the years of World War I or the Civil War, the decade of the Eisenhower presidency, the number of U.S. Senators per state, or identities of their representatives in Congress. The identity of Thomas Jefferson, the date of the Declaration of Independence, and the content of the Bill of Rights were equally obscure. And, as an extreme example, reported Stein, none had ever heard of Vladimir Ilyich Lenin. ("Was he the drummer with the Beatles before Ringo Starr?") Stein's conclusion does not speak with optimism about the future if, indeed, it is a requisite of a democratic society that there be a close link between an informed citizenry and political survival: "The kids I saw (and there may be lots of others who are different) are not mentally prepared to continue the society because they basically do not understand the society enough to value it" (p. 19).

Sources of Popular Ignorance

Prompted in part by puzzlement over the knowledge gap, that is, the contrast between the requirements and realities of democratic citizenry, but more by a desire to justify its performance (Witt 1983), the television industry in 1959 commissioned the Roper Organization to survey public perceptions of TV. Among the questions posed to a cross-section of Americans was one asking where respondents acquired their political information—from newspapers, radio, television, or news magazines. Newpapers led as a source (59 percent), TV was second (51 percent); radio and news magazines trailed far behind. The surveys became an annual affair. But since 1963, television has been the preferred news choice. Thus, in the 1982 survey 65 percent of respondents selected TV as the source of their news, 44 percent named newspapers (Witt 1983).

Until relatively recently television news has been restricted in the amount of time available for coverage of public affairs. It was, therefore, but a small inferential leap to the conclusion that, if people relied upon the medium for information and little information was actually carried in the "headline" service, television was the principal source of Americans' low levels of political information. But, as other studies have pointed out, fingering television news as the culprit is misleading. For one thing, there have been challenges to the conventional wisdom that TV is the principal information source of the American citizenry. Lawrence Lichty (1982), for example, reports that 68 percent of repondents in annual surveys conducted by the Simmons Market Research Bureau report reading a newspaper each day, less than a third watch local TV news, and less than a third watch national television news (i.e., 56 percent watch TV news of some type). As Witt (1983, 48) concludes in his comprehensive review of where most Americans receive their information about public affairs, they "get their news from a variety of sources" and "many regularly rely on both television and newspapers."

The relatively low levels of information about politics regularly reported in nationwide surveys, thus, cannot be attributed simply to the fact that Americans rely upon any particular news source. Perhaps, therefore, it is from the nature of news itself that popular ignorance derives. Sociologist Robert Park (1940), himself once a journalist, suggested as much more than four decades ago. News, he wrote, does not give people "knowledge about" what is going on, only "acquaintance with" things. Acquaintance involves simple, superficial familiarity with persons, places, ideas, or objects; knowledge implies detailed factual and contextual information and a grasp of origins, implications, and consequences. One may be no more knowledgeable about the presidency or congressional committees, although acquainted with their existence, than about one's hairdresser, busdriver, or check-out clerk at the supermarket. Although news—be it from whatever source—is admirably suited to acquaint people with what might be going on in the political world, it is not a vehicle of informed political knowledge.

Park's observations are well taken and have been made in other forms before and since. Lippmann's (1922) distinction between "the world outside" and "pictures in our heads" made much the same point, i.e., stereotypes may be accurate or inaccurate, but the news that feeds them is not truth, just acquaintance with an "ocean" of possible truth. Other writers have likened political news less to verifiable information than to fantasy (Bormann 1972, 1982; Nimmo and Combs 1983), real-fictions (Fisher 1970, 1980), and the "politics of illusion" (Bennett 1983).

Robinson and Levy (1983) have given empirical credence to such views in a study designed to discover precisely what information people derive from prominent stories carried on the wire services, television news, and in newspapers. Drawing upon representative samples of 526 respondents in Washington, D.C., and 544 respondents across the United States, they probed levels of acquaintance with and information about news events. They found that "the public often picks up garbled messages and images, and that news about similar events, officials, or terms may run all together in kind of a 'melt down' in the public's mind" (p. 39). Public acquaintance with events varied considerably; detailed information in all areas was rare. Higher proportions of respondents were acquainted with foreign affairs than with domestic. For example, about two-thirds had "heard or read" about national elections in Great Britain but only two-fifths had "heard or read" of a recent Supreme Court decision striking down legislative vetoes and "half of them did not know that ruling affected the balance of power between the legislative and judicial branches of government" (p. 40).

A key finding of the Robinson and Levy study suggests another possible source of popular ignorance or misinformation about politics. Although only 28 percent of respondents had heard or read about U.S. policy in Central America, almost everyone polled knew of a recent space shuttle flight carrying the first American woman astronaut. Observed Robinson and Levy, "In short, our survey showed how quickly Americans pick up on stories with a 'human interest' angle, but how difficult it is for the press to convey successfully the gist of complicated, technical and important stories" (p. 38).

What may be at issue is a phenomenon Converse (1964) unearthed in his seminal investigations into the nature of mass belief systems in the American electorate. Relying upon open-end responses to questions asked in representative nationwide surveys in the 1956 and 1960 presidential elections, Converse classified respondents in accordance with the sophistication of their knowledge of the concepts of *liberal* and *conservative*, and their ability to match Democratic and Republican parties correctly to such labels. Converse's five categories reflect an implicit "acquaintance with" and "knowledge about" distinction: (1) Ideologue—recognition of the labels, broad understanding of the terms, and proper matching to parties; (2) Near Ideologue—recognition and proper matching but a narrow understanding of the terms; (3) Group Interest—recognition of the terms but errors in matching; (4) Nature of the Times—recognition, attempt at matching, but no understanding of terms; (5) No Issue Content—no recognition of terms.

Converse found that less than one-fifth of respondents were ideologues (individuals one could say possessed both acquaintance and knowledge), more than one-third oriented themselves to the nature of the time or no issues at all (neither acquaintance nor knowledge), and the remainder were near ideologues or concerned with group interests (individuals with acquaintance, but minimum knowledge).

Although Converse's research spawned a host of critical and confirming studies, the interest here is not in his specific findings but in a more general conclusion. Noting that the distribution of sophisticated belief systems among the mass public was limited, Converse went on to say, however, that there might yet be "folk ideologies" among the populace. One could presume that these folk ideologies contain an ample supply of stereotypes, fantasies, real-fictions, illusions, and misinformation. In any case, they perhaps direct popular attention toward information with precisely the "human interest angle" that Robinson and Levy found so apparent.

Michael McGee (1985) makes this point in his far-ranging critique of research in the general areas of political communication and information. McGee argues that researchers fail to honor "popular culture" (i.e., to take seriously the folk ideologies alluded to by Converse). If they were to do so they might, writes McGee, "come to significant conclusions about the nature of political communication"—hence, political information—in Anglo-America. Three of McGee's postulates are relevant here: (1) From the standpoint of acquiring political information people regard personality (that is a "human interest angle") as more important than technical competence in reasoned discourse (communicating "knowledge about"); (2) people accept that leaders must "justify" policies—use selective information, perhaps misinformation—as evidence that officials considered all options before acting; and, (3) ideological values are, and should remain, so equivocal that their meaning must be negotiated in every critical circumstance of usage, hence, ideological sophistication is not only unlikely but inappropriate.

Reassessing the Information Requisites of Democracy

Implicit in McGee's critique is the understanding that philosophers and researchers may have missed the point in stressing that the possession of factual and contextual information is a requirement of democratic citizenship. Perhaps a more general, ambiguous, even equivocal "handle" on what is going on in the world is sufficient. Traditionally in American politics one such handle has been provided by the symbolic presence of the two major political parties. Although the point had been made earlier, Campbell, Converse, Miller, and Stokes (1960) underscored it in their influential study of voting behavior in presidential elections. They found that there is a strong affective tie of voters to one or the other of the two major political parties. This tie—party identification—consists of which party the voter considers himself or herself close to emotionally. Depending upon the strength of that identification, partisanship provides a terministic screen through which the voter filters information and evaluations of candidates, issues, and parties.

Voting studies pinpointed a specific relationship between the flow of political information and voting behavior, a relationship built upon the strength of partisan ties (Campbell et al. 1966). Research indicated that voters most intensely identified with a political party also tended to be those most interested and involved in elections. Consequently, they were the most likely to be exposed to information about a political campaign. Yet, because of their close partisan tie, they were voters likely to have their partisan choice shaped by the information acquired—either because they gathered only information supportive of their partisan choice or because they were unmoved by contradictory information. Partisans of weaker intensity, or those independent of party identification, were determined by researchers to be those most likely to succumb to the influence of political information in reaching voting decisions. Yet, these were also voters least likely to be involved and interested in the campaign, hence least likely to be exposed to precisely that political information which might influence their voting behavior.

This posited relationship between information and voting behavior has been questioned by later researchers. Dreyer (1972), for instance, noted that the growing availability of the mass media, including television, and its use by candidates for campaigning penetrates all segments of the electorate—strong and weak partisans, and independents—both during and be-tween elections. Hence, the relationship between strength of partisan ties and information exposure could no longer be taken for granted. Moreover, widespread distribution of partisan loyalties may well be a thing of the past (O'Keefe and Atwood 1981). And, DeVries and Tarrance (1972) argued that a new form of independent voter, one neither bound by party loyalty nor apathetic to politics, had been born. This nonaligned voter was more likely to be exposed to and influenced by political information garnered from both conflicting and nonpartisan sources and to sift it through terministic screens that have nothing to do with political parties.

Such observations suggest that as partisanship declines voters will have no overriding anchor for electoral choices, hence, will seek out and acquire information in order to reach decisions. However, Downs's (1957) early but still insightful economic analysis of voting in democratic politics raises doubts about this and, in the process, continues to suggest why a demo-cratic citizenry is unlikely to be politically informed. Downs argues that a rational voter votes for the political party that one believes will provide the voter with a higher utility income than any other party. This is the voter's party differential. Utilizing "free" information (that readily available so that the time and energy in acquiring it is minimal), the voter estimates a party differential, the cost of voting, and the value of voting at all. Having made that estimate the voter reaches a decision based upon whether chang-ing the party in power will improve one's lot, hurt it, or make no difference. Once having made that decision it will take a considerable body of infor-mation to produce a likely reversal of it.

Acquiring new information, whether free or non–free, does take time and effort. In sum, all information has a price. Hence, as costs of being informed exceed expected payoffs, the voter is less likely to seek out infor-mation. Those indifferent to which party wins the election (for they see no difference in payoffs to them) have no motivation to pay the costs of being informed, hence they refrain from becoming so. Thus, an economic analy-sis of voting concludes that information has a role to play in political behavior only to the point that the benefits of becoming informed outweigh the costs.

Conceived in terms of the costs it takes to accumulate it, it is not surpris-ing that, as Downs notes, some members of society are more capable of paying those costs than others. Those that do, he argues, are also those more likely to be taken into account by elected governments, for they are likely to bear the costs of being sufficiently informed to be influential in at least selected policy areas. For most people, however, Downs concludes that "it is irrational to be politically well-informed because the low returns from data simply do not justify their cost in time and other scarce resources." Here, then, is an explanation as to why detailed information plays such a minor role in the political behavior of so many, why "many voters do not bother to discover their true views before voting, and most citizens are not well enough informed to influence directly the formulation of those policies that affect them" (p. 259).

Analyzing the Information Polity

To say that no homogeneous set of "ideal," politically informed democratic citizens has been uncovered by research is not to say that the politically informed do not exist. Nor can one conclude that information has no role in politics. Indeed, Berelson and his co-workers (1954)—accepting that there was no homogeneous ideal citizenry—restated the requirements of democracy in terms of realities that might apply to the entire political system rather than to individual citizens. This restatement reflected a shift of emphasis away from the informed citizen to the information polity.

The basic thrust of the restatement was that democracy demands not uniformly high levels of interest, information, principle, and rationality among citizens but a "distribution of qualities along important dimensions" (Berelson, Lazarsfeld, and McPhee 1954, 315). For example, since highly involved voters know more about campaigns and participate more, but are also more partisan, they are less likely to change their minds. This, Berelson and his colleagues maintained could easily culminate in "rigid fanaticism" if carried to an extreme (p. 315). Persons with low affect toward elections—citizens who do not care very much—provide a requisite balance between strong commitment, on the one hand, and maneuverability, compromise, and controlled change on the other.

Similarly, Berelson and his colleagues speculated that low levels of political involvement and knowledge among portions of the citizenry might also make for a useful balance between stability and flexibility, progress and conservation, consensus and cleavage for the polity as a whole. Thus, voters "least admirable when measured against individual requirements" of democracy lend a "curious" flexibility to what might otherwise be a fixed, static electorate if made up solely of strongly involved, committed, and partisan voters (p. 317).

A campaign reaffirms partisans in their convictions, yet is a creative force for the unsettled and uncommitted. Whereas the campaign provides an opportunity for partisans to recommit themselves to their heritage and conserve the past, the "campaign citizens" (what Berelson and colleagues liken to "Sunday churchgoers") constitute a potential group that can be rallied in the name of progressive change. The strongly involved, well-informed, partisan voter is also likely to be the bloc voter, the voter that helps align the electorate into sharp cleavages. The less involved, relatively uninformed, and nonpartisan voter who shifts political loyalties is open to the kind of compromise of principle so often vital to producing social and political consensus.

This general restatement of the requirements of democracy which emphasized the positive side of low levels of political involvement and information among the masses, high levels among elites, served as conventional wisdom for at least two decades. Democracy, it was said, did not require an informed, knowledgeable citizenry, but instead an "attentive public" (Almond 1960), "carriers of the creed" (Key 1961), and mass acceptance of the general principles of a liberal political culture rather than information about specifics (Prothro and Grigg 1960; Devine 1972). Convention, however, did not go unchallenged. Many political observers—especially during the political ferment of the 1960s—argued that singing the praises of an uninformed citizenry, claiming that ignorance and apathy were indicators of stability, was both incorrect and little more than a rationalized acquiescence to elite rule (e.g., Bachrach and Baratz 1962; Walker 1966).

This is not the place to review the arguments supporting and refuting alternative theories of democracy. Rather, the point is that social scientists were moved by empirical evidence generated by studies of voting to recast classical assumptions about the ideal democratic citizenry, with an attendant shift in focus from concern about the role information plays in orienting individuals to politics, to the role information plays in the polity as a whole. That emphasis, however, gradually moved away from debates over normative democratic requirements. Descriptive and analytical accounts of information transmission throughout the polity and information-processing by individuals began to occupy communication and political scientists as topics of theoretical and empirical inquiry.

Information Diffusion in the Polity

As Savage (1981) notes, an interest in the process of information diffusion has guided inquiry in various areas of the physical, biological, and social sciences. It found its way into the communication sciences principally through the works of Everett Rogers (1962; see also Rogers and Shoemaker 1971). The character of the diffusion model has been reviewed so frequently (Rogers 1962; Chaffee 1975; Savage 1981) that a summary statement will suffice to relate it to the linkage of information and political behavior. Diffusion consists of the process whereby an idea, innovation, or other message once transmitted through the mass media penetrates an audience and receives acceptance, rejection, or indifference as a result of intervening factors, stages, and filters in that process. Although on the surface the process appears linear, mechanistic, almost deterministic (Krause and Davis 1976) in that messages flow from sources through intervening agents, then penetrate audiences, research indicates that diffusion involves a host of complex, reciprocal transactions between sources, media, and receivers.

Chaffee (1975) has hinted at the relative complexity/simplicity of the diffusion of political information as it relates to political behavior. In so doing he poses a provocative insight into the nature of the modern information polity. Diffusion studies, he notes, have uncovered a recurring pattern in the adoption of ideas, innovations, and messages. In the early stages of diffusion, relatively few people hear or know of an idea—say, for example, a presidential political candidate such as George McGovern in 1972 or Jimmy Carter in 1976. As transmission continues, or as the campaign unfolds, diffusion accelerates and larger numbers of people learn about and, perhaps, adopt the idea (for example, favor the former unknown for a party's presidential nomination). The process, however, decelerates and relatively fewer people learn of or adopt the idea as the campaign progresses, for instance. If plotted on a graph so that the horizontal axis consists of time and the vertical of the number of persons learning of the innovation, the pattern takes the form of an S-curve.

Chaffee stresses that such a curve is not inevitable. Rather, when it shows up it reveals evidence of an absence of constraints, of a randomness of interaction, in the process. Political institutions—for example, political parties in a campaign—are types of constraints that can work to modify the S-curve pattern and, in effect, organize the flow of political information. The mass media, hypothesizes Chaffee, may override the influence of such constraining political institutions, enhancing the likelihood of a random pattern of diffusion—an S-pattern.

What is intriguing about Chaffee's hypothesis is that it says, in effect, that as a polity relies more on the mass media and less on political institutions as principal diffusion agents the more likely an S-curve in the diffusion and distribution of political information. This comes close to being precisely the curve that Berelson and his colleagues (1954) sketched as being typically required of a democratic polity. "Happily for the system," they noted (p. 322), democratic voters distribute themselves along a continuum. At one end is "Ideological Man" consisting of the relatively few who are absorbed in public affairs, highly partisan, highly informed, and so on. They can be likened to the relatively few who acquire information about an innovation in the early stages of diffusion. At the other end of the continuum is "Sociable Man" (indifferent to public affairs, nonpartisan, uninformed). This category is akin to those who never learn about an innovation and are at the decelerating end of diffusion's S-curve. In the

middle is "Political Man," the bulge in the S-curve, that set of heterogeneous citizens of contrasting involvements and information who learn
about politics, and may or may not act on the basis of such information.

Research employing diffusion models of political information has not
explored the implications of Chaffee's hypothesis and has thus not probed
the degree that the distribution of political information and knowl-
edgeability is a function of institutional and/or random factors. There
have, however, been two principal areas where the link between information and political behavior has been made via diffusion research. One
concerns the diffusion of policy innovation among the American states. To
cite but one example, Walker (1969) relied on the years when U.S. state
legislatures had adopted a vast array of policies to compute an index of
policy innovativeness. He correlated index scores with social, economic,
and political factors that might explain differences in adoption rates.
Walker found that such differences are partially attributable to variations
in the size, wealth, and urbanization of states and to regional blocs. He
hypothesized decreasing time lags between states due largely to the active
role being played by the national government and to national organizations at the period of his study—thus, perhaps, hinting at a decreasing S-
curve pattern flowing from the constraints supplied by national
institutions.

A second albeit much less researched area of the diffusion of political
information involves voting behavior. The basic diffusion model posits that
certain factors associated with an innovation partially explain its rate of
acceptance among adopters. If an innovation is better than the idea it
supercedes, it has relative *advantage.* If it is consistent with the prevailing
values, past experiences, and needs of potential adopters, it possesses *compatibility.* If it is difficult to understand, it is *complex*—(the assumption
being that the more simple the new idea, the less adopters must invest in
learning it, hence, the greater the likelihood of adoption). And, if it is easy
for people to see the results—as with the cash-value of adopting a new
idea—the innovation has *observability.* Zukin (1976) operationalized these
four characteristics—relative advantage, compatibility, complexity, and
observability—using political indicators to examine voting patterns in the
1972 U.S. presidential campaign. His diffusion of information model explained almost 60 percent of the variance in voting behavior, a level equal
or superior to alternatives to the diffusion of information model.

A Polity of Information Networks

It is but a small step from describing the polity as a system of information diffusion to that of considering it as a set of overlapping networks of
political information. As Rogers and Cavalcanti (1980) assert, individuals

are members of a wide variety of social groups—families, peer groups, religious groups, work groups, and so on. Although persons in such groups may have equal opportunities to communicate with one another, they do not. They regularly give preference to communicating with some individuals more than others. Cliques form; some clique members who communicate within them join other cliques as well (members called "bridges"); others do not join other cliques but do communicate with members of other cliques (such people serve as "liaisons" between cliques). Clique members communicate in variable ways within their own groups and across groups as a function of personal intimacy, subject matter, etc. All such concepts constitute the basic terms of network analysis. To analyze the polity via such an approach would be to discover the variety of structures of political communication within it and their linkages.

Although early studies of voting behavior (Lazarsfeld, Berelson, and Gaudet, 1948; Berelson, Lazarsfeld, and McPhee 1954) did not have networks of political information as a primary focus, they did uncover the important role such networks play in mediating between channels of mass communication and voters' interest, knowledge, and choices. Later generations of voting studies (see Nimmo 1978) have addressed the role of social groups in voting behavior in a more oblique fashion, preferring instead to partial out the relative influence of partisan loyalties, candidate images, issue positions, and the mass media. Network analysis, as such, has not played as prominent a part in charting the relationships between information and political behavior.

Network analysis, however, does offer a promising avenue for exploring such relationships (Rogers and Kincaid 1981). To cite but one case in point, in 1973 Granovetter argued that the strength of ties between people influences the kinds of networks that link them. By strength, Granovetter referred to the amount of time, emotional intensity, mutual confidences (intimacy), and reciprocity between people. Within groups the ties between members tend to be strong—for example, group members talk to one another directly. Between people of differing groups, however, ties are weaker. For instance, group members represent their groups in relating to one another, talking on behalf of others rather than all group members speaking with all members of another group or groups. Thus, there are bridges between groups rather than strong ties. Granovetter suggests how this might relate to politics and to political information. Citizens do not deal with political leaders directly but through representatives who do so on their behalf. That is, ties to political leaders are weak. It is, therefore, on the basis of weak ties that citizens are informed or not, trust or mistrust their leaders, and so on. An obvious question, similar to that raised by Chaffee's account of the diffusion of political information, is what happens when the mass media rather than institutional leaders (party officials, interest group leaders) constitute the weak ties between citizenry and government. Consequences for the role information plays in political behavior flow from such varying networks, but as yet students of political communication have not clarified their precise nature.

The Polity as Informed Steering

In considering system–wide approaches to the link between information and political behavior—rather than solely those that focus upon variable levels of information among citizens—one other approach demands consideration. Political scientist Karl Deutsch (1963) has gone far beyond an interest in individual differences, systemic information requirements, information diffusion, or information networks. Deriving his posture from cybernetic theories of communication and control, Deutsch analyzes governments as "learning nets," by which he means self-modifying communication networks. Networks process information, information essential for a polity to make continuous adaptation to its environment. *Information* in such a model has a more specific meaning than that employed heretofore. Information consists of a patterned relationship between events. That pattern teaches a mechanism or organism (in politics, a government) about itself in relationship to its environment. Moreover, on the basis of information governments "learn," that is, they modify their behavior through and in response to communication.

Deutsch's analysis is far too rich and replete with implied testable hypotheses to review here. Concepts such as complementarity, memory and pattern recognition, feedback, goal, and purpose are all brought to bear upon problems of political self-awareness, autonomy, self-closure, and growth. Deutsch's analogue, in sum, is government as a servomechanism—i.e., a communication network that "produces action in response to an input of information, and includes the results of its own action in the new information by which it modifies its subsequent behavior" (p. 88).

Politics is, thus, an information-processing activity, not only processing information directly but processing information about information which, inevitably, modifies the nature of information itself. This possibility, pursued by Deutsch and others on the macro level (see especially Galnoor 1982) is a viable perspective from which to return to the link between information and political behavior at the micro level as well.

The Citizen as Information Processor

Viewed from the evaluative perspective provided by democratic requirements for individual citizenship, any observer is forced to measure (or at least ask about) the quantity and quality of political information possessed by citizens. Considered, instead, from the viewpoint of the polity as an overall system, questions of quantity and quality are also key, whether those questions be of the distribution of political information among elites and masses, the diffusion of political information, networks of political information, or the intelligence available to the polity as a steering, information servomechanism. However, taking note of the citizen as an information processor changes the emphasis. The amounts and kinds of information possessed by citizens, although still relevant matters, are of lesser concern than the question of what it is that people do with whatever quantities and types of political information or misinformation they may have—how they process it. Models of information processing vary along several dimensions. An important one for politics is how simple or complex is the task of processing political information.

The (Not So) Simple Act of Voting

The most straightforward of political information models have derived from efforts to explain voting behavior. For example, as noted previously, Downs (1957) in his economic model of rational voting posited that an individual compares utilities of retaining or changing the party in power. If the citizen establishes a preference between parties, that is the key factor in voting choice. But, if no distinctions can be made, the citizen abstains. Downs went on to add qualifying factors to the model, but, nonetheless, the overall picture of information processing is devoid of major complexities.

In a roughly similar fashion, Kelley and Mirer (1974) developed an account of voting choice, one that has become standard in much of recent literature (see Conway and Wyckoff 1980; Hinckley 1981; Wattier 1983). Kelley and Mirer derived their voting rule, which is also an information-processing rule, from concepts derived from the principal voting studies of Campbell and associates (1960; 1966). For Kelley and Mirer the "simple act of voting" occurs as follows:

> The voter canvasses his likes and dislikes of the leading candidates and major parties involved in an election. Weighing each like and dislike equally, he votes for the candidate toward which he has the greatest net number of favorable attitudes, if there is such a candidate. If no candidate has such an advantage, the voter votes consistently with his party affiliation, if he has one. If his attitudes do not incline him toward one candidate more than toward another, and if he does not identify with one of the major parties, the voter reaches a null decision (p. 574).

There are parallels in the Downs and Kelley-Mirer formulations, as is apparent. Downs, however, is interested in rational calculations based upon voters' self-interests. Although this does not escape Kelley and Mirer, there is another consideration implicit in their information-processing rule. Whereas Downs conception of voters' party differentials is based upon their estimates of utilities—"a measure of benefits in a citizen's mind which he uses to decide among alternative courses of action" (1957, 36)— Kelley and Mirer, since they draw upon the tradition of empirical voting studies, have something else in mind. Partisan differentials in the Kelley-Mirer rule—although they do not employ the term—have affective, non-rational derivations. They stem from citizens' long-term party identifications, the affective and emotion-laden ties citizens have for one political party or another. Moreover, "likes and dislikes" of candidates constitute short-term factors specific to the election (Campbell et al. 1966) that are frequently emotionally based (Nimmo and Savage 1976).

As with the underlying assumptions from which the voting studies derive, the Kelley-Mirer rule, thus, relies in part upon ideas borrowed from various attitude-change theories (Kirkpatrick 1975) including balance models (Heider 1946), congruity models (Osgood 1960), cognitive dissonance (Festinger 1957), and discrepancy models (Sherif and Sherif 1965). Those theories have in common the view that humans strive for consistency rather than inconsistency in processing information.

The assumption of these theories is that one seeks to reach an equilibrium between one's views of the world and new information challenging those views. Depending upon the attitude-change theory employed, individuals experiencing attitudinal conflict reduce the perceived discomfort by changing values (the balance model), bringing values and beliefs into line with one another (the congruity model), changing beliefs (the dissonance model), or accepting/rejecting new information (the discrepancy model).

The "simple act of voting" is itself a striving for consistency, through calculating net numbers of favorable attitudes toward candidates, behaving consistently with partisan loyalties—ideological identifications in Wattier's (1983) application of the rule to primary elections—or a combination thereof. But, as Donohew and Tipton (1973) stress, a review of a host of attitude-change studies questions the facile notion that consistency is a normal state for individuals that, when interrupted, they labor to reduce. There is strong evidence that people are variety-seeking as well as consistency-seeking. Whereas people in accordance with the tenets of consistency theory are activated by an effort to reduce inconsistency, in variety theory people seek to maintain a level of activation to which they are accustomed, hence covet variety in order to keep from falling below that level. In such a view, variety, in the form of information, is a prized commodity regardless of whether it is consistent with prior beliefs and/or values.

It is beyond the intent of this review to resolve, if it were even possible, the conflict between proponents of consistency- and variety-seeking assumptions. Rather, the point here is to suggest that to the degree that information-seeking flows from other than the urge to be consistent, information-processing models must take on an accordant degree of complexity. Such may especially be the case when considering the processing of political information. For to most people political information itself may appear to be inconsistent. The common-sense view that politicians seem to say one thing today, another tomorrow, that political statements and acts are not consonant, cannot be summarily dismissed. As Edelman (1980) points out, political rituals provide a mystifying illusion of continuity that permits people to make sense of the abundant variety of discordant political information that flows through networks of political communication.

Processing a variety of sometimes parallel, sometimes consistent, and often overlapping but contradictory versions of reality is, as Watzlawick (1976) has stressed, no easy task. It extends far beyond the classic democratic requirements of being interested, informed, principled, and rational. Indeed, faced with a flow of perceived political inconsistencies and contradictions, guidelines of what constitute interest, information, principle, and reason change, blur, even vanish. The simple act of processing political information, as recent investigations demonstrate, turns out to be unique and complex.

Information Processing and Politics

Current scholarship emphasizes the multi-stage, transactional nature of information processing (Bennett 1981; Kirkpatrick 1975; Ruben 1984). Uppermost are questions about how persons sort out and select particular bits from the ongoing flow of information, how they interpret and code what they select, how they remember or forget, how they organize what they retain, how they learn, and how they apply and/or transmit information. It is generally accepted that these activities are not linear, but mutually reciprocal, and that information processed influences future processing. Moreover, the very acts of processing influence information previously selected, interpreted, and retained.

Students of information processing have not confined themselves just to indentifying activities comprising its stages. Factors influencing processing have also been explored. As Ruben (1984) notes, these include those associated with the individual (needs; attitudes, beliefs, and values; goals; capabilities; uses and gratifications; style; experience and habit); the type, nature, character, organization, and novelty of the information; characteristics associated with the information source; and environmental factors including context, repetition, consistency, and competition between sources and data sets.

It is apparent that the nature, stages, and influences on information processing in general can apply to the processing of political information as well. However, relatively few political scientists (Kirkpatrick 1975 and Bennett 1981 are examples of notable exceptions) have explored the overall character of political information processing at the individual level. Instead the focus has been upon aspects and phases of the total process. For example, conclusions from studies of voting behavior led to a conventional wisdom that exposure to and selection of political information obeyed a principle of selective perception, a postulate that has since come under criticism (Sears and Freedman 1967); and Edelman (1964) provided seminal instruction in how persons interpret and respond to political symbols. The nature of political learning has also been the basis of a plethora of studies in political socialization (see Renshon 1977 and Atkin 1981 for reviews). As might be expected these and other studies touching upon the selection, interpretation, and retention of political information have identified a variety of influences over each, most notably the role of partisan identification, social alignments, and belief systems on selective exposure, perception, and retention (Kirkpatrick 1975).

Since the study of the processing of political information has been conducted in essentially a piecemeal fashion, it is necessary to stand back as Bennett (1981) has done and ask what an information processing approach to understanding political behavior might include. Bennett focuses upon two key processes, perception and cognition. Although he recognizes that no absolute distinction between the two activities exists, he does build a framework for examining political behavior upon an analytical difference.

Bennett defines perception as including "the selection and reception of sensory inputs and the transmission of these inputs to various perception centers in the brain." Cognition includes "the transformation of neural signals into recognized codes that can become the basis for various symbolic operations" (p. 83). So regarded, perception involves what many information-processing models, be they macro-level (Deutsch 1963) or micro-level (Ruben 1984) in nature, label as in the realm of information seeking, or information selection. Cognition pertains to information interpretation, retention, and use. Of course, perception and cognition may be regarded as joint information-processing activities that underlie such political tasks as decision making, learning, the formation of belief-systems, and so on.

Whatever the political task, some degree of information is required to perform it. What that information is, and how much there is, depends in large measure upon information available to processors. This in turn is influenced by how information is controlled: "Perhaps the most universal and politically important distinction to be drawn about political information is the manner in which it is controlled" (Bennett 1981, 95). Control over the flow of information lies at the heart of the possession and exercise of political power (Lasswell and Kaplan 1950). Bennett argues that implicit in any discussion of information control is a distinction between how governments and/or individuals control information via "selection" (through secrecy, management, public relations, audience targeting, and so on) and "symbolic transformation," through symbols employed, propaganda and advertising devices, rhetorical techniques, and so on.

In Bennett's scheme, then, it is important to distinguish these two kinds of information-processing tasks, namely, information selection and symbolic transformation—processes obviously closely matching perception and interpretation. Equally important is a second distinction, whether the "information" process is newly introduced to a person in a situation (in the form of new facts, ideas, arguments, concepts, and so forth) or is brought to a situation by the individual (through concepts, identifications, loyalties, and other notions stored in memory). Drawing upon Lindsay and Norman (1977) Bennett labels these "data-driven" and "conceptually driven" processes, respectively. The former involves a person's response to immediate sensory data, the latter a response on the basis of expectations the individual uses to define the situation.

The data vs. conceptually driven distinction in processing modes appears frequently in accounts of how people deal with political information. Do, for example, voters perceive candidates on the basis of the images candidates project—that is, are voters data-driven—or on the basis of screening perceptions through party loyalties—a conceptually driven view? This distinction is at the heart of the "stimulus-determined" vs. "perceiver-determined" debate in the study of political imagery (McGrath and McGrath 1962; Sigel 1964). Or, does campaign information (data-driven) change votes by overriding partisan loyalties (conceptually driven)?

Bennett stresses that distinctions between data-driven and conceptually driven modes of information processing, and between information selection and symbolic transformation processing tasks, are not hard and fast. One mode does not operate in isolation and exclusive of the other; one task is not performed totally apart from the other. However, the distinctions serve him well in constructing a four-fold typology of political information processing that applies equally well to individuals, organizations, or politics. It reminds researchers of the complexity of political information processing—and, hence, how problematic and conditional the role that newly presented data may play in changing political behavior—and that in any single situation or across time there may be shifts in modes and tasks of processing.

Bennett cites political socialization as an example. Socialization studies take different approaches to political learning. Some employ classical conditioning models, others models of operant conditioning, still others cognitive development approaches, and there are studies of direct learning through civic training. Bennett's typology suggests that at points in the socialization process there may be data-driven information selection (an awareness of civic information being directed at the individual). Classical conditioning involves data-driven processing of symbolic transformation—the pairing of conditioned and unconditioned stimulus and the production of an invariant, learned response. But when conceptually-driven, information is akin to operant conditioning (a person imposes a conceptual scheme on reinforced behavior to learn correctly). And, when symbolic transformation is conceptually driven, learning follows a cognitive development pattern—generalizing, deducing new inferences, and seeking new information.

Bennett's framework for the analysis of the processing of political information is far more intricate and rich in its content and implications than can be captured in a capsule review. Anyone interested in the tasks and contexts involved in linking information to people's political behavior would do well to consult his lengthy discussion.

One clear point that his scheme should remind researchers of, for instance, is that the "simple act of voting" consists of an individual undertaking a variety of different information-selection and interpretation tasks across the conduct of a political campaign. At some points new information will inform those tasks—the voter's processing will be data-driven. At other points the voter's candidate orientation, issue positions, and party identification will come into play in acquiring and interpreting campaign messages—hence, conceptually driven. One should not be surprised, then, to see party identification enter and depart information processing during a campaign, becoming more or less useful to the voter depending upon the flow of information about issues, candidates, and events; the character of that information; and the utility of party identification in condensing parallel, contradictory, and supportive information in a meaningful way. Considered from the viewpoint of the processing of political information, then, the concept that has been the anchor of so many voting studies relates information to political behavior in ways considerably more complex than frequently hypothesized.

Information Processing or Information Overload: The Future of Information in Politics?

The tasks and contexts involved in processing political information do not go unchanged. If for no other reason than rapidly developing technologies of communication, the quality and quantity of political information potentially available to citizens increases—through expanded news programming by television networks, penetration of cable television to households (37 percent in 1983), the emergence of national daily newspapers, and so on. Nor can one ignore the growth of the number of personnel that make up the broadly defined information industry—political media consultants, pollsters, direct-mail specialists, and many others.

It is precisely this potential for an increase in the information flow that has led some political observers to prophesy an information overload. Sociologist Orrin Klapp (1982) is typical. Klapp argues that, in spite of a "flood of information" there is a "crisis of meaning" (p. 57). He develops that argument by, first, introducing Ogburn's (1922) concept of cultural lag. Ogburn believed that material culture changes much more rapidly than nonmaterial culture. Technology, for instance, develops at a more rapid pace than human beliefs, values, and habits. As a result the nonmaterial culture of ideas fails to keep up with the shifting world of material realities. Toffler's more recent (1970) work suggests much the same problem.

Klapp in his analysis substitutes *mere information* for *material culture, meaning* for *nonmaterial culture.* Klapp sees *mere information* as "bits" of data characterized by being additive, digital, analytical, accumulative, and easily measured in some fashion. *Meaning* involves the relation of something to a pattern or a scheme of knowing. One is reminded of a distinction introduced earlier, namely, Klapp treating *information* as Robert Park (1940) did *acquaintance with* while viewing *meaning* as Park did *knowledge about.* Klapp's position is that the information culture, the flow of information, is rapidly exceeding the meaning culture, i.e., the processing of mere information.

The reason for cultural lag in Ogburn's mind was that humans are biologically limited in their capacities for adjustment. Klapp proposes a different limitation, one related to the capacity of people to process information. As the flow of information increases into limited human processing channels, people impose less and less meaning on more and more information. The result is information overload leading to a meaning lag. Contributing to this state of affairs is noise; that is, information of no use to receivers as contrasted with useful information. Noise consists not only of nonsense but of "information getting in its own way" (p. 59). Since noisy information must be processed along with useful, it merely adds to the information overload such that the more noise the less meaning.

Klapp spells out several reasons why *meaning formation,* another term for information processing, lags behind the flow and accumulation of mere information. For one thing information processing is inherently slow. Making sense of things demands time-consuming pondering, contemplation, meditation, deliberation, even brooding. In sum, meaning formation, since it is information processing, consists of that complex series of tasks involved in information selection and symbolic transformation and of modes, both data-driven and conceptually driven, of which Bennett (1981) writes. Allied with the requirement of reflection, meaning formation demands discussion, debate, exchanges, and so on. Meaning formation, as symbolic interactionists argue (Blumer 1969), thus relies upon the formation of significant symbols, symbols that strike responsive chords.

Reflection and talk, however, have been replaced by lesser activities that demean the meaning of information, political information as well as other forms. As Downs (1957) argued, the rational voter unwilling to pay the costs of information seeking and selection can transfer those costs to others who share the voter's "selection principles" (p. 214). But the current tendency is to turn to celebrities (TV talk shows, pundits, alleged experts, and others) for "instant analysis." This is no substitute for careful deliberation in meaning formation as far as Klapp is concerned.

Ritual and ceremony serves as another substitute, but be they political inaugurations, conventions, debates, or dramas they are, for Klapp, so banal as to fail to bestow meaning. Vicarious experience via the persona of political candidates, television anchors, and packaged personalities replaces first-hand political participation.

Finally, whereas social networks—families, friends, and peers—once were forums of deliberation and discussion, that role has been taken over by the mass media, which moves from event to event, crisis to crisis, in rapid succession. Particularly noteworthy here is the demise of political party organization. No longer do parties serve as vehicles of political interpretation and education. Instead they grow sparse and diffuse, unable to compete with the noise generated by countless political action committees making specialized pleas on behalf of interests by flooding airwaves, telephone lines, and mailboxes with mere information.

Klapp's thesis, to the degree that it is plausible, poses a problem for all concerned with pinpointing the relationship between information and citizens' political behavior. Viewed from the perspective of the requirements of a democratic citizenry, information plays a minimal role in political behavior among the masses because citizens possess so little of it. Viewed from the perspective of requirements for information processing and meaning formation, information plays a minimal role in political behavior among the masses because there is too much of it. Given these possibilities, it may well be that the ultimate crisis in an era of "crisis politics" is not foreign entanglements, war, energy, big government, inflation, unemployment, or public confidence. It has been and remains an information crisis.

References

Almond, G. *The American people and foreign policy,* New York: Praeger, 1960.

Atkin, C.K. Communication and political socialization. In D. Nimmo and K.R. Sanders (eds.), *The handbook of political communication.* Beverly Hills, Calif.: Sage, 1981.

Bachrach, P. and M.S. Baratz. Two faces of power. *American Political Science Review,* 1962, 56, 947-962.

Bennett, W.L. Perception and cognition: An information-processing framework for politics. In S. Long (ed.), *The handbook of political behavior I.* New York, Plenum, 1981.

_____. *News: The politics of illusion.* New York: Longman, 1983.

Berelson, B., P. Lazarsfeld, and W. McPhee. *Voting.* Chicago: University of Chicago Press, 1954.

Blumer, H. *Symbolic interactionism.* Englewood Cliffs, N.J.: Prentice-Hall, 1969.

Bormann, E.G. Fantasy and rhetorical vision: The rhetorical criticism of social reality. *Quarterly Journal of Speech,* 1972, 58, 396-407.

_____. Fantasy and rhetorical vision: Ten years later. *Quarterly Journal of Speech,* 1982, 68, 288-305.

Campbell, A., et al. *The American voter.* New York: John Wiley & Sons, 1960.

_____. *Elections and the political order.* New York: Wiley, 1966.

Chaffee, S.H. The diffusion of political information. In S.H. Chaffee (ed.), *Political communication.* Beverly Hills, Calif.: Sage, 1975.

Converse, P. The nature of belief systems in mass publics. In D. Apter (ed.), *Ideology and discontent.* New York: Free Press, 1964.

Conway, M.M. and M.L. Wyckoff. The Kelley-Mirer rule and prediction of voter choice in the 1974 senate elections. *Journal of Politics,* 1980, 42, 1146-1152.

Deutsch, K.W. *The nerves of government.* New York: Free Press of Glencoe, 1963.

Devine, D. *The political culture of the United States.* Boston: Little, Brown, 1972.

Devries, W. and V.L. Tarrance. *The ticket-splitter.* Grand Rapids, Mich.: William B. Eerdmans, 1972.

Donohew, L. and L. Tipton. A conceptual model of information-seeking, avoiding, and processing. In P. Clarke (ed.), *New models for mass communication research.* Beverly Hills, Calif.: Sage, 1973.

Downs, A. *An economic theory of democracy.* New York: Harper & Row, 1957.

Dreyer, E.C. Mass media use and electoral choices: Some political consequences of information exposure. *Public Opinion Quarterly,* 1972, 35, 544-553.

Edelman, M. *The symbolic uses of politics.* Urbana: University of Illinois Press, 1964.

_____. Restricted and general political communication networks. *Communication,* 1980, 5, 205-217.

Festinger, L. *A theory of cognitive dissonance.* Evanston, Ill.: Row, Peterson, 1957.

Fisher, W.R. A motive view of communication. *Quarterly Journal of Speech,* 1970, 56, 132-139.

_____. Rhetorical fiction and the presidency. *Quarterly Journal of Speech,* 1980, 66, 119-126.

Galnoor, I. *Steering the polity.* Beverly Hills, Calif.: Sage, 1982.

Granovetter, M.S. The strength of weak ties. *American Journal of Sociology,* 1973, 78, 1360-1380.

Harris, L. *Confidence and concern: Citizens view American government.* Cleveland: Regal Books/King's Court, 1974.

Heider, F. Attitudes and cognitive organization. *Journal of Psychology,* 1946, 21, 107-112.

Hinckley, B. *Congressional elections.* Washington, D.C.: Congressional Quarterly Press, 1981.

Kelly, S., Jr. and T.W. Mirer. The simple act of voting. *American Political Science Review,* 1974, 68, 572-591.

Key, V.O., Jr. *Public opinion and American democracy.* New York: Knopf, 1961.

Kirkpatrick, S.A. Psychological views of decision-making. In C.P. Carter (ed.), *Political science annual VI.* Indianapolis: Bobbs-Merrill, 1975.

Klapp, O.E. Meaning lag in the information society. *Journal of Communication,* 1982, 32, 56-66.

Krause, S. and D. Davis. *The effects of mass communication on political behavior.* University Park: Pennsylvania State University Press, 1976.

Lasswell, H. and A. Kaplan. *Power and society.* New Haven: Yale University Press, 1950.

Lazarsfeld, P.F., B. Berelson, and H. Gaudet. *The people's choice.* New York: Columbia University Press, 1948.

Lichty, L.W. The news media. *The Wilson Quarterly,* 1982, Special Issue, 49-57.

Lindsay, P.H. and D.A. Norman. *Human information processing.* New York: Academic Press, 1977.

Lippmann, W. *Public opinion.* New York: Macmillan, 1922.

McGee, M.C. 1985: Some issues in the rhetorical study of political communication. In K.R. Sanders, L.L. Kaid, and D. Nimmo (eds.), *Political communication yearbook 1984.* Carbondale: Southern Illinois University Press, 1985.

McGrath, J.E. and M.F. McGrath. Effects of partisanship on perceptions of political figures. *Public Opinion Quarterly,* 1962, 26, 236-248.

Nimmo, D. *Political communication and public opinion in America.* Santa Monica, Calif.: Goodyear, 1978.

———, and R.L. Savage. *Candidates and their images.* Pacific Palisades, Calif.: Goodyear, 1976.

———, and J.E. Combs. *Mediated political realities.* New York: Longman, 1983.

Ogburn, W.F. *Social change.* New York: B.W. Huebsch, 1922.

O'Keefe, G.J. and L.E. Atwood. Communication and election campaigns. In D. Nimmo and K.R. Sanders (eds.), *Handbook of political communication.* Beverly Hills, Calif.: Sage, 1981.

Osgood, C.E. Cognitive dynamics in the conduct of human affairs. *Public Opinion Quarterly,* 1960, 24, 341-365.

Park, R.E. News as a form of knowledge. *American Journal of Sociology,* 1940, 45, 669-686.

Pierce, N. Political illiteracy among the young. *Manchester Guardian Weekly,* February 4, 1979, p. 19.

Prothro, J.W. and C.M. Grigg. Fundamental principles of democracy: Bases of agreement and disagreement. *Journal of Politics,* 1960, 22, 276-294.

Renshon, S. (ed.). *Handbook of political socialization.* New York: Free Press, 1977.

Robinson, J., and M. Levy. What do readers digest? *Washington Journalism Quarterly,* 1983, 5, 38-40.

Rogers, E.M. *Diffusion of innovations.* New York: Free Press, 1962.

———, and C.P.B. Cavalcanti. Communication networks and political behavior. *Communication,* 1980, 5, 161-167.

———, and L.D. Kincaid. *Communication networks: A new paradigm for research.* New York: Free Press, 1981.

———, and F.F. Shoemaker. *Communication of innovations.* New York: Free Press, 1971.

Ruben, B.D. *Communication and human behavior.* New York: Macmillan, 1984.

Savage, R.L. The diffusion of information approach. In D. Nimmo and K.P. Sanders (eds.), *The handbook of political communication.* Beverly Hills, Calif.: Sage, 1981.

Sears, D.O. and J.L. Freedman. Selective exposure to information: A critical review. *Public Opinion Quarterly,* 1967, 31, 194-213.

Sherif, C., and M. Sherif. *Attitude and attitude change.* Philadelphia: W.B. Saunders, 1965.

Sigel, R.S. Effect of partisanship on perception of political candidates. *Public Opinion Quarterly,* 1964, 28, 483-496.

Stein, B.J. Valley girls view the world. *Public Opinion,* 1983, 6, 18-19.

Toffler, A. *Future shock.* New York: Random House, 1970.

Walker, J.L. A critique of the elitist theory of democracy. *American Political Science Review,* 1966, 60, 285-295.

_____. The diffusion of innovations among the American states. *American Political Science Review,* 1969, 63, 880-889.

Wattier, M.J. The simple act of voting in 1980 Democratic presidential primaries. *American Politics Quarterly,* 1983, 11, 267-291.

Watzlawick, P. *How real is real?* New York: Random House, 1976.

Witt, E. Here, there and everywhere: Where Americans get their news. *Public Opinion,* 1983, 6, 45-48.

Zukin, C. Voting in 1972: An application of an innovation diffusion model to the political system. Paper presented at the annual conference of the International Communication Association, April, 1976, Portland, Oregon.

18

Metaphors, Information, and Power

Stanley Deetz and Dennis Mumby

This chapter explores the nature of power and information within organizations. Power is not fundamentally a property of individuals or groups, but of systems of meaning which constitute information favoring certain individual and group interests. Systematically distorted cultural formation can result from the uncritical acceptance of the political conditions of information production. Metaphors are part of the "deep" political structure producing particular forms of information. Metaphor analysis and criticism is proposed as a means of examining the role of power in the construction of organizatonal information and of developing more representative and adaptive forms of information.

The purpose of this article is to explore the relationships among metaphors, information and power in organizations, and to suggest the value of metaphor analysis as a productive means of investigating power relations within organizational cultures. We will begin by briefly exploring five basic premises—*The Five "P's"*—upon which the subsequent analysis is based. These premises could be developed out of a number of theoretical traditions. We will use the language of modern phenomenology and critical theory to give them expression (see Deetz 1983; Merleau-Ponty 1962; Habermas 1971).

The Frame of Reference: The Five "P's"

Perception Is Primary

No matter how mediated, all human knowledge is ultimately grounded in perceptual experience. All knowledge verification and change ultimately rest in further perceptions. To some extent this assumption is counterintuitive since we experience ourselves in direct contact with the world. We are often in direct contact. That contact is perception. On the basis of perceptions human beings make judgements, have feelings, decide courses of action, and make claims about the nature of reality. All information is perceived information.

Fundamental Perception Is Participatory

Experience of the world is formed interactively between subjects and objects. Human concepts and object attributes initially arise together in perception. The perceived qualities of an object require both a way of looking and an object capable of being perceived in that manner. Neither is primary. Such a claim is in opposition to empiricists who privilege the transcendent world and, hence, take for granted or forget the constructivist activities of data production. It is also in opposition to subjectivists who fail to recall the materiality of the world and the subject's placement in it when they examine interpretive processes.

In this sense, information is always produced information. Information is not objective and value-neutral. It only exists in the perceptual and expressive activities of cultural members.

Perception Is Positional

The human side of perception production is a way of "looking," an orientation, or set of activities. Positionality is much more basic than personal interpretive frames or psychological states. Positionality is based most fundamentally on the human possession of certain types of sense equipment or a manner of comportment, but is extended by institutions, social practices, tools, and testing instruments.

Perhaps the most central human institution is language. Language participates with other institutions to provide humans with a way of being in the world. Like other institutions it positions cultural actors to make certain distinctions, to highlight certain aspects, and make other parts of the world into background. Idiosyncratic, personal, and sectional positions intersect and compete with general social institutions to provide a relatively stable but incompletely shared cultural perception. Every perception, every piece of information implicitly carries with it the conditions of its formation.

Perception Is Political

Owing to the socio-historical development of positionality—of institutions, technology, and sensual extentions—certain human groups, interests, and possible futures tend to be more fully represented in perceptual formation. Institutional arrangements and technological developments enhance the possibility of certain perceptions and ease the expression of them in social interaction. Since time and timing make a difference in open organizational systems, such enhancement and easing have considerable effect on decision making. Certain informational frames and contents are privileged and provide nonrational advantage to certain groups and interests. Disadvantaged cultural members' own perceptions support advantaged groups and interests. The most important forces of politics lie, not in changing attitudes and providing information, but in changing the conditions under which perception is formed. Power and privilege are realized in each perceptual act. Contrary perceptions, if found at all, have an unreal and irrational quality about them.

Perception Becomes Protected Opinion

What arises initially as interactively formed perception in a practical context can become reified, reproduced, and protected from examination in further experience and expression. In these cases culture becomes ideology. Dominant opinions substitute for interactive perceptions in experience. Social/institutional arrangements tend to normalize experience and preclude the continual formation of new perceptions. The experienced world and information regarding it become artifactual.

The natural corrections from alternative positions and further experience are frustrated by the development of secondary institutional mechanisms. Such mechanisms principally work to make the current state of affairs seem natural and necessary. For example, historical narratives in the form of stories and claims about the "real" objective nature of people and things pull experience together in a particular fashion, making heroes, villains, responsibilities, and certainty (Jameson 1981). The constitutive conditions of experience are thereby closed to rational assessment and historically embedded power arrangements dominate experience.

Applying the Framework to Organizational Analysis

Using the above assumptions to conceptualize the relations among metaphor, information, and power, we can suggest the following themes to be developed in the remainder of the article.

The language of organization members provides an aspect of positionality of perception. The organization's conceptual and operational paradigm positions perception, what is taken as real, and what is left problematic (Kuhn 1970; Morgan 1980). The paradigm becomes the "interpretive frame" through which organizational information is created and makes sense.

The primary origin of power is not with the individuals who possess information, but in the system of meaning which constitutes information (Clegg and Dunkerley 1980). Organizational paradigms are partially ideological in nature. That is, they articulate a particular organizational reality for the individual, while simultaneously foreclosing the possibility of other, alternative realities. Representation of differing interests and open participation in information construction is arbitrarily foreclosed.

Based on recent work by Lakoff and Johnson (1980a,1980b), linguistic metaphors can be shown to be intrinsically connected to human cognition. An examination of metaphors in organization language thus provides a way to explicate the organizational paradigm and members' perceived realities (Koch and Deetz 1981; Deetz forthcoming).

Metaphor analysis encourages critical examination of organizational ideology by showing how organizational information is a human construct and not merely representative of "the way things are." The "protection" of "protected opinion" is removed. Organizational metaphors are not arbitrarily imposed. They emerge out of the material infrastructure of the organization. As such they support the particular power interests within the organization, serving both to produce and reproduce the existing systems of domination. A description of metaphor structures not only allows for examination and discussion of organizational "reality" but also provides opportunity for the creation of alternative metaphor structures (see Pondy 1983).

Ideology and Power

The metaphor structure of language can serve to direct, constrain, and often distort members' perception of reality by articulating information within the framework of a particular ideology. As such, certain power relations are maintained and reproduced, resulting in the perpetuation of sectional interests.

Language and Social Reality

Language is the principal medium through which social reality is produced and reproduced. In this context, Berger and Luckmann (1971, 85-86) state:

> Language objectivates . . . shared experiences and makes them available to all within the linguistic community, thus becoming both the basis and the instrument of the collective stock of knowledge. Furthermore, language provides the means for objectifying new experiences, allowing their incorporation into the already existing stock of knowledge, and it is the important

means by which the objectivated and objectified sedimentations are transmitted in the tradition of the sedimentation in question.

Berger and Luckmann's claim—that language is the principal medium through which social reality is shared—is a useful beginning to cultural analysis but fails to make explicit the issue of power. Discourse is a product of—and reproduces—the dominant power interests in social formations. In other words, language is not simply a neutral transmitter of information-as-social-knowledge, as Berger and Luckmann suggest. Rather, it must be seen as the principal means by which the dominant ideological structures in a social system perpetuate themselves. As such, metaphor analysis provides a way of uncovering these ideological structures as they are reflected in language use.

The relationship between language, ideology, and power is manifest in the relationship between the social actor and the institutions in which he or she conducts day-to-day activities. Language systems (and metaphors specifically), reflect ideological meaning formations. They produce and reproduce the subjectivity of the individual vis-à-vis social institutions. That is, language positions the social actor to look at and respond to the world in a particular way. This "positioning" is more than just having a set of beliefs or attitudes.

Information and Ideology

Following contemporary interpretations of Marx (Althusser 1971; Gramsci 1971; Coward and Ellis 1977; Giddens 1979; Therborn 1980), "ideology" and "beliefs" are not the same thing. Ideology is not something that a person has along with other things. Ideology constitutes the way that social actors think and act in the world. An examination of ideology addresses questions of power and domination, since particular ideologies generally sustain and reproduce certain world views at the expense of others. From an information perspective, ideology is that which determines, for a particular community, what is to be considered information, and what is mere noise. Ideology is, thus, a material force. It is a construction of human subjectivity, and is represented in the practices and meaning systems of concrete institutions.

Functions of Ideology

In this context, we wish to suggest that organizational cultures are intrinsically ideological, and that the dominant metaphors in an organization serve to provide selective information that produces and reproduces certain ideological meaning formations. The link between metaphor structures and the creation of a particular subjectivity ("organizational consciousness") among organization members can be examined using Therborn's (1980, 18) three modes of ideological analysis:

Ideologies subject and qualify subjects by telling them, relating them to, and making them recognize:

1. What exists, and its corollary, what does not exist: that is, who we are, what the world is, what nature, society, men and women are like. In this way we acquire a sense of identity, becoming conscious of what is real and true; the visibility of the world is thereby structured by the distribution of spotlights, shadows and darkness.
2. What is good, right, just, beautiful, attractive, enjoyable, and its opposites. In this way our desires become structured and normalized.
3. What is possible, and impossible; our sense of the mutability of our being-in-the-world and the consequences of change are hereby patterned, and our hopes, ambitions, and fears given shape.

These three modes serve to "qualify" the subject as a participant in a culture or subculture. Therborn uses the term *subjection-qualification* to point out the dialectical nature of the relationship between the subject and ideology. *Subject* can mean both subject (subjugated) *to* a particular meaning system, as well as "subject" in the sense of the locus of creative practices. *Qualification* designates both the ability to perform certain activities and the setting of restrictive parameters on those activities. These terms capture the sense in which subjects are socialized through ideological practices both in a limiting sense and in the sense of being qualified to participate fully in a structured, meaningful world.

Organizational metaphors inform members about what exists, what is good, and what is possible—they shape and delimit social reality within organizations. In this sense, metaphor structures, as vehicles of information in an organizational context, can be viewed as manifestations of organizational ideology.

The role of metaphors in the ideological structuring of information can be illuminated by examining the relationship between ideology and material reality. First, as indicated above, ideology does not exist as a set of dislocated and autonomous ideas and values, but is grounded in material practices. Therborn (1980) suggests that ideologies reproduce themselves through a system of affirmations and sanctions. Adherence to, or contravention of, ideological specifications are rewarded and punished, respectively. Without this parallel between material conditions and ideology, the latter would have little or no effect on the social reality of organizational members. For example, the use of time clocks is a way of distinguishing workers' "wages" from managers' "salaries." Workers' failure to comply with this regimentation of the work day results in loss of pay. The ideological distinction between workers and management is thus maintained by an actual material sanction.

Second, although ideology is grounded in material reality, it is not simply a reflection of it. The power of ideology lies in its ability to produce a selective and distorted view of material reality, while at the same time presenting that "reality" in factual and objective terms. Althusser (1971, 164-65) presents a radical version of this position when he states, "What is presented in ideology is . . . not the system of real relations which govern the existence of individuals, but the imaginary relations of those individuals to the real conditions in which they live. Not only does ideology deny the real conditions of existence, it also systematically excludes them from thought and hence from possibility."

Third, ideology functions in various ways to maintain and reproduce certain nonarbitrary forms of social reality (Giddens 1979). One function involves the presentations of sectional interests as universal ones. Any dominant interest must be perceived as more than specific to a particular group if it is to be accepted as anything more than one interest among many competing points of view. For example, in organizations there are vested interests by management to maintain a structural hierarchy. These interests are presented as universal by connecting the notion of managerial success with that of economic wealth, which, in turn, is viewed as beneficial for everyone, especially the workers, who benefit through wage increases, bonuses, and so forth.

Another function of ideology rests in its power to structure social reality such that system contradictions are masked, excluded from thought, or transformed into a more acceptable form. The primary contradiction in capitalist society—that between private appropriation and socialized production—is legitimated by the ideological separation of the social sphere from the economic sphere, embodied in the notion of "keeping politics out of the workplace." The nondemocratic authoritarian structure of an organization can thus remain intact. If contradictions become visible, they are made more acceptable by their justification through a higher authority. Thus, the authoritarian nature of many organizations is made tolerable by arguing that democracy can limit efficiency, production, and profit, and is therefore ultimately detrimental to everyone.

Ideology thus functions to mystify and naturalize social reality by making essentially human constructs appear fixed and external to those who created them. Social relations and meaning structures are viewed as static and immutable—"the way things are." The reification of day-to-day experience forecloses the possibility of envisioning alternative social realities. When alternatives become visible, they are often considered "strange" and "unnatural." In organizations, for example, the concept of *hierarchy* is not seen as an inherently political (metaphoric) construct which serves the interests of a small, elite group. It is, rather, perceived and experienced as a tangible, physically existing structure which guides and regulates organizational activity.

Finally, ideology is an effective means of control. It functions primarily not as a means of coercion, but acts to produce a dominant, shared meaning system about which there is a consensus, and to which everyone "freely" consents (Althusser 1971; Gramsci 1971; Habermas 1975).

Williams (1977, 18) describes this condition of ideological dominance as "effective *self-identification* with the hegemonic forms: a specific and internalized 'socialization' which is expected to be positive but which, if that is not possible, will rest on a (resigned) recognition of the inevitable and the necessary." This function of "control through consent" not only maintains the existing power relations but simultaneously obscures the fact that power is being exercised, causing certain interests to predominate over others.

Generally, dominance is manifested not in significant political acts but rather in the day-to-day, taken-for-granted nature of organizational life. As such, the exercise of power and domination exists at a routine level, further protecting certain interests and allowing the order of organizational life to go largely unquestioned by its members.

Giddens (1979, 188-92) emphasizes the role of language in the functioning of ideology when he states:

> To analyze the ideological aspects of symbolic orders . . . is to examine how structures of signification are mobilized to legitimate the sectional interests of hegemonic groups. . . . To examine ideology institutionally is to show how symbolic orders sustain forms of domination in the everyday context of "lived experience." . . . To study ideology from this aspect is to seek to identify the most basic structural elements which connect signification and legitimation in such a way as to favour dominant interests.

From this perspective, language and the information it provides is not ideological per se, but can be used by dominant groups as a means of reproducing the existing social relations.

One of the problems associated with the analysis of ideological meaning structures, however, is the development of an adequate descriptive scheme. In the section that follows, we will argue that metaphor analysis provides a powerful tool for explicating the means by which a particular "symbolic order" maintains and legitimates certain forms of social reality. Following Giddens, metaphor itself will not be treated as inherently ideological. Rather, metaphor structures are appropriated ideologically and thus function to reproduce a certain ideology, serving particular, sectional interests.

Metaphors and Organizational Reality

In *Metaphors We Live By* Lakoff and Johnson (1980a, 6) state, "Metaphor is not just a matter of language, that is, of mere words. . . . On the contrary, human *thought processes* are largely metaphorical. This is what we mean when we say that the human conceptual system is metaphorically structured and defined. Metaphors as linguistic expressions are possible precisely because there are metaphors in a person's conceptual system." From this perspective, metaphors are not simply persuasive devices which present in a particular way what is already understood. Rather, the act of conceptualizing and understanding is achieved through the metaphoric structure of language.

Metaphors are a vehicle for understanding by virtue of their grounding in experience. A metaphor makes sense only because we are able to overlay one aspect of our experience on another. For example, many people conceptualize an argument in terms of *war* insofar as they defend their positions, attack opponents, use strategies, and try to win. Such a metaphor works only because we are able to apply our experience of war to arguments, and thus experience arguments *as* war.

Functions of Metaphors

Although Lakoff and Johnson's theory of metaphor is grounded in experience, they present no conception of the politics of experience, or the ideology which mediates it. We, therefore, wish to build on Lakoff and Johnson's position by moving beyond individual experience per se, and looking at the way in which metaphor functions to produce social formations as a whole. Metaphors do not simply provide a neutral description of reality for the individual. Rather, they are an instance of a particular, ideological social reality, which is the expression of certain sectional interests within the community. Metaphoric structures function to universalize these interests through the commonality of the experience which grounds them.

Metaphor functions to provide individuals with information about the environment by linguistically structuring events in a particular way. Metaphor structures present an experiential gestalt, fixing the relationship between figure and ground. This conceptualization process is only partial, however, since other competing metaphor structurings are usually present. The individual's thinking is directed and limited by these structures.

To the extent that the organization's discourse is restricted to particular metaphors, the member is provided with the information necessary to participate fully in the organizational structure, while simultaneously restricting thought and behavior that falls outside the reality delimited by the organization. To this extent, metaphors can be analyzed in terms of the way in which they produce and reproduce particular organizational ideologies.

Types of Metaphors

Metaphors are grounded in material reality and, as such, provide the subject with the information necessary to make sense of his or her environment. Lakoff and Johnson (1980a,1980b) argue that all metaphor structures are reducible to three basic types, each of which is tied to three corresponding and pervasive nonmetaphorical features of human life. The three types of metaphor are termed *orientational, ontological,* and *structural.*

Orientational metaphors reflect the fact that our bodies serve as the center of our world with objects, people, and events perceived in this context. Orientations such as up-down, front-back, in-out, and so forth, can structure concepts linearly. In each case, abstract concepts are grounded and conceptualized in terms of our bodily experience of the physical world.

In Western culture generally, "up" is positive while "down" is negative. For example, *more* is up ("His spirits rose"), *control* is up ("He had power over her"), *good* is up ("Things are looking up"), *happiness* is up ("I'm on a high"). In the Western world such basic conceptions assure that hierarchy will mean stratification and not simply differentiation. Hierarchy and power become intrinsically connected and associated with goodness, expansion, and happiness. In the absence of competing structures (e.g., "small is beautiful") the imaginary relation holds sway.

Ontological metaphors involve the projection of entity characteristics onto areas of experience which have a nonphysical status. For example, *ideas* become food ("That's hard to swallow," "Chew on the idea for a while," "That's a theory you can really sink your teeth into"). Lakoff and Johnson stress that ontological metaphors "are so natural and persuasive in our thought that they are usually taken as self-evident, direct descriptions of mental phenomena" (1980a, 28). Further, such shifts are not as innocent as they might seem. When non-things like ideas become things, linear and mechanical conceptions of communication are not far behind, and possession and ownership become issues. All processes in which phenomena are reified or given status as a commodity are made possible by ontological metaphors.

Structural metaphors perform the function of projecting the characteristics of one structured experience or activity onto another. In Western culture, for example, "time is money" is a dominant metaphor in which one structured experience—monetary transactions—is projected onto the concept of *time.* Thus, conceptualizing time as money ("My time was well spent") makes it a limited resource ("I don't have much time") and hence a valuable commodity ("My time is precious—I mustn't waste it").

These three generic metaphors help to map out the prereflective social reality of the indivdual. They provide the basis upon which members of a given culture can develop shared meaning systems, allowing them to distinguish what is meaningful from what is nonsense. The metaphor structures that are built on these three basic metaphors provide a frame and context for the interpretation of information, and ground the taken-for-granted reality out of which individuals act towards the world.

In an organizational context there are a number of possible ways these general metaphors may be specified and actualized. For example, the prevalence of an organization-as-machine metaphor will predispose members to watch for breakdowns, make necessary repairs, be sure that everything runs smoothly and with minimal friction, and so forth. Information that is sought will thus tend to relate to the mechanical conditions in the organization, while information concerning member's quality-of-life, for example, will be considered less important and more peripheral.

In contrast, conceiving an organization as a family entails the use of a different interpretive frame, in which structural information becomes more peripheral and members place greater importance on interpersonal issues. Efficiency and smooth running may still be important, but these concerns are approached in terms of whether organization members can work well together and develop close working relationships. In the case of a fall in profits, for example, information may be sought on the ability of managers and subordinates to "get along" together, while the question of whether organization members are "working to capacity" may be considered less important.

The Double-Structuring Function of Metaphors

Metaphor use produces and reproduces certain forms of organizational reality by directing members toward particular, commonly shared perceptions of their environment. This continual reproduction of reality is achieved through what can be called the double-structuring function of metaphor. Here, a reciprocal relationship exists between the metaphors used to describe an organization and the type and structure of information that is disseminated there. This reflexivity between metaphor and information is premised on the assumption that organizational information may take the form of certain dominant metaphors. These metaphors themselves mediate information flow, determining both what is to be considered information and the frame in which the information is to be interpreted. Metaphors are therefore both the medium and product of information flow (see Giddens 1979, 69-95). Such a double-structuring function is intimately linked with the question of power interests in an organization.

Metaphors, thus, on the one hand allow this possibility for shared meaning and consensus, while on the other hand such consensus (achieved through the channeling of perception) leads to systematically distorted communication (Habermas 1971, 1975; Frost 1980)—that is, communication that considers the highlighted aspects of events without enabling competing perceptions and interpretations. As noted previously, such distortion is not arbitrary, but is largely determined by the interests that it serves. At any one time the interests of a particular group will be held as privileged—for example, the managerial interest in organizations (see Deetz and Kersten 1983).

As another illustration, the predominance of an organization-as-military metaphor usually indicates the existence of a fairly formal, strict hierarchy, an emphasis on the importance of following orders, and little opportunity for participation in decision making by "subordinates" (Weick 1979). In this context, not only is information formulated using military metaphors, but the use of military metaphors itself functions to mediate in the sensemaking process of organization members. Thus, the use of a military metaphor gives legitimacy to the interests of the managerial group, both by the ongoing description of the organization as hierarchically structured, and by providing information within this perceptual context. Efficiency is facilitated by following the chain of command and difficulties are attributed to the breakdown of order. While the notions of participation and quality of life align well with other conceptual structures (become thinkable) in an organization where a family metaphor prevails, in a military organization they do not. In a military context, such metaphors may be viewed as subversive and a threat to the "natural order of things."

The concept of an organization as a military unit or a family is essentially a human, social construct. The "naturalness" of dominant metaphor structures, however, is such that the subjective nature of organizational structure is largely obscured. As indicated previously, metaphors perform a primary task of ideology by producing and reproducing reified experiences, and produced experiences which are taken as the "way things are."

The Conduit Metaphor

The very concept of information flow in organizations is almost exclusively processed in terms of what Reddy (1979) calls the conduit metaphor. He has observed that this metaphor is particularly prevalent in talk about ideas and information. It is structured such that ideas or meanings are objects, the linguistic expressions used are containers, and the communication process involves some kind of linear movement ("sending"). Thus a speaker puts ideas and meanings (objects) into words (containers), which are sent along a conduit (the communication channel) to the recipient, who then takes the ideas out of the words. A few examples are:

- I need to get my ideas across more effectively.
- It's hard to put the idea into words.
- Our communication channel is blocked.
- Send the information as soon as possible.

The use of this metaphor has consequences in terms of the way in which information is conceptualized. For example, the notion that words and expressions are containers for meanings entails that words have *a* meaning independent of origin and context. Furthermore, ideas attain an object-like, fixed quality which they do not actually possess—they are conceptualized as "things" that can be passed back and forth between people when, in effect, they do not actually materially "exist" anywhere.

As a consequence of this view of information, organizational policy and philosophy seems more tangible and permanent than it actually is; it is seen as natural and self-evident, rather than the product of certain interests. Here, the notion that information must be both tangible and sent through the "correct channels" reduces the possiblity for alternative, more flexible means of communication which encourages the development of innovative and original ideas and the possible representation of different interests.

Despite the many discussions of the limits of such a conception and of process views of communication, both talk in organizations and management texts continue to be totally dominated by such a conception (Axley 1983). The development of new information technologies would appear both to be stimulated by this and to enhance the metaphor structure in place (Hiemstra 1983), even developing with it new mechanical terminology (e.g., *interface, boilerplate*).

The persistence of this metaphor complex, despite opposition and theoretical writings, is not surprising. The extensive use of the conduit metaphor in information flow also fits well with the more mechanistic views of organizations (such as machine and military metaphors, in which the effective transportation of messages through the organizational structure becomes the primary measure against which efficiency is judged. In such instances, maintenance of the information networks becomes more important than the nature of the information itself.

Metaphor and Organizational Control

The ability of metaphor to control organizational reality emerges as a necessary consequence of the other ideological functions. By channeling perception in certain directions, metaphor structures exclude other potential cultures or social realities. This not only produces a commonly shared consciousness amongst organization members, but also diminishes the possibility of challenges to the existing meaning structures. Once a particular taken-for-granted worldview has been established within an organization, its reproduction is assured as long as new information is mediated by the dominant metaphors used to describe the organization.

The degree to which metaphors can maintain and reproduce a particular (ideological) social reality, however, is largely dependent on what Lakoff and Johnson (1980a, 1980b) term *internal* and *external systematicity.* Internal systematicity relates to the coherence of a metaphor structure, for instance, the way that hierarchy, following orders, and subordinates fit into the complex gestalt that makes up the organization-as-military metaphor. External systematicity emerges when different metaphors overlap to form more complex metaphoric groupings as, for example, in the case of the conduit, machine, and military metaphors. Such groupings allow organization members to tie together different aspects of organizational experience into a structured whole. Consequently, the systemic ramifications of changing any component conception is great. Further, such groupings tend to protect component systems from change and enhance conceptual stability.

The ability of a group of metaphors to present a particular ideological structuring of organizational reality is, therefore, most seriously threatened when an alternative and noncoherent metaphoric structuring emerges, threatening to displace the dominant metaphors. It is only through the emergence of such alternatives, however, that other organizational realities can evolve. In this sense, organizational change is intimately linked to the ways in which members talk about, and hence conceive of, organizational functions. It is here, also, that the importance of members' "discursive penetration" of organizational meaning structures becomes clear—that is, the extent to which they can reflect on and examine the construction of their "reality."

Values of Metaphor Research

The values of metaphor analysis are clear in light of this theoretical discussion. These values may be summed up in terms of the goals of a "critical" research program (Deetz 1982)—understanding, critique, and education. Metaphor analysis provides understanding by enabling members to reflect on the nature and construction of information present. Critique is accomplished by the demonstration of the "systematic distortion of communication" owing to unexamined power interests supported by particular conceptions and making difficult the perception and expression of legitimate competing interests. Education refers to the possibility of the enrichment of the "natural" language of the organization (Daft and Wiginton 1979). Such enrichment, stimulated by the exposure of power interests embedded in particular conceptions, enhances the expression of the full range of interests. Space here does not allow a demonstration of this analysis as applied to a particular organization (for example, see Deetz, forthcoming), but we wish to discuss the implication of these values briefly here.

The development of alternative metaphors opens the possibility of self-determined change, in the sense that the structure of organizations is revealed as a human, social construct which is by no means fixed and immutable. Providing members with other ways to make sense of information in an organization allows for a more critical stance vis-à-vis the nature of power interests in that organization. Power interests manifest themselves in metaphor by directing members' attention towards certain aspects of organizational life; alternative metaphors focus attention more directly on those areas which would otherwise remain perceptually obscure.

An important and related question involves the issue of whether novel metaphors can be introduced into organizational talk. Much of the power of metaphor, as an expression of organizational ideology, rests in its ability to "sustain the sectional interests of dominated groups" by making them appear natural, beyond change and independent of human action. In order to become part of the natural language in an organization, new metaphors must therefore overcome the deeply rooted patterns of thinking that the prevailing metaphors create and overcome the material arrangements in the organization such as reward systems and technology which support them.

The chances of success of a metaphor training program are probably limited, but the possibility of providing new ways of expression seems within reach, especially when the new metaphors are in some sense compatible with the dominant metaphor system, or when major structural changes are occurring. For example, many organizations operate within the framework of a world-as-jungle metaphor (see Maxwell 1982). Here, the organization is viewed as existing in a predatory, competitive environment, in which only the fittest (economically most viable) prevail. A focus on survival stresses the importance of qualities such as adaptation, defending territory, aggressiveness, defeating rivals, and so forth. Similar perspectives may exist within a single organization; individual members may perceive advancement as contingent on being ruthless, "looking out for number one," developing a "killer instinct," and so on.

Alternatively, an organization-as-organism metaphor, although related to the jungle metaphor, highlights different facets of organizational behavior. There is still an emphasis on survival, but this is achieved through the promotion of overall growth and health. As such, rugged individualism becomes subordinated to the need for a more holistic perspective of the organization, in which success is based on an "ecological balance" both at the organizational and wider societal levels.

Conclusion

Metaphors are an instance of organizational ideology insofar as their usage operates in the production and reproduction of information and human subjectivity. Organizational reality is constructed through the metaphoric channeling of perception in particular directions. The informing of the organizational member by certain metaphor structures positions the individual to make sense of the world in a particular manner. The sense-making process is not arbitrary, but represents and supports hidden power interests. The pervasive use of certain metaphors, wittingly or not, reflects a selective social reality and excludes others which might challenge or contradict the existing power relations.

Information is not a neutral product of organizational activity, but is a result of an inherently political activity—a political activity often hidden from those engaging in it largely due to presumed neutrality. The construction and character of information is as critical to issues of power as questions of information flow and distribution. Analysis of power and information construction must get beyond the "surface" of organizational structure to its infrastructure (Benson 1977). In this context, metaphor analysis emerges as a way of examining the way in which the deep-structure power relations manifest themselves at the surface level.

Our objective here has been to explore the issues that are important in considering the relationships among information, metaphor, and power. Our main concern has been to move beyond conceptions of information as symbolic representations of reality distributed in organizations, to concentrate, instead, on ways information and reality are constructed. The importance of this move rests on the need to explicitly recognize the link between the structuring of information in an organization and the nature and center of power interests in that organization.

Information as structured by particular metaphors serves hidden power interests and ultimately certain groups and individuals by presenting a social reality which preserves the prevailing system of domination which arose at particular times reflecting certain historical conditions. Such constructions are taken as the "natural order of things" by those who live them—metaphors become strangely literal. In this way, the current conditions of existence are obscured and their human construction is closed to self-reflection. Metaphor analysis opens up the possibility of reflection by exposing the constructed nature of organization reality and providing the means to conceptualize more representative and adaptive ways of perceiving the world.

References

Althusser, L. Ideology and ideological state apparatuses. In B. Brewster (trans.), *Lenin and philosophy*. New York: Monthly Review Press, 1971, pp. 127-186.

Axley, S. The conduit metaphor in organizational texts and talk. Paper presented at the third annual Interpretive Approaches to Organizational Communication Conference, Alta, Utah, 1983.

Benson, J. Organizations: A dialectic view. *Administrative Science Quarterly*, 1977, 22, 1-21.

Berger, P. and T. Luckmann. *The social construction of reality*. Harmondsworth, England: Penguin, 1971.

Clegg, S. and D. Dunkerley. *Organization, class, and control*. London: Routledge & Kegan Paul, 1980.

Coward, R. and J. Ellis. *Language and materialism: Developments in semiology and the theory of the subject*. London: Routledge & Kegan Paul, 1977.

Daft, R. and J. Wiginton. Language and organization. *The Academy of Management Review*, 1979, 4, 179-192.

Deetz, S. Critical interpretive research in organizational communication. *The Western Journal of Speech Communication*, 1982, 46, 131-149.

_____. Negation and the political function of rhetoric. *Quarterly Journal of Speech*, 1983, 69, 434-441.

_____. Metaphors and the discursive production and reproduction of organization. In L. Thayer (ed.), *Organizations and communication: emerging perspectives I*. Norwood, N.J.: Ablex, forthcoming.

_____, and A. Kersten, Critical models of interpretive research. In L. Putman and M. Pacanowsky (eds.), *Communication and organizations: An interpretive approach*. Beverly Hills, Calif.: Sage, 1983.

Frost, P. Toward a radical framework for practicing organization science. *The Academy of Management Review*, 1980, 5, 501-508.

Giddens, A. *Central problems in social theory: Action, structure and contradiction in social analysis*. Berkeley: University of California Press, 1979.

Goldhaber, G., et al. *Information strategies: New pathways to corporate power*. Englewood Cliffs, N.J.: Prentice-Hall, 1979.

Gramsci, A. *Selections from the prison notebooks*. Ed. and trans. Q. Hoare and G. Nowell Smith. London: Lawrence & Wishart, 1971.

Habermas, J. *Knowledge and human interests*. Trans. J. Shapiro. Boston: Beacon Press, 1971.

Habermas, J. *Legitimation crisis*. Trans. T. McCarthy. Boston: Beacon, 1975.

Hiemstra, G. You say you want a revolution: Information technology as a cultural category in four American corporations. Paper presented at the annual conference of the International Communication Association, Dallas, 1983.

Jameson, F. *The political unconscious: Narrative as a socially symbolic act*. Ithaca, N.Y.: Cornell University Press, 1981.

Koch, S. and S. Deetz. Metaphor analysis of social reality in organizations. *Journal of Applied Communication Research,* 1981, 9, 1-15.

Kuhn,T. *The structure of scientific revolutions.* 2d ed. Chicago: University of Chicago Press, 1970.

Lakoff, G. and M. Johnson. *Metaphors we live by.* Chicago: University of Chicago Press, 1980a.

_____. The metaphorical structure of the human conceptual system. *Cognitive Science,* 1980b, 4, 195-208.

Maxwell, M. Brawlers, military men, and savages: Images of violence in a paralyzed executive group. Paper presented at the second annual Interpretive Approaches to Organizational Communication Conference, Alta, Utah, 1982.

Merleau-Ponty, M. *Phenomenology of perception.* Trans. C. Smith. London: Routledge & Kegan Paul, 1962.

Morgan, G. Paradigms, metaphors, and problem solving in organization theory. *Administrative Science Quarterly,* 1980, 20, 605-622.

Pondy, L. The role of metaphors and myths in organization and in the facilitation of change. In L. Pondy et al. (eds.), *Organizational symbolism.* Greenwich, Conn.: JAI Press, 1983.

_____, et al., eds. *Organizational symbolism.* Greenwich, Conn.: JAI Press,1983.

Reddy, M. The conduit metaphor. In A. Ortony (ed.), *Metaphor and thought.* Cambridge, England: Cambridge University Press, 1979.

Shannon, C. and W. Weaver. *The mathematical theory of communication.* Urbana: University of Illinois Press, 1949.

Therborn, G. *The ideology of power and the power of ideology.* London: Verso, 1980.

Weick, K. *The social psychology of organizing.* 2d ed. Reading, Mass.: Addison-Wesley, 1979.

Williams, R. *Marxism and literature.* Oxford: Oxford University Press, 1977.

19

Privatizing the Public Sector:
The Information Connection

Herbert I. Schiller

Twenty-five years ago, interventionism on behalf of the weaker groups in the nation was regarded as a necessary and acceptable response to the instabilities of the economic order. More recently, and especially observable in the United Kingdom and the United States, there has been a marked erosion in public sector activities. A corporate sphere, grown powerful during and after World War II, intensifying competition internationally, and the availability of new information technologies have been major factors in promoting privatization and rationalization throughout the economy. Information, increasingly a saleable good, is now a basic determinant in the growing privatization of education, science, government information and other erstwhile public functions. One consequence of privatization is to put all information-related services on an ability-to-pay basis. A significant social effect is to extend further the influence of corporate thinking and to weaken social perspectives. At this time, it is not clear how these developments may be reversed. If anything, deepening crisis may accelerate them.

Twenty-five years ago, Gunnar Myrdal, an internationally-renowned Swedish social scientist, delivered the Storrs Lectures on Jurisprudence at the Yale Law School. In his lectures, published two years later, Myrdal (1960) surveyed developments in the Western market economies. He came to some optimistic but seemingly quite reasonable conclusions.

Declaring that state intervention in the practices of the economic systems of the West was largely a response to twentieth century economic crises, Myrdal considered the Welfare State a progressive development. It was acting, he wrote, to protect the social and economic needs of the people. He believed that it furthered the democratization of the political process.

Additionally, Myrdal was convinced that no Western country would ever again tolerate severe unemployment. He regarded this as the crowning achievement of the Welfare State. More enthusiastic still about the direction of this kind of society, Myrdal (1960, 77) declared that "a created harmony has come to exist in the advanced Welfare State" and that most people feel freer in this environment (p.86).

Recalling these views, expressed a quarter of a century ago, is not intended to denigrate a scholar whose inclinations have been humane and whose analyses often better grounded than most in the international social science community. Rather they are useful for appreciating how far thinking has moved in a different direction in the last few years.

What has been happening at an accelerating pace is the undoing of the "created harmony." A rapid dismantling is underway of the public sector that had emerged from the state's social interventions over the last one hundred years, and in the first half of this century in particular.

Myrdal's analysis calls attention also to what today would appear to be a striking paradox. According to his and others' understanding, it was the imperative of economic crisis that made necessary the interventionist state and the growth of the public sector in pre-World War II years. In the 1980s, crisis is again a central preoccupation in the West, but the means of coping with it are entirely different, or so it would seem. Instead of an expansion, the public sector is being contracted. Privatization proceeds apace.

What accounts for the very different means by which economic crisis is being handled today in the Western economies, especially in the United Kingdom and in the United States? It is the contention here that information and information technology play crucial, if not determining, roles in this latest stage of capitalist development and crisis.

What follows is a brief overview of the extent to which traditionally public functions in the United States are being transferred to private enterprise, an explanation of why this is happening, and, finally, a discussion of what already are—and what are likely to be—some of the effects of the push toward privatization.

Undercutting the Public Sector

We do not need to be futurists to be able to recognize the patterns that are being established now for the years ahead. One, hard-to-miss feature of the current institutional landscape is the wave of privatization sweeping over society, pushing back and often eliminating spheres of activity that historically have been public and noncommercial. Most heavily affected are the cultural, educational, health, and economic security of large groups of people.

Indigent municipal governments, for example, petition private companies to build or maintain their parks and other social services (*New York Times*, April 1, 1982.) Shopping centers offer occasional resting places to weary shoppers, but their interior malls are private enclaves whose owners are insistent on their right to deny public access at their discretion. Financially-stricken public institutions are enabled to remain open on selected days or evenings through the generosity of private patrons—who sometimes inscribe their corporate logos, in gilt letters, on the entrance doors.

In St. Louis, advertising space on parking meters is being sold to raise money for the municipality (*New York Times*, June 25, 1983). Gold medallions struck by the federal government are being sold by a private company (*New York Times*, April 13, 1983). Proposals have been made to introduce Coast Guard charges for rescues at sea (*New York Times*, April 24, 1983).

The private envelopment of what was once public space and activity reaches further. Some of the most prestigious museums are establishing joint housekeeping with corporate partners. In Stamford, Connecticut, for example, a branch of the Whitney Museum has moved into the Champion International Corporation complex. The same trendsetter, the Whitney, will be opening another branch—almost like a transnational banking corporation—in the Philip Morris Corporation's new headquarters on Park Avenue in New York City.

These are admittedly small markers but there are more substantive indicators of the decline of the public sphere. Public school systems are shut down weeks or months ahead of scheduled closing because of exhausted finances. Public libraries across the country are open on a part-time basis only and unable often to maintain even this partial service. Mass public transit in a number of major urban areas is near collapse. The public mails—about which more will be said later along—are in no better shape, and private delivery systems are flourishing.

Until there was an outcry from some parts of the private sector iteslf—from those whose interests felt threatened—the Reagan administration proposed to sell the national weather service to private enterprise (*New York Times*, May 14, 1983).

The Public Sector Defined

These conditions would seem to describe a decaying social order on the point of breakdown. Yet it is an incomplete picture, partly because it is based on a misleading understanding of what constitutes the public sector.

Also, it ignores other developments in the economy that demonstrate substantial growth and vitality.

The public sector is no monolith with a single social function. It is, in practice, a combination of operational machinery. Included is the bureaucracy, an assortment of social functions, and a large apparatus of coercive force. The blend, at any one time, depends on the historical moment and the contending strengths of the underlying social forces, the property-holding and the employed classes. Consequently, the public sector is an historically changing category, immediately and continuously the outcome of social struggles and political and economic crises.

America, of course, has been capitalistic from the outset with a few exceptions in specific locales. Activities were left to the state either because they were unprofitable to private enterprises or because they were deemed important to the development and protection of the new nation. In some instances both reasons applied. For example, the national mails were essential for national growth, but providing service to the outlying, vast continental settlements was hardly a profitable activity.

In the early decades of the republic, it was possible and indeed likely that the growth of the nation and expanded governmental functions corresponded to the interests of the governing class. At the same time, the services that were developed were of undeniable value and comfort to the entire population. Some of these included the mails, the promotion of transportation and commerce, the extension of literacy with schools and libraries, and the establishment of municipal and state utilities.

Later came telegraph and telephone and broadcasting. But these were private undertakings from their inception. They were regulated by the government as monopolies. Regulation of the telephone and, subsequently, broadcasting is illustrative of a public sector function that basically favored the ownership class. The regulation, more than not, accorded with the needs and well-being of the industry. At the same time, however, the regulation did confer general utility and satisfaction. For example, eventually universal telephone service became an accepted standard, and broadcast programming, for which no charge was levied, also was institutionalized.

In this sense, the bureaucratic-regulatory role of government, an important constituent of the public sector, can be regarded as an ambiguous function. It is of indisputable advantage to capital, but it is also capable, under certain historical circumstances, of affording benefits to the general public, including the least advantaged groupings in the nation.

There is, however, no ambiguity about the coercive component in the public sector. The state, from its earliest formation, is organized by and for its most powerful constituents and groupings. These groups supervise and command the coercive instruments, the police, the armed forces and the intelligence services. Officers holding the command posts in these services have been selected for their dedication and reliability to the prevailing order. This sector is public only insofar as it derives its financial support from general tax levies. The public pays for, but the governors depend on, the police and the rest of the control machinery.

There is, finally, the social welfare component of the public sector. Customarily, it is this sphere which is considered to represent what is called the "public sector." And it is true that this area comes closest to the ideal of a public sphere. It is concerned directly, to some extent, with education, health, old age, and general security.

Yet this part of the government especially is a socially determined category, representing at any one time the historical and contemporary balance of social forces. The welfare features in this component, protestations notwithstanding, have never been fully accepted or acceptable to the affluent governing class. Bitter battles continue to be waged in the United States and elsewhere for job security, workers' protection on the job, and for overall guarantees against the cyclical fluctuations of capitalism. It is precisely here that the current attack of capital is concentrated.

From all this, one can conclude that the public sector is a heterogeneous sphere of governmental nonprofit activity, embedded in a private, capitalist economy. It is never absolute, never unchanging, and frequently ambiguous insofar as the benefits and protection it offers to the general population.

Surrounding the public sector is the corporate sphere, benignly labelled the "private sector." Though there are actually millions of private businesses in America, it is the huge corporations that account for the greatest share of the nation's production, employment, profits, and investment. A few thousand companies, at most, hold the levers of national economic power and thereby largely direct the political system. The same enterprises undertake their operations on a global scale. A majority of them derive substantial and increasing shares of their revenues from their international activity. The accretion of wealth and power in the concentrated corporate sphere in recent decades has been staggering.

Information for Crisis Management and the Destruction of the Social Component of the Public Sector

What has happened to disrupt the "created harmony," the faint outlines of which Myrdal thought he saw and wrote about a quarter of a century ago? What forces have cut into the public sector and drained it of its social character while simultaneously promoting the growth of its coercive side? What explains the movement to privatize—turn into profit-making activities—so many functions until recently managed by public authority and regarded as socially desirable, non-profitmaking areas?

Decisive, though not an exclusive factor in these developments, is the tremendous power, wealth, and authority of the core, corporate sector, which has grown monstrously in the last forty years. Then, there is the reappearance of sharp economic crisis and the resulting intensifying competition between the Western, U.S., and Japanese market economies. Last, and coincidental with these developments, are the invention and deployment—also since the end of World War II—of new information technologies and processes.

Big Business after the War

The welfare state was one of the outcomes of the great crisis and long depression of the 1930s, though its origins can be traced further back still. The American economy did not climb out of its slump until the war abroad made enormous demands on American industries for maximum output. After the war, American industries found the domestic markets of the exhausted European powers—as well as the Europeans' colonial holdings—thrown open to their exports and investments. American companies grew rapidly in the postwar boom. Their assets climbed astronomically on the profits flowing in from international and domestic demand.

While the boom based on postwar reconstruction abroad and expansion at home lasted, the corporate order contributed to a growing social welfare sector. Also, there was no urgent need to check the advances of its working force. Deals were struck which allowed organized labor in the major industries to share, modestly, in the great revenue stream that flowed into corporate treasuries. In return, organized labor supported unquestioningly, indeed enthusiastically, American foreign policy, from cold war expansionism to antisocialist interventions.

When the boom finally began to subside and markets began to become far more competitive, American corporate controllers, not surprisingly, also began to see things differently. Speedily, economic concessions were sought from the state, and company obligations to social welfare were reduced to the barest minimum.

It is noted occasionally, though hardly given the attention it deserves, that the corporate income tax has practically been repealed through the practice of widespread exemptions and allowances. And, while corporate strategies for withholding funds from general social services achieve striking success, many individual property holders are no less adept in adopting tactics of tax avoidance. It is reported, for example, that "44 percent of all capital gains—profits on sale of securities, real estate, commodities and other forms of wealth—went unreported (to the Internal Revenue Service in 1981)" (Cowan 1982). And this is probably an understatement! A significant source of the public sector's economic woes, therefore, is the ability of property to determine national resource allocation to suit itself. Property holders, especially the major corporate ones, do not contribute to national housekeeping in proportion to their ability to pay, or to the benefits they receive.

The unwillingness of Congress to enact legislation in 1983, to insure the collection of taxes on interest and dividend income, is yet another reminder of the ability of the investing class to avoid what is, for the population at large, a national legal obligation (*New York Times,* July 29, 1983).

The corporate order, besides withdrawing its economic support for the social needs of the nation, has returned to its prewar, open and aggressive opposition to organized labor. Now, however, it possesses, along with the traditional weapons, modern means for combating labor. These include using consultants, public relations, advertising and general access to its own as well as the national informational system.

While availing itself fully of the legal, political and modern ideological means at its disposal, American big business continues to place its greatest reliance on familiar and traditional capitalist methods to maintain profits and retain markets. To overcome crisis, according to longstanding entrepreneurial wisdom, demands ruthless cost-cutting, cut wages, and sweeping rationalization of plant and equipment—labor displacement.

Information, Technology, and Business

Each national market economy proceeds from these same assumptions. It is for this reason that the new information technologies and processes are seen by managers and decision makers in all market economies as providential means of delivery from economic slump. The new equipment and systems, according to this logic, should increase productivity, cut labor costs, and intimidate the work force so that it will accept cutbacks and "give backs."

Still, according to this reasoning, labor costs throughout the economy will be lowered while entirely new lines of goods and services—computers, peripherals, programming, entertainment, etc—can be produced. These, it is hoped, may take up some of the slack created by the continuing slump in the older industries.

Those with a more daring vision see the new technologies as a means of reorganizing the global division of labor. In this view, the already industrialized countries will move to a new and higher rung on the international production ladder and maintain world dominance through the seizure of the information-handling machinery (Eger 1981).

Information as Commodity

There is more than mere fantasy involved here. Whether it will work itself out in the manner its promoters envision is still an open question, but the capability of the new technologies to change significantly the means of production seems no longer an unproved assertion. Consequently, the information sphere is becoming the pivotal point in the American economy. And, as the uses of information multiply exponentially by virtue of its greatly enhanced refinement and flexibility—through computer processing, storing, retrieving and transmitting data—information itself becomes a primary item for sale.

The emergence of information as a valuable good, applicable to a wide range of uses, is certainly one of the primary factors in the sweeping changes occurring in the economy. Along with the current drive of capital to cut costs and rationalize production to meet the crisis—and assisting in those objectives—information in its commodity role has become a basic determinant in the growing privatization in the economy-at-large.[1]

Activities, functions, and services that until recently were limited, relatively unchanging, and unprofitable, all at once have become potential and actual profit centers with the assistance of the new information technologies. Health, education, municipal services, information itself, suddenly have emerged as sites for private investment and profit-making activity. Banking, insurance, communication, advertising, travel, and entertainment are now dependent on massive information flows and vast amounts of data processing. These are the developments that are packaged attractively in the so-called information society.

As the possession of, or at least access to, information, is now a means to profit-making, information stockpilers and stockpiles are being swept up by private enterprisers searching out new areas for investment. On all sides, functions that rarely were regarded as revenue makers are being eyed, taken over, and reorganized with the assistance of improved information handling. This is the source of privatization in much of the economy today. It is especially visible in the information field itself, where longstanding nonprofit arrangements are being pushed aside to incorporate information production, processing, and dissemination into money-making activities.

The Commercialization of Information

The consequences of the changing importance and growing commercialization of information are extraordinary. Once information is a saleable item, as it now is, the public institutions that customarily have produced, preserved, and disseminated it—universities, libraries, and the government itself—are themselves forced to become privatized or lose their function in the information process. Accordingly, the observable changes affecting public informational and cultural institutions across the country, to a large extent, are attributable to treating information as a commodity.

As recently as 1979, a museum director could call upon the example of libraries and universities to defend the principle of public financing. The director of the Cleveland Museum of Art, for example, declared that museums "should be like libraries and universities and get their money with as few strings attached as possible" (Metz 1979).

This is no longer either an appropriate comparison or recommendation. Those units of the nation's universities which are sources of potentially profitable information—microbiology, microelectronics, artificial intelligence, etc.—are being integrated, financially or structurally, with corporate enterprise. In March 1982, for example, a private meeting—the public and the press were excluded—brought together the "presidents of five major universities with the leaders of ten high-technology companies and a group of prominent scientists . . . to try to work out guidelines to govern the growing commercialization of scientific research, especially in biotechnology"(Butterfield 1982).

Science magazine reported that the Pajaro Dunes Conference—named after the meeting site—was unable to resolve such issues as the exclusive rights that industry secures from the university research it sponsors (Culliton 1982). The search for agreement continues.

Meanwhile, the national laboratories of the federal government are being forced to yield to the same influence. There are 755 laboratories in the federal system "ranging from tiny facilities with a handful of employees to enormous installations the size of small towns" (Boffey 1983). The budget for these laboratories in fiscal 1984 was more than $15 billion, and the research carried out is on military, energy, health, space, and agricultural problems. *Business Week* (April 18,1983) reports that "the labs are opening themselves to industry. More and more companies, from Exxon Corp. to 3M Co., are enthusiastically forging collaborative ties with the labs on projects . . . ."

The account notes further that, in the past, the main obstacle to industry collaborating with the labs was "the difficulty of conducting proprietary research," that is, keeping the research findings private for company commercial utilization and application. This barrier has been swept away and the road is clear for the largest corporations to do just that. As current "partners" with various federal laboratories, *Business Week* lists, among others, Dupont, Exxon, 3M, Bell Labs, Gulf, IBM, Xerox, Rockwell, Allied Chemical, Hewlett-Packard, Ford, Babcock and Wilcox, Westinghouse and GM. "Many in government," *Business Week* adds, "find it difficult to justify granting exclusive patent licenses for work conducted at the tax-payers' expense."

Troubling to some or not, the private appropriation of public property has become the general pattern and it is being applied to the entire infor-mation generating sector. It is not only universities and governmental labs that are beginning to be enmeshed in a widening net of commercial enterprise.

Public and university libraries are experiencing similar pressures from the pull and tug of private information suppliers and vendors. Holdings and acquisitions are being put into machine readable formats, while librar-ies themselves are being obliged to link up their facilities with the databases offered by commercial vendors—Lockheed, SDC, BRS, and others.

Library information capability is greatly enhanced but accompanied by the abandonment of libraries' historical free-access policy. User charges are introduced. The public character of the library is weakening as its commer-cial connection deepens. No less important, the composition and character of its holdings change as the clientele shifts from the general public to the ability-to-pay user (Schiller 1981).

Similar powerful forces are at work in the nation's largest storehouse of publicly-generated information, the national government's information supply. Increasingly, the heavy output and great holdings of scientific and social information of the federal government are being organized, pack-aged, and sold by private information companies.

Information which was produced originally from the expenditures of public funds is being acquired by private information companies for com-mercial sale. The privatization of the governmental information supply is a complex process which includes the transfer of public data to corporate ownership and a combination of policies that are reducing or eliminating full public access to publicly compiled information (Schiller and Schiller 1982). The Information Industry Association wages a relentless campaign on behalf of private information vendors and packagers against public preparation and dissemination of governmentally acquired information (Kirchner 1983).

The privatization of public information in the governmental sector is undertaken with careful semantic protection. "In a relatively short time," a library publication notes, "the phrase 'national information policy' has had its meaning drastically narrowed from designation of the collective needs and rights of all Americans to a kind of code word expressing the concerns of the private sector, and especially its claims to the riches it perceives in the bureaucratic wilderness preserves of government-produced and distributed information "(*LJ/SLJ Hotline* 1982).

Commercialization and Privatization of the First Amendment

The new information technology, in addition to being used to create an ability-to-pay standard for information which used to be socially available, facilitates the privatization of information and human values in still more direct ways. The flexibility and penetrability of the new instrumentation, and the greatly expanded number of channels available for message transmission, enable nonmedia corporations—enterprises whose main economic activity is *not* media production—to engage heavily in message making and dissemination.

The television production facilities at the disposal of the big, nonmedia firms already surpass the installations owned by the commercial national media networks (Hurwitz 1979). The outpouring of corporate messages directly to the general public is still at an early stage, but already it is considerable. Cable TV and video cassettes are favored means of distribution. Programs for in-plant audiences ordinarily are transmitted through closed-circuit systems. But much of this corporate media production is offered without cost to the commercial outlets and may be played in entirety or in part, often without attribution to the original producer.

Some suggestion of the possible future significance of private corporate media outputs is provided by the rulings of the Supreme Court in recent years. In one important case in 1978, the Court ruled that "corporate speech" is entitled to the same protection as individual speech under the First Amendment (First National Bank 1978). This interpretation is startling, but not inconsistent with earlier opinions in which the Bill of Rights is made an accomplice to privatization and corporate power.

Ideological Privatization: Advertising

Alongside these new forms of private message creation and dissemination is the ongoing and continuously expanding volume of commercial advertising, penetrating deeper and extinguishing further what remains of individual and public space.

For decades, the United States has had the dubious distinction of leadership of the world in advertising expenditures, calculated either absolutely or on a per capita basis. With the growth of the transnational corporation, many nations are being elevated—if that is the appropriate way of putting it—to the American level.

The development of direct satellite broadcasting—signals from a communication satellite beamed directly into a home receiver without passing through an intermediary—produces euphoria in the advertising community. The chairman of J. Walter Thompson (Britain), one of the top ten transnational ad agencies, has this vision of the future for pan-European television: "For the first time, it would seem, we can fulfill the multinational's corporate dream: to establish at a single moment in time, eyeball to eyeball interface with the man in the street on a global scale. What power, what savings, what consistency, what an opportunity" (Newman 1982).

To those who thought the point of saturation had been reached in the United States, the projections of the advertising industry for the next twenty years and into the 21st century, indicate how mistaken they are. A tidal wave of commercialization, if the industry's expectations are to be taken at face value, is still to come (Coen 1980).

It seems to be so. Advertising outlays, in all media, rise from year to year. Though difficult to prove empirically, the impact of the flood of commercial messages on human values and behavior should not be underestimated. How much they contribute to the reinforcement and extension of already powerful patterns of individual acquisitive and consumerist behavior cannot be specified. What is clear is that the weakening of publicly supported cultural activity proceeds apace with minimal opposition.

Internationally, the same pressures are evident. Public systems of broadcasting in Europe and elsewhere are being weakened and diminished. Publicly financed systems are unable to withstand the assault of corporate-financed, advertising-supported media corporations.

Effects of Privatization

In sum, the massive privatization of the public sector, and especially its cultural/communication component, is the outgrowth of a combination of factors at work over recent decades. There is, to begin with, the enormously expanded wealth and power of corporate business in America and in a few other advanced industrialized countries. Multi-billion dollar transnational companies utilize the media and communication circuits for their direct and indirect messages. They saturate the commercial information systems of the world, while at the same time manufacture in-house growing numbers of messages.

The phenomenal growth of the transnational system is facilitated by the development of new information technologies which allow the tight coordination of transnational global operations. Additionally, the new instrumentation and processes provide expanded opportunities for profit making in the system overall, particularly in the field of information generation and data processing and dissemination. No less important, the new technologies are utilized increasingly in the entire production system to rationalize, cut costs, increase worker productivity, and, intentionally or not, decrease worker autonomy.

Under these powerful systemic forces, the public domain shrinks, its activities and functions increasingly relocated in the profit-making private sector.

Does it matter? If basic social functions are being performed, sometimes even more comprehensively with the new machinery and techniques, does it make any difference where the support comes from and how it is delivered? Does the source of the financing affect the product or service? Is there anything inherently harmful in the shift from public to privately-paid-for or assisted communication, education, and community services?

Economic Consequences: Division of Society

Some direct economic effects are immediately detectable. When information becomes exclusively a commercial product, it is information that is produced for profit. Who can pay for it and how much it will cost are questions that affect everyone. When an ability-to-pay criterion becomes the standard for information access—which is precisely what occurs when information provision and dissemination are turned over to market enterprises—the divisions in the society deepen. The poor become poorer still because they are excluded from the means by which their condition could be improved. The rich become more affluent than ever because they have the means whereby to consolidate and extend their power base.

Some of these developments are already evident. A local situation, for example, that has its national and international counterparts, is reported by Winerip (1983). Another example is government information, produced with public money, turned over to private information handlers and providers. These organizations then make the processed data available but often at prices far beyond the reach of ordinary people. A report on federal spending for information technology in 1983, using government statistics but produced by a private research firm, bore a $995 price tag (*Computerworld,* September 19, 1983).

Other examples abound. As the public mail service yields more and more of its functions to private businesses, the losers are the groups who cannot pay for the high-class private services—the rural communities, inner-city neighborhoods, and small suburbs (Califano 1983).

Similarly, as the national telephone system is "deregulated" and is withdrawn from public accountability, local telephone rates escalate, and the end may be in sight for a universal telephone service standard. Again, library user fees are becoming standard as libraries are compelled to computerize and tie into privately-owned data banks. Finally, universities that link up with the high-tech private sector overcome the funding troubles of those schools left out. In the dynamic tension developed in this increasingly popular relationship, courses, curricula, and research supporting the association of business and academia are privileged. Faculties that are outside the university-industrial nexus begin to ressemble a proletarian work force in higher education.

Political and Ideological Consequences:
Control of Culture and Communication

These economic consequences of privatization of the cultural/information sphere of the public sector are only beginning to be experienced. Still it is apparent that national social divisions are being widened and hardened. The social damage resulting from privatization of the information sector, surprisingly enough, may be deeper still and longer lasting, though extremely difficult to measure precisely.

Once withdrawn from its social context and made into an item for sale, necessary information may just not be available. This would occur not because of censorship, though this is no small concern, but for the reason that it will be controlled by the marketplace. Information we should have or might need may never be gathered, much less organized and transmitted. If it is produced, it will have to be purchased. With the destruction of public information, the basis of democracy disappears.

There are still other perils inherent in an information/cultural order administered and financed largely by a few thousand super corporations. One is that a people's social consciousness is made the prime target for ideological attack.

A recent study, generally favorable to corporate philanthropy in education and the arts, suggests the direction that may be expected in such an order. The question is asked: "Should business support those who don't support the system?" The reply:

> Corporations have no obligations to provide funds to organizations or individuals that are devoted to the replacement of our economic, political and social system by some other system. This would be foolhardy. Looking at it from another viewpoint, it is appropriate for corporations to provide support to worthwhile organizations that are proponents of the free enterprise system. . . . On the other hand, it is unwise for corporations to fund those who, while well motivated, would damage or destroy the system they nominally support with reforms that are unworkable or unnecessary. (Koch 1979, 131)

There is evidence enough to conclude that this is precisely the outlook of prevailing corporate authority. Gulf Oil, one of the patrons of public television, calls attention to the programming it supports with newspaper and magazine headlines such as these: "We Believe Television Should Command Your Attention," and "We Believe Television Should Explore Vast Wastelands. Not Be One." To generate what it terms "provocative" programming, Gulf collaborates with the National Geographic Society and offers the viewing public such items as *Polar Bear Alert, The Thames,* the world of wild chimpanzees, and efforts to save the whooping cranes, an endangered species.

All of these themes are surely unobjectionable and hardly ideological—except insofar as they are presented as provocative material and actually preempt time and resources that could be applied to genuinely thought-provoking work.

In the same vein, Mobil, also a benefactor of public television and other cultural projects, found it necessary to request the return of its contribution to the American Writers' Congress—a sum of $1,000—in the fall of 1981. According to Mobil's Robert P. Maxon:

> The meeting was not directed at preserving the character and quality of our literary culture—it was a political platform to advance causes contrary to the fundamental ideals upon which America is based. The political positions taken by the Congress are an anathema, and we disavow any association with them (*Publishers Weekly* 1982, 22).

The record could be added to at will. The decisive issue is not overt censorship by corporate free enterprise bluenoses, though of course this occurs regularly. It is rather the creation of a cultural-communication atmosphere in which the dialogue is purged at the outset of critical discourse, and significant, alternative formulations are avoided. Self-censorship becomes a built-in component in intellectual and creative work. Eventually, it enters into the mode of social existence.

This is the production side of the process. On the consumption side, that of the national public, impacts are no less great, and no less easily measurable. Navigating in a media-pervasive environment, almost all of which is commercially saturated, the viewer/listener/reader is overwhelmed with corporate ideology and is instructed day and night on the benefits of privatized behavior.

One prominent corporate supporter of the arts emphasizes that "bringing the arts to the people is the core of the capitalist system" (Hanes 1979). Perhaps. More persuasive is the acknowledged corporate effort to create for itself an aura of social responsibility and good citizenship by contributing to national cultural activity. David Rockefeller, in 1966, calling for a much larger role for business corporations in the arts, put it this way:

It can provide a company with extensive publicity and advertising, a brighter public reputation, and an improved corporate image. It can build better customer relations, a readier acceptance of company products, and a superior appraisal of their quality. Promotion of the arts can improve the morale of employees and help attract qualified personnel (Koch 1979, 131).

Philip Morris Corporation, a major contributor to the nation's health problem, as well as a generator of cultural pollution with its advertising, at least has a forthright board chairman. He has said: "We are in an unpopular industry. [While] our support of the arts is not directed toward that [problem], it has given us a better image in the financial and general community than had we not done this" (Metz 1979).

IBM's charitable art contributions are considered a "great insurance policy." And few failed to miss the greatly increased support, from the "seven sisters" who dominate the international oil industry, for public TV and related cultural events that coincided with the price-gouging and profit-taking accompanying the oil crisis in the 1970s.

The Gulf between Popular Understanding and Global Realities

These initiatives are all well within familiar traditions of the American historical experience. However, the current practices and institutional changes are unique and especially dangerous because they are deepening, at an alarming rate, the gulf between popular understanding and global realities.

An increasing number of Americans are being informed, educated, and entertained by corporate-created or sponsored media and cultural programs and materials. Often, these programs exclude or minimize or misrepresent the great social conflicts of our time and the near global insistence on fundamental change in economic and cultural relationships.

In this process, the informational apparatus, the arts, and the educational enterprise are being locked into supporting—or at least not confronting critically—a system of transnational business and culture which is coming under scrutiny and challenge almost everywhere else around the world. Communication and creativity in America, in sharp contrast, are being identified positively with corporate production and management.

The many ways in which the information structure and the creative material that passes through it are linked directly to the private sector may be of assistance in the near term in relieving pressure and diverting attention from the social issues that trouble people everywhere. In the long term, the efforts and outputs of the privatized cultural communication sphere will lose their credibility by their unwillingness or inability to illuminate the problems that affect Americans and everyone else. But the dangers are present now and nondeferrable.

For a very long time, we have known but not admitted that what is needed desperately in the country is an expansion, actually an explosion, of human services: educational, health, cultural, social and so on. It is a feat of a disordered imagination to see these created in a private context. Human services are the quintessential social expressions of community.

The fact that medical and health services are now provided increasingly by privately-owned and administered systems and that there are chains of for-profit hospitals, are startling indicators of how far privatization has entered into the fabric of American life (Butterfield 1983; Starr 1982).

Yet it is the private envelopment of the communication-cultural sector that severely limits and practically excludes public knowledge and open debate of these developments. The agendas of the major political parties avoid these subjects. Public discussion, for the most part, is nonexistent. The proposals for coping with the current, still relatively manageable, economic crisis—leaving aside the crises looming—are pathetic in their narrow vision. The most elementary and basic questions of political economy are ignored for fear of contradicting the tenets of privatism. Perhaps social solutions hardly can be expected to be formulated, much less implemented, when people have been systematically privatized in their physical, spiritual, spatial, and cultural lives.

But while privatization accelerates, so do the social breakdowns, at home and abroad. These intrude despite considerable efforts to contain and screen them out of popular awareness. For the fact of the matter is, the private sector, comprised of divergent interests, is unable to administer, much less coordinate, the national economy, especially in a period of increasing general turbulence.

The situation already has become bizarre. There is a crumbling public sector with dispirited supporters, under siege from a private sector, which is itself in continuous battle among its own groupings. Actually, the antagonistic and conflicting activity inside the private corporate sector limits, for the moment at least, the full force of privatization from being experienced throughout the social realm.

Whether private sector internal disagreements can be relied upon to protect the common interest in the long term, is quite another matter. The pressure of intensifying economic and social crisis more likely will compel the beleaguered system to find increasing attraction in a "command economy," in which the orders come from the most powerful private commanders. If indeed this is the direction in which affairs go, the communication/cultural sphere would be one of the first areas of attention and direct intervention.

It is moving to the rim of the abyss to put hope in the intensification of crisis. But it seems there are few, if any, alternatives. The sense and pursuit of community have been so badly undermined that their recapture requires a momentous reversal in social-political direction. In the years immediately ahead this seems doubtful.

The special lesson that the Thatcher government in Britain provides is that even a stricken, advanced capitalism still possesses a great capacity to smother its opposition, divert and disorient popular resistance, and push back living standards of the weaker sectors of society to pre-World War II levels.

Under what and whose auspices an alternative reconstitution of our own social order will occur, if indeed it does, are at this time unknown and perhaps unpredictable.

Note

1. Some see other explanations for the shift to privatization. Hirschman (1982) suggests a nonmaterial factor. In his reading of history as well as recent developments, he regards the shift from public to private involvements essentially occurring as the result of disappointment. The high expectations that individuals had for public sector activities were not fulfilled. In reaction, there is a pronounced turn to private preoccupations. The changes are expressed in terms of individual preferences and subjective valuations.

References

Boffey, P.M. Experts criticize U.S. laboratories for "deficiencies." *New York Times*, July 16, 1983, p.1.

Business Week. Industry finds a new ally in the national labs. *Business Week*, April 18, 1983, pp.44E-44K.

Butterfield, F. Town and gown of high tech seeking guidelines. *New York Times*, March 25, 1982, p. 47.

_____. Proposed sale of a hospital by Harvard is raising fears. *New York Times*, September 4, 1983, p.17.

Califano, J.A. The little guy will get hurt. *New York Times*, April 17, 1983, sec. 3, p.F2.

Coen, R.J. Vast U.S. and worldwide ad expenditures expected. *Advertising Age*, November 13, 1980, pp.1-16.

Cowan, E. Dole bill to ask brokers to file gains by clients. *New York Times*, March 6, 1982, p.1.

Culliton, B.J. Pajaro Dunes: The search for consensus. *Science*, 1982, 216, 155-158.

Eger, J.M. The international information war. *Computerworld*, 1981, 15(11a), 103-119.

First National Bank of Boston et al. v. Bellotti, 435 U.S. 765(1978).

Hanes, R.P. *Art News*. May 1979, p.61.

Hirschman, A.D. *Shifting involvements: Private interest and public action.* Princeton: Princeton University Press, 1982.

Hurwitz, S. On the road to wired city. *Harvard Magazine,* 1979 (September/October), pp. 18-19.

Kirchner, J. IIA statement calls for reliance on private sector. *Computerworld,* September 5, 1983, p.1.

Koch, F. *The new corporate philanthropy,* New York: Plenum, 1979.

LJ/SLJ Hotline. National information policy forum. 1982, 11(12),5.

Metz, R. The corporation as art patron: A growth stock. *Art News,* May 1979.

Myrdal, G. *Beyond the welfare state.* New Haven: Yale University Press, 1960.

Newman, B. European states face problem of controlling their neighbors' TV. *Wall Street Journal,* March 22, 1982, p.1.

Publishers Weekly. Mobil has second thoughts about writers' congress. *Publishers Weekly,* 1982, 221(5), 22.

Schiller, A. Instruction or information: What's changed? *The Reference Librarian* 1981, 1/2, 3-11.

_____, and H. Schiller. The privatizing of information: Who can own what America knows? *Nation,* April 7 1982, pp. 461-463.

Starr, P. *Social transformation of American medicine.* New York: Basic Books, 1982.

Winerip, M. In computer education, rich districts get richer. *New York Times,* June 24, 1983, p.16.

20

Videotex and American Politics:
The More Things Change. . .

David Blomquist

It is widely argued that videotex—two-way, computer-based information networks—will become an important medium of mass communication in the 1980s and 1990s. This article explores the potential effects of videotex on American politics, focusing on those effects that seem especially likely to materialize early in the medium's growth. It is argued that the critical difference between videotex and traditional mass media is the ability of videotex to personalize information as it shares it—to change its message to suit the particular needs and interests of each member of its audience. Based on inferences from the limited empirical data currently available, it is observed that this power of personalization could have considerable impact on political communication. But after reviewing the industry structure and technical constraints likely to evolve in the medium's first 20 years, it is concluded that most of this potential will probably remain untapped for decades. Nevertheless, several policy changes are suggested that could insure that videotex remains a benign influence on politics as it matures.

Nearly 40 years ago, at the dawn of television, sociologists Paul Lazarsfeld and Robert Merton looked back on what they knew about the media's role in society—and together they probably knew more than anyone else—and concluded that the purported power of the press was considerably overrated. "It is not unlikely," they asserted, "that the invention of the automobile and its development into a mass-owned commodity has had a significantly greater effect upon society than the invention of the radio and its development into a medium of mass communication." They explained:

Consider the social complexes into which the automobile has entered. Its sheer existence has exerted pressure for vastly improved roads, and, with these, mobility has increased enormously. The shape of metropolitan agglomerations has been significantly affected by the automobile. And, it may be submitted, the inventions which enlarge the radius of movement and action exert a greater influence upon social outlook and daily routines than inventions which provide avenues for ideas—ideas which can be avoided by withdrawal, deflected by resistance, and transformed by assimilation (1971 [1948], 558-559).

If Lazarsfeld and Merton foresaw that television would prove to be a more potent engine of social action than radio, they did not say so. But, then, how many people realized in 1948 that television would mature into arguably the most important new social force in mid-century America?

One can debate Marshall McLuhan's (1964, 25) notion that television became more influential than radio because TV "speaks, and yet says nothing," obliging the viewer to participate with it to fill in the blanks. Nevertheless, McLuhan's point that the television medium contains in itself the seeds of considerable influence is well taken. Edward R. Murrow, Walter Cronkite, Lucille Ball, and Norman Lear certainly assisted, yet television ultimately became a critical social force because, like the automobile, it is a technology that cannot help but expand its users' field of vision.

As they took possession of their first Model T's, Americans gained the ability to roam about relatively large distances entirely at their discretion; distance and time ceased to be such constraining factors. Similarly, the first DuMont or Philco in the living room became a window with an ever-changing view, a window that gave distant people, places, issues, events, and ideas compelling form and substance. Radio may have been cheaper and just as immediate as television, but seeing—unlike hearing—is believing.

The dawn of yet another mass medium, the computer-based, two-way hybrid of text and electronic drawings called videotex, is just around the corner. And though videotex systems aren't expected to be widely marketed until the end of the decade, comparisons to television and the automobile have already begun. For the technology driving videotex has amply demonstrated that it can dramatically change the ways in which people learn about their world, manage their money, conduct business, and amuse themselves.

Because videotex uses the extraordinary processing ability of computers to format and transfer messages, it communicates to mass audiences, yet personalizes its response to the particular needs, interests, and desires of each consumer. One might say videotex can talk to millions one at a time.

Consequently, much as the automobile gave people more power to decide for themselves when and where to travel, videotex gives consumers more power to decide for themselves what, when, and how to learn about their world.

Will videotex grow into a significant political force in the 1990s, as television did in the 1950s? The British media historian Anthony Smith, for one, thinks so. His historical perspective is typical: "The switch from a scribal society to a printing one changed the whole focus of knowledge in the West . . . . The transition of paper to telecommunications systems can hardly prove to be less important" (1980, 323). But John Carey reads an opposite lesson in history. After studying parallels between videotex and the early history of the newspaper and the telephone, Carey (1982, 83) concluded that "videotex in this century will more likely become a service for businesses and a small, elite group of consumers, not a mass audience"—thereby becoming important in some limited spheres, but not attaining anywhere near the influence of television.

There are very few empirical findings to support either perspective. At the end of 1983, no more than a half-dozen field trials of mass-market videotex systems had been conducted in the United States, and most of the research gathered in these experiments was still proprietary to the companies which sponsored them. Still, enough material has become available about the potential and reality of videotex use in the home to support some tentative hypotheses about the ultimate impact of changing communication technology.

This article, then, is—to borrow Daniel Bell's phrase—a venture in social forecasting. The view expressed here is neither alarmed nor euphoric. Videotex technology has the capacity to exert considerable influence in politics. Nevertheless, for reasons to be discussed later, it is most unlikely that videotex will fulfill that potential, certainly for the next generation, and perhaps beyond. However, the personal control which videotex affords each member of its audience is addictive, and over time may generate demands for similar personalization from other media that traditionally have been less permeable.

A Brief History of Videotex

Just defining videotex is itself a venture in forecasting. An informal international consensus takes the term *videotex* as a generic label for any electronic system that uses a modified TV set to send and receive information on demand from a distant computer (Tydeman et al. 1982; Gecsei 1983). But since there are a multitude of such technologies, *videotex* in practice means very different things to different people.

Videotex originated as a simple extension of conversational computer time-sharing, the now-familiar practice of rigging up a large mainframe processor to telephone lines and permitting several dozen people to run programs on it simultaneously. Such systems ordinarily allow customers to share more than the processor's time: because they are using the same data storage devices, customers can also swap programs and other information with each other.

In the mid-1970s, Britain's publicly-owned telephone network, seeking to raise revenues by inducing people to make more local telephone calls—unlike the former AT&T operating companies, the British have long charged customers for each local call—came up with the idea of a time-sharing system in which the focus was on sharing information rather than computing. The storage disks of the central computer would be loaded with millions of words about thousands of subjects—everything from the latest news headlines to encyclopedia articles, thoroughly indexed and immediately available. Instead of rummaging through books or newspapers for information, the British reasoned, people could let the time-sharing computer do the walking.

The British Telecom system, dubbed Prestel, began public operation in 1979. To join the Prestel network, customers lease a decoder and typewriter-like keyboard, which hooks up to their telphone lines and to any color TV set. A 3-pence local call connects them to the Prestel computer. It receives their typed requests, fetches the desired information from the files, and sends it back to the decoder, which displays it on the color TV. Each request takes between five and fifteen seconds to process.

There is not much glamor and flash to Prestel. For the most part, all that appears on the TV screen is text. There are no pictures, at least not the sharp, vivid pictures we associate today with color television. The only respites from text Prestel can produce are crude, fuzzy drawings—line drawings painfully assembled from squares of colored light, much like mosaic designs in a tile floor.

The earliest American systems didn't even go that far. Beginning in the late 1970s, computer hobbyists took over two existing commercial time-sharing networks after business hours to exchange news. There were no drawings whatsoever, and the networks made few changes at night in the stilted, clipped command language used by programmers during the day. Yet two such networks, The Source and Compuserve, survived and grew—grew enormously, to more than 100,000 subscribers by the end of 1983. Soon thereafter, AT&T, IBM, and several other major American communication and computing firms became interested in Prestel's information-sharing concept. These companies developed decoders with much better drawing capabilities than Prestel. Nevertheless, the emphasis remained on fetching and displaying text.

This meaning of videotex—an electronic system for retrieving text and simple pictures upon request from a central computer—predominates in the field today. Outside the computer industry, most of the commercial interest in videotex in the United States now comes from newspapers, magazine publishers, and broadcasters, who think of videotex as an "electronic newspaper." But text isn't the only sort of information computers can share. Indeed, the convergence of computer, telecommunications, and video technologies is rapidly making text the least likely candidate.

The designers of Prestel assumed, like the designers of any time-sharing system in the mid–1970s, that the customer's decoder would be no more dextrous than an ancient Western Union teletypewriter; the Prestel decoder serves only to convert the customer's typed instructions into telephone signals and to convert the computer's telephone signals into text on the customer's TV screen. The first information-sharing services tested in the United States by AT&T and IBM made the same assumption.

By the beginning of 1984, however, at least a dozen companies in the United States operated small systems that assume the customer's decoder is a "smart" personal computer rather than a "dumb" Teletype-like terminal. Like Prestel and its American cousins, these systems contain vast amounts of textual matter, and can retrieve any piece of it upon request in a few seconds. But they can also do much, much more. They can monitor a home's burglar and fire alarms, automatically signaling authorities if alarms are tripped while the owner is away. They can play Pac-Man, balance checkbooks, and chart biorhythms. They can even display real color television pictures, not stubby mosaic drawings, with stereo sound.

This more intelligent model of videotex appears all but certain to supercede Prestel look-alikes in the United States by the end of the decade. The difference in cost between a "dumb" terminal and a midrange personal computer is now trivial, and several proprietary industry studies have found that consumers prefer videotex packages which offer the additional capabilities of a personal computer. Moreover, the tremendous growth in sales of home computers—industry analysts project between 26 and 70 million households will own at least one personal computer by 1988 (*VideoPrint,* December 8,1982; CSP International 1983b)—will undoubtedly spur further development of videotex networks to serve that market.

Videotex and Traditional Mass Media

With a personal computer as the home decoding device, videotex becomes vastly different from traditional mass media. In the Prestel model, videotex is little more than an automated library, a reading machine; it works much as if hundreds of printed books and newspapers had been photographed on slides and arranged in a huge carousel, available to be projected for viewing at will.[1] "Smart" decoders free videotex from this dependence on the written word. When delivered through a personal computer, videotex is no longer just print carried over a telephone line but a medium in its own right—a medium with the depth and breadth of print, the immediacy and intensity of broadcasting, and an extraordinary quality all its own: the power to personalize information as it shares it.

In traditional mass media, every member of the audience receives the same message; different people may perceive it in different ways, but the message itself is the same. If you and I watched the CBS Evening News last night, we saw the same stories presented in the same order. If you and I picked up the *Wall Street Journal* this morning, we read the same front page.

Advanced videotex systems, on the other hand, put the audience in charge of the message. They use the processing capabilities of the central and home computers to permit every member of the audience to decide for himself or herself what the message ought to be. One subscriber might ask the computer just for news about the Middle East and toxic waste; another might inquire about taxes and the MX missile; a third might not ask for news at all. And each would come away from videotex with very different pictures of the world—the many pictures that each wanted to see, not the single picture that some editor or producer thought they all ought to see.

In theoretical terms, the difference between traditional mass media and videotex is that videotex technology provides mass audiences with an instantaneous and highly effective feedback loop. No other form of communication provides such immediate, personal response—not even one-on-one conversation. When I meet a friend on the street, I can try to steer talk toward my needs and interests, but I succeed only insofar as my friend will cooperate. For the most part, however, the computers driving a videotex network will fetch me whatever I want whenever I want—and do so for hundreds and hundreds of people simultaneously.

A simple matrix captures this distinction (see Figure 20.1). Face-to-face conversation conveys a personalized message, yet only to an audience of one. Lectures and other group discussions reach somewhat larger audiences, but the message is no longer so personal; it must be aimed at the group's least common denominator. Traditional mass media are completely impersonal, but can reach vast audiences over great distances. Yet videotex can reach vast audiences while adapting its message to the particular needs of each audience member.

FIGURE 20.1
Videotex, unlike Traditional Mass Media, Talks to Millions One at a Time

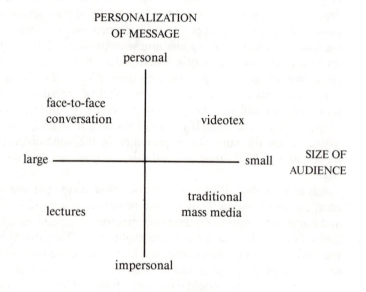

This does not mean that videotex eliminates gatekeeping. Someone still has to decide what information goes into the computer and what programs are available for getting at it (Graber 1982). But the dimensions of the gatekeeping problem in videotex are considerably different from traditional media. Television and newspapers choke on space limitations, yet in videotex practically nothing needs to be excluded on grounds of space alone; computer storage is simple, efficient, and dirt cheap.

Indeed, videotex's ability to personalize information effectively brings audiences into the gatekeeping process—and not just to help watch their own gate but to contribute to the information available to others, too. For information-sharing can work in both directions; not only can the central computer share its database with distant decoders but the "smart" decoders—the home computers—can share their information with the central machine. In other words, each consumer of information through videotex can also be a producer of information for videotex—an opportunity that no traditional mass medium easily affords.

Videotex and Political Involvement

Some political scientists, most notably Ted Becker, are excited by the prospect of all this new power in the hands of voters. Becker believes videotex will bring new vigor to American government by making direct democracy—the active, informed, universal involvement in politics venerated in the New England town meeting—feasible in a large republic. He writes: "Teledemocracy—. . . electronically-aided, rapid, two-way political communication—could offer the means to help educate voters on issues, to facilitate discussion of important decisions, to register instantaneous polls, and even to allow people to vote directly on public policy" (1981, 6).

That capacity horrifies others. Michael Malbin (1982) recalls Madison's warnings in the Federalist Papers about the tyrannical potential of direct democracy. Not only does the deliberative process of a representative legislature help thwart mob rule and tyranny of an ignorant majority, says Malbin, but it promotes the refinement and enlargement of political views by forcing participants to compromise, to think of the needs of others. Gerald Benjamin (1982) concurs, noting that previous advances in communication technology have been accompanied by less political participation, not more.

But the possible political impacts of videotex extend well beyond mechanization of voting. Our research at CBS suggests that bringing consumers into the gatekeeping process makes them feel more intimately involved with the information it produces; information that one helps create seems to be much more compelling, even when it comes from the same sources and is conveyed in the same language as messages in traditional media. Consequently, the potential effects of videotex encompass not only new tasks its powers can perform, but the state of mind these powers can instill in the videotex audience.

If any empirical studies of videotex's impact on politics or political beliefs have been conducted—either in the United States or abroad—they are not well publicized. In Europe, videotex services are operated by government-owned utilities, which have a vested interest in overlooking the medium's political significance. At the end of 1983, the only videotex services no longer considered experiments in the United States still couldn't display graphics, and their subscribers were mostly computer hobbyists—hardly a representative population.

Therefore, to anticipate how videotex might affect politics, one can for now only call on educated guesses drawn from the little evidence that is available. Naturally, this process goes just so far, since it is impossible to foresee in abstract all the ramifications new technology may have for social systems, especially when the technology is as different from its predecessors as videotex. Yet many crucial impacts do seem immediately apparent—enough to imply that the full picture will be well worth pursuing.

It has been suggested that the impact of mass communication on politics varies with three elements: (1) the technology through which communication takes place; (2) the environment in which it transpires; and (3) the content of its message (Blomquist 1981, 1982).[2] Let us consider how the differences between videotex and traditional mass media in each of these areas could generate distinctive effects.

The Role of Technology

Television technology puts a premium on political messages that are highly visual. The more the message stimulates our eyesight, the more it is likely to be absorbed. Videotex technology favors short messages that prompt audience involvement. The ideal videotex message must be short because many home computer screens hold only about 75 words at once, and each additional chunk of text requested entails a significant delay. It must encourage audience involvement because individual choice is the essence of videotex technology—and, indeed, its most important potential effect on political communication.

Our experience with radio and television demonstrates the impact on political discourse of communication technology which favors brief messages: events, concepts, and problems tend to be oversimplified or portrayed in extremes. Videotex may be no better. The central database can file items of any length, but the existing telephone network can only distribute so much so fast; each 75-word screen with a small drawing may take fifteen seconds to travel from the host computer to subscribing homes. Tests by CBS and other videotex system operators have found that few consumers have the attention span to wade through screen after screen of text on one topic at that rate; most stop after one or two screenfuls (that is, no more than 150 words) and continue browsing for something else.

For candidates and other political actors accustomed to traditional media, the close, continuing relationship videotex maintains with its audience will be strange, maybe even frightening. Print and broadcast, which must deliver the same message to every consumer, favor masters of generalities—that is, those who can spin a single tale that appeals to large numbers of people. Videotex, which enables each consumer to pursue his or her individual interests, favors those who can devise multiple messages appealing to the idiosyncracies of dozens of small subaudiences. Of course, this is not a new art in American politics; candidates for major offices routinely concoct several stock speeches for use before different audiences—one for farmers, one for labor groups, one for chambers of commerce, and so forth. But videotex will exacerbate this tendency for politicians to build coalitions by appealing to interest groups rather than the collective welfare.

Audiences seem to give unusual credence to information received through videotex. Perhaps this should be no surprise. One of the recurring findings in the social sciences is that participation reduces alienation. People are more likely to endorse the outcome of a process if they have been involved in it. The experience with evolving media technology has followed this pattern. As each technological advance establishes a more vivid and realistic stimulus—a stimulus that intensifies the consumer's feeling of "you are there"—audience involvement with and trust in the media increases as well. Television is but the latest, though most extreme, example. The halftone liberated newspaper readers from dependence on reporters' prose to visualize the scene of events, thereby lending new urgency to columns of gray type. And radio enabled distant audiences to share for the first time the nuances of meaning that only voice inflection can convey (would FDR's "Martin, Barton, and . . . Fish" have been so effective in print alone?).

The opportunity to share in the gatekeeping process apparently adds to an audience's sensation of vicarious participation, thereby imparting an extra measure of believability. In the winter and spring of 1983, CBS and AT&T jointly sponsored two in-home trials of a prototype videotex system. Each trial placed videotex terminals for three months in 100 upscale households in a northern New Jersey suburb. The terminals accessed a database of more than 20,000 screens of information prepared by CBS journalists, updated around the clock during the trial. At the end of the test period, the male and female household heads completed a battery of questionnaires probing their response to videotex technology and the prototype service. Our analysis of these questionnaires, reported in Table 20.1, found that videotex was rated significantly more trustworthy, objective, and accurate than traditional mass media—even though the content presented in the prototype largely duplicated wire services, local newspapers, and CBS broadcasts word-for-word. If commercial videotex services maintain this advantage in credibility, videotex may supplant television as the critical source of information about politics in households that subscribe to it.

On one hand, it is tantalizing to think of audiences becoming more actively involved in the flow of information. Though the notion of a totally passive audience is long dead, it is nevertheless fair to argue that traditional

TABLE 20.1
The CBS/AT&T Videotex Experiment

Question: Using a scale from 1 to 5, please indicate how well, in your opinion, each
of the five media listed across the top of the table is described by the
characteristics listed on the left. The better a characteristic describes a
medium, the higher the number you should give to the medium for that
characteristic.

Characteristics:	News-papers	Tele-vision	Maga-zines	Radio	Videotex[1]
Trustworthy	3.1	2.9	3.2	3.2	3.6
Involving	3.1	3.2	3.3	2.8	3.4
Objective	3.0	2.8	3.0	3.0	3.6
Accurate	3.4	3.1	3.3	3.3	3.7

Mean Ratings: 5 = Highest, 1 = Lowest
☐ = significantly smaller than videotex
(95% level of confidence, two-tailed test)

[1] Participants rated videotex the most credible mass medium.
Source: Postplacement questionnaires, CBS/AT&T videotex field trial, Ridgewood, New Jersey,
Wave 2, 1983 (N = 170).

media—especially television—inculcate passivity and escape more than
they encourage activity and involvement. But videotex compels audiences
to action; after all, the machine just sits there unless the consumer gives it
instructions. Writes John Wicklein (1981, 260): "Rather than depositing
people in front of their sets for hours, . . . an interactive system has the
potential for whetting their appetites for human interchange and pulling
them out of their homes to make that happen."

On the other hand, one is aghast at the thought of audiences having final
say over what information they receive. To be sure, audiences of traditional
media also have the final say about which pictures they assimilate and
which they reject. Yet is is one thing to reject an issue or idea upon reading
or viewing it, and another to have a machine unobtrusively cull it out for
you. Selective perception from print or broadcast is an active process—
subconcious, perhaps, but at least at some level one actively decides to
ignore or reinterpret new information. Selective perception from videotex,
on the other hand, can be deceptively passive. The computer displays only
what it is instructed to retrieve; everything else stays out of sight and thus
out of mind. At least traditional media offer some hope that an errant

reader or viewer will bump into political information enroute to the weather forecast. With videotex, people can avoid politics altogether—and one suspects many will.

And how valid is this sensation of vicarious participation, anyway? To be sure, interactive technology permits audiences to have more of a say about what they see and read. Yet the final authority still rests with the system's designers and operators. Audiences can't retrieve information that isn't loaded into the computer or that isn't "flagged" by the program which responds to their requests. The videotex consumer may play a more active role in gatekeeping than the consumer of traditional mass media but is still far from equal to the gatekeepers who maintain the central computer.

Consequently, the power to "get what you want when you want it" that videotex claims to provide can be rather an illusion. Videotex encourages audiences to think that they alone are in command; when a request for news about a certain subject comes up empty, the consumer usually cannot tell whether it failed because in fact nothing newsworthy has happened or because the system programmer omitted that information. It is a dangerous paradox: audiences come to videotex significantly less suspicious than they approach traditional mass media, and yet can be almost undetectably misled.[3] As Wicklein (1981) points out, the videotex system operator is in a strong position to impose its peculiar tastes of what is and isn't fit for public viewing unseen and unchallenged. And unlike human editors, computers don't have second thoughts: if the system programmer leaves an interest, issue, or point of view off the agenda—intentionally or unintentionally—the machine will never notice.

As the differences between the mediated world and political reality grow less and less obvious, the likelihood of what Lazarsfeld and Merton called the "narcotizing dysfunction" of the media increases—the possibility that citizens will confuse absorbing information through mass communication with doing something about it in the political system. Without leaving home, the videotex subscriber can survey the environment, focus in on issues of particular interest, record his or her opinion, and ask to be kept abreast if the situation changes. But politics as we now know it demands more than pushing buttons on a box in the living room. At the very least, someone has to decide what questions the machine should ask.

Moreover, as Malbin and Benjamin observe, such livingroom referenda threaten the political institutions which nurture long-term decision making. If every issue faces public review every day, what chance is there to develop coherent strategy, to think about the linkages between problems, to replace rhetoric with ideology? The ability to pass up quick band-aids for more complicated but enduring policies depends upon an "information float" (Naisbitt 1982, 14)—a gap between the implementation of decisions and citizen feedback about government's performance. Yet the "float" is already too small; policy choices are already too affected by last month's Gallup poll or the prospect of next year's congressional election.

Finally, the videotex system's capacity to record individual interests and generate mailing lists of subscribers likely to respond favorably to particular messages has enormous potential for single-issue organizations and candidates without cohesive constituencies. Television advertising and direct mail have already enabled interest groups and politicians to reach voters without the help of political parties. Direct video—electronic mail appeals sent instantly by lobbies and candidates to voters identified by the computer as likely supporters—might similarly reinforce the declining significance of party organizations.[4]

Naturally, this capacity poses a considerable threat to personal privacy. If the central computer keeps a diary of every screen requested by each subscriber, should it be available to politicians, interest groups, and parties—or anyone?

The Role of Environment

Political communication through videotex is likely to take place in a somewhat different context than conventional mass communication. Even the simplest videotex networks require a decoder and some familiarity with computers (or at a minimum the ability to be at ease around them). As a result, the videotex audience will probably be better-off financially than audiences for television and some print media, at least for the next decade. And though videotex will develop strong local roots—after all, the industry won't get off the ground if people must call long-distance to reach the central computer—system ownership will likely be concentrated in a few national companies.

Though subscription fees and hardware costs are declining, videotex will remain sufficiently expensive through 1990 that many families will be priced out of the market. Charter subscribers to Knight-Ridder's Viewtron, the first commercial videotex system in the United States to incorporate high-resolution graphics, paid $600 in 1983 for an AT&T decoder, $12 per month for basic service, and $1 per hour for the telephone line. Cheaper decoders that plug into existing home computers should become available in the mid-1980s, but computer and decoder together probably will still run at least $500.

Even if lower-income families could afford a home computer and videotex decoder, most wouldn't know how to use it. Computer literacy is largely restricted to the middle-class, and likely to remain so for years to come. White-collar workers deal with computers every day in the office;

their children attend suburban school systems that purchase microcomputers and program libraries for every classroom. Blue-collar workers get
laid off when computers come into the factory; their children attend big-
city and rural schools barely able to meet the payroll each week.

Of course, no mass medium begins life with a truly mass audience. It
took eleven years before 75 percent of American families owned a television set (Tydeman et al. 1982). But during those years before widespread
acceptance, those who own a new medium have an advantage—in terms of
both the information they possess and in sheer social status—over those
that don't. In its early years, then, videotex may add to the gap in political
knowledge between rich and poor.

With the ascendance of network news, television became a force for
nationalizing politics. The networks naturally concentrated on national
officeholders and issues. So far, videotex has been predominantly national
as well; every public videotex system operating in the United States at the
beginning of 1984 worked out of a single computer center, with distant
subscribers reached by long distance. By the end of the decade, however,
system operators will eliminate the telecommunication overhead by establishing remote computer centers in major markets. No doubt these remote
affiliates will develop local editorial content to support local advertising,
shopping, and banking.

Television fostered local news, too, but has remained primarily a national medium. Local TV stations cannot hope to cover well all two or
three score municipalities within range of their signals—not in a couple
hours of air time each day. The videotex affiliate, blessed with unlimited
storage capacity and the ability to deliver neighborhood news and advertisements just to residents of that neighborhood, may evolve a more robust
local presence. Indeed, the initial Viewtron package, targeted to residents
of metropolitan Miami, included news from each high school and Little
League team in south Florida.

Nevertheless, the industry will probably be dominated by a few nationwide system operators—the videotex equivalents of television networks. In
the first place, videotex consumes enormous quantities of capital—perhaps as much as $250 per subscriber. Not many companies have pockets
that deep. Second, big national systems can take advantage of favorable
economies of scale. Once written, electronic encyclopedias, cookbooks, or
travel guides can serve a hundred markets as well as one. Finally, most
industry observers think households will subscribe to only one videotex
service; consequently, the handful of operators offering the most services
for the least price will carry away most of the business.[5]

Centralization is already more the rule than the exception in traditional
mass media. Network programs account for 79 percent of prime-time tele-

vision viewing; chains control two-thirds of the nation's newspapers.[6] Beginning with the Hutchins Commission (Commission on Freedom of the Press 1947), critics of the press have often maintained that concentrated control inhibits the marketplace of ideas by replacing dozens of independent voices with cookie-cutter copies of one or two slick formulas. Videotex certainly won't make matters easier if it turns out to be an oligopolistic business as well.

The Role of Content

Videotex started out as merely an electronic newspaper, a simple transfer of print journalism to a computer so that news and features could be retrieved more easily and refreshed constantly with the latest details. But few still envision videotex as just the news. Today, the greatest appeal of videotex to consumers and advertisers is its power to act on information by performing services: its abilities to manage bank accounts, play video games, organize shopping, and to carry out dozens of other household tasks. As we depend on videotex more and more for these services, our life and leisure time will focus ever more intently on a sleek box in the living room. That may have significant impact on our evaluations and expectations of each other.

Harold Lasswell (1971, 85-86), writing for the same 1948 symposium as Lazarfeld and Merton, identified three basic functions of communication that have become touchstones for media research. Communication, Lasswell said, gathers information (the surveillance function), analyzes information (the correlation function), and transmits the social heritage from one generation to the next (the socialization function). A dozen years later, Charles Wright (1960) added another function Lasswell overlooked: entertainment.

The evolution of videotex from a high-tech carrier of information to a vehicle for acting on information suggests that Lasswell's list needs to be expanded once more. For it seems increasingly likely that people will use videotex most often not to read news stories (surveillance), editorials (correlation), or educational materials (socialization), but for entertainment and for personal service—for checking a bank balance, shopping for the lowest price on Levi's, ordering tickets for the Boston Symphony, or countless other personal tasks.

At the conclusion of a nine-month market test in southern California in 1982, Times Mirror Videotex Services asked heads of its 350 sample households to choose the content areas they felt were "essential" to have on videotex. The top three choices were entertainment or service functions—video games, rated essential by 79 percent of the respondents; shopping and product information, 72 percent; and electronic bill paying, 71 percent. News headlines placed fourth; only 60 percent of the respondents said news was essential (*VideoPrint*, April 8, 1983).

And though American field trials committed substantial resources to news coverage, the positive consumer response to and profit opportunities in service activities probably will cause services to take the lead in full-scale videotex businesses. CSP International, a telecommunications consulting group, estimates that videotex system operators will divide between $850 million and $1.1 billion in revenues from electronic shopping transactions alone in 1990, or about $4 per month per subscriber (CSP International 1983a).

If videotex had remained solely an electronic newspaper, one might be able to dismiss its other potential effects with the same argument Lazersfeld and Merton used against radio: an institution that merely conveys information "can be avoided by withdrawal, deflected by resistance, and transformed by assimilation." But when it began to offer personal service—the ability to do something with information as well as to consume it—videotex "enlarged the radius and movement of action" of its subscribers. Suddenly people could accomplish practically anything from a living-room easy chair.

The commercial possibilities of this function are almost unlimited. These days, the two-worker or single-parent family, for which time is at a premium, is increasingly the norm. Videotex can provide busy people an effective, more convenient, and often less expensive option for coping with mundane needs.

But one wonders if this convenience might be too much of a good thing. If videotex subscribers can complete most everyday transactions without leaving home, will they lose interest in the outside world? If I can do most of my shopping in a few minutes through videotex, will I ever realize if downtown stores are in decline? If my computer always points me straight to the best buy, why should I care that groceries cost more in Harlem than in Scarsdale?

It is by no means clear that videotex will have such a dehumanizing effect. Indeed, it is just as plausible that videotex will broaden its subscribers' social contacts. After video games, the most frequently-used features of videotex systems—especially among women—are the communication services, through which subscribers can talk back and forth to one another. In interviews after the CBS/AT&T market trials, several consumers reported becoming friends with other test participants that they met by exchanging notes on an electronic bulletin board.

As the videotex audience becomes more heterogeneous, subscribers might find themselves striking up relationships through electronic mail with people they would never address face-to-face. Electronic mail gives no hint of the correspondent's gender, race, or social class; it introduces people anonymously and without prejudice. Over time, it might help break down the invisible barriers that still separate Americans from different backgrounds.

Obstacles to Videotex

It isn't difficult to devise any number of chilling "worst cases" from the potential effects described above. Imagine this 1990 scenario: levels of political awareness plummet still further as 7 to 10 million videotex subscribers turn the power of a hundred-million-dollar information network loose to keep track of Wayne Gretzky; the president loses a critical budget vote on Capitol Hill after videotex telepolls taken the night before show resounding opposition to higher taxes; a huge media conglomerate buys a videotex network and rearranges the master control program for its entertainment database so that movies produced by the conglomerate's studio always print out first.

The worst is possible, to be sure—but it is very, very unlikely. A variety of obstacles—perceptual, technical, and organizational barriers—will almost certainly control and constrain the growth of videotex through the 1980s, limiting the medium's actual political influence.

It is conceivable that videotex won't even make it out of the starting gate. The medium utterly depends upon the success of home computers, and industry pessimists are having doubts about the extraordinary growth in microcomputer sales widely projected in 1982 and 1983. The source of doubt, as Marcian E. Hoff of Atari recently put it, is that home computers have become "a wonderful solution looking for a problem" (*Time,* February 6, 1984). There are plenty of obvious applications for microcomputers in the office, but not many for around the house aside from video games, which are swiftly losing sales momentum. How many homes write enough letters to justify a word processor? Who needs VisiCalc to balance a checkbook?

The devotees of interactive technology think consumers will be so captured by videotex that it will sell home computers all by itself. Well, maybe. Most consumers who try videotex do seem enthralled by its convenience and personalized control. Nevertheless, John Carey (1982) argues that Americans are so accustomed to media which do all the work for them that videotex will seem too intense and demanding. Videotex, he writes, "proceeds from an assumption that users need or want specific pieces of information," and want that information badly enough to master how to get it out of the computer.

Yet "the existing patterns of information consumption by the general public emphasize habitual behavior." In most cases, people "will not make a decision to buy the newspaper, but will do it as part of a regular pattern, like brushing . . . teeth." Indeed, he asserts, the traditional media became large and influential only when they fostered habitual usage. If videotex encounters the same behavioral tendencies—and Carey believes it will— the record suggests "a very slow growth for this new medium as an infor- mation utility in the decades ahead" (pp.81-86).

If videotex does find a niche in the marketplace, it is doubtful that it will become the primary source of political news for any large number of people. This is simply the flip side of the preference for personal services noted above. The convenience of banking and shopping at home sells videotex to more consumers than the prospect of automated news. From mid-1980 to mid-1982, the Associated Press and eleven member news- papers prepared daily news summaries for the Compuserve text-only vid- eotex system, and issued terminals without charge to 100 households scattered across the country. The AP found that news accounted for less than 10 percent of the average test family's time on the Compuserve net- work. Moreover, this usage was heavily skewed toward a few respondents; one out of ten households accounted for half of all news reading (Blasko 1982).

It seems most plausible that people will use the information retrieval capabilities of videotex as a backstop for traditional media—as a quick means to follow up a specific headline heard first somewhere else. This is consistent with the CBS/AT&T market test experience, in which more than a third of the time consumers requested fewer than ten screens (750 words maximum) of information at a sitting. This suggests they often knew what they were looking for before they signed in. As Carey observes, videotex is not well suited to browsing: telecommunications delays make idle scanning of pages impossible; reading quantities of text off a screen soon tires the eyes; people on the go can't carry the computer with them on the train. And even with video-game graphics, videotex just isn't as visually satisfy- ing as television.

The upper and upper-middle class people who will be the first videotex subscribers are already the most intense consumers of mass media. The University of Michigan's quadrennial studies of the presidential electorate ask citizens which of the traditional media—newspapers, magazines, radio, and television—they use to follow politics. In 1976, respondents with the highest incomes were 44 percent more likely to use all four media than one or more, while respondents with the lowest incomes were 30 percent more likely to use just one medium or none at all than to use all four (Miller, Miller and Schneider 1980, 311). Even if videotex establishes itself as the most credible medium, it will still be only one information channel among many, facing tough competition for this audience's time and attention.

Many possible applications of interactive technology will likely be closed off by consumer concerns about privacy. Fear of Big Brother is deeply felt in this country; videotex companies that peddle personal data gathered by their computers will encounter sharp backlash. As an executive of Warner Communications' Qube interactive cable system told Wicklein (1981,27): "We have to be very careful and set up very strict rules. If we abuse them, we're fools." Certainly some system operaters will attempt to sell whatever diary data they can without killing the golden goose, but open access to consumers' reading habits is improbable.

The large corporations poised to take the lion's share of the videotex business are more likely to cultivate conservative uses of new technology that are variations on things they already know rather than radically departing from traditional media fare. Big organizations distrust the unknown and venture into it only by little steps; they foster gradual change, not sudden innovation.

In addition, many of these companies are already successful publishers or broadcasters. They enter videotex with well-entrenched expectations and operating practices forged in their traditional businesses—expectations which may not work in a very different new medium. Broadcasters who cut their teeth on radio took twenty years to realize that television was a visual medium and to produce TV news with that in mind; videotex will require a similarly long and painful adjustment.

These major system operators will surely avoid anything that might tempt government regulation. Even though broadcasters and the Federal Communications Commission have had a fairly cozy coexistence, station owners still felt encumbered by Washington bureaucrats. So far, videotex is virtually unregulated, and videotex companies—some of whom have first-hand experience with the FCC as broadcasters or telecommunications utilities—want to keep it that way. Since the more political influence videotex acquires, the more public officials are likely to seek to regulate it, system operators feel a stong incentive to maintain a low profile.

Finally, as Malbin (1982) points out, there is no reason to believe that the sudden appearance of videotex in American livingrooms will increase the electorate's interest or propensity to participate in politics. Many people, maybe even most, have a political attention span of six weeks every four years. Teledemocracy could be fun for a while, but one suspects that soon enough another diversion would come along, and teledemocracy would join Pong, Asteroids, and Space Invaders in computer heaven.

Conclusion

Videotex, then, will not abruptly alter the face of politics when it becomes generally available in the late 1980s. Like television, which took a generation to evolve from John Cameron Swayze seated in front of a wall map to Walter Cronkite ankle-deep in South Vietnamese rice paddies, videotex will mature very slowly, becoming a medium in its own right and a critical social and political institution only after years of experimentation.

But that does not mean we can forget about videotex for a decade or two. We should use this growth period to take steps which will help insure that videotex remains a positive force in politics as it matures.

First, we should make certain that all would-be operators of videotex systems have equitable access to local telephone exchanges. This will promote maximum diversity within the marketplace. Videotex depends upon local lines like magazines depend on the mails. And just as the mails are open to any periodical, local lines must be equally available to any videotex system able to pay for them. Telephone companies shouldn't be able to decide who can and cannot launch a videotex service.

Second, we should establish a legal right for videotex subscribers to know what records are being kept of their activity, to examine copies of those records, and to restrain distribution of this information as they choose.

Third, we should evaluate whether a public-owned videotex network ought to be established to ensure that groups, ideas, and opinions that don't make a profit have an outlet in this medium.

Fourth, we should take every opportunity to encourage computer literacy in elementary and secondary schools. The sooner children understand computers, the sooner they (and perhaps by osmosis their parents) will comprehend the advantages and limitations of videotex as an information source.

Finally, editors and publishers should start thinking now about the ramifications for traditional media of a competitor that treats audiences as a partner rather than a subordinate. Newspapers, magazines, radio, and television are not especially permeable institutions. Traditional news organizations, obsessed with professional pride, often strike outsiders as narcissistic and aloof. Yet even when its choices are illusory, videotex denies the carefully cultivated fiction that only professional journalists can decide what's fit to print. Conventional media need to consider how they can be more participatory, more receptive to audience feedback, before the public discovers how few are the clothes that the emperors wear.

Notes

An earlier version of this paper was presented to the Block Foundation conference on communications technology and American politics, Yale University, in April 1983. I am grateful for the comments of participants in that conference and Henry Heilbrunn, Carol Landers, and David Shnaider of CBS. The views expressed herein remain entirely my own.

1. This metaphor originated with my colleague David Waks.
2. I employ somewhat different category labels here than previously to reflect some minor improvements to the scheme.
3. This conflict may well be part and parcel of evolutions in media technology. Each new medium appears more realistic than its predecessors—and is, to some extent. A television camera seems to capture a football game with less artifice than a sportswriter; videotex systems, unlike print or broadcast, seem to speak with the give-and-take of real conversation. Yet each technological advance merely makes the deception more difficult to find; the mirage grows more lifelike.
4. National party committees have found some new life recently as clearinghouses for mailing lists and technical assistance. No doubt they could fulfill the same role for videotex, though personal computer technology tends to decentralize expertise: these days, a clever 12-year-old with an Apple in her bedroom often can accomplish as much as a fair-sized Washington consulting group.
5. One can quarrel with this collective wisdom, which is based largely on the industry's experience with cable TV. The thinking goes that people will spend about as much on videotex per month as cable. The average cable subscriber spends around $20 monthly for basic service and extra channels like Home Box Office and Showtime; therefore the monthly videotex ante will be around $20—or enough, at current market rates, to buy one basic service with a couple frills. But it is entirely possible that videotex subscription fees will decline substantially as the installed base increases and advertising, shopping, and banking revenues foot more of the bill. This would enable households to buy into several videotex services for the same $20 monthly charge.
6. These figures were provided by, respectively, the CBS Television Network research staff and Paul Jess of the University of Kansas.

References

Becker, T. Teledemocracy: Bringing power back to the people. *The Futurist,* 1981, 15, 6-9.

Benjamin, G. Innovations in telecommunications and politics. In G. Benjamin (ed.), *The communications revolution in politics. Proceedings of the Academy of Political Science,* 1982,34(4).

Blasko, L. Untitled summary of AP/newspaper/Compuserve experiment. New York: Associated Press, 1982.

Blomquist, D. The mass media and politics: a systems approach. Paper presented at the annual meeting of the Northeastern Political Science Association, 1981.

Blomquist, D. *Elections and the mass media.* Washington, D.C.: American Political Science Association, 1982.

Carey, J. Videotex: the past as prologue. *Journal of Communication,* 1982, (Spring), 32, 81-86.

Commission on Freedom of the Press. *A free and responsible press.* Chicago: University of Chicago Press, 1947.

CSP International. Electronic financial services in the home. In *inContext Profit Opportunity Series.* New York: CSP International, 1983a.

_____. Personal computers: market trends and quantitative projections. *inContext Update,* 1983b, (November), 3 (4).

Gecsei, J. *The architecture of videotex systems.* Englewood Cliffs, N.J.: Prentice-Hall, 1983.

Graber, D. Executive decision-making. In G. Benjamin (ed.), *The communications revolution in politics. Proceedings of the Academy of Political Science,* 1982, 34(4).

Lasswell, H. The structure and function of communication in society. In W. Schramm and D. F. Roberts (eds.). *The process and effects of mass communication.* Urbana: University of Illinois Press, 1971.

Lazarsfeld, P. F. and R. K. Merton. Mass communication, popular taste, and organized social action. In W. Schramm and D. F. Roberts, (eds.) *The process and effects of mass communication.* Urbana: University of Illinois Press, 1971.

Malbin, M. Teledemocracy and its discontents. *Public Opinion,* 1982, (June/July), 5(3), 58-59.

McLuhan, M. *Understanding media.* New York: New American Library, 1964.

Miller, W. E., A. H. Miller and E. J. Schneider. *American national election studies sourcebook, 1952-1978.* Cambridge: Harvard University Press, 1980.

Naisbitt, J. *Megatrends.* New York: Warner Books, 1982.

Smith, A. *Goodbye Gutenberg.* New York: Oxford University Press, 1980.

Time. The zinger of Silicon Valley. *Time,* February 6, 1984, 50-51.

Tydeman, J., et al. *Teletext and videotex in the United States.* New York: McGraw-Hill, 1982.

VideoPrint. Norwalk, Conn.: International Resource Development, 1982, 1983, various issues.

Wicklein, J. *Electronic nightmare.* Boston: Beacon, 1981.

Wright, C. R. Functional analysis and mass communication. *Public Opinion Quarterly,* 1960, (Winter), 24, 605-620.

21

Information Technology and
Social Problems:
Four International Models

Jerry L. Salvaggio

*This chapter argues that: (1) The relationship between information tech-
nology and social problems is not a simple cause and effect relationship
but is a complex process; (2) for purposes of understanding the complexity
of this process at least four different models representing four different
types of societies can be studied; (3) the difference in each society is the
manner in which information technology is "developed and used"; (4) the
manner in which information technology is developed and used can be
attributed to at least six variables; (5) the key variable in all six models is
ideology; (6) the key variable which determines how information technology will be developed and used is variable x; and finally (7) ideology
and variable x will largely determine the nature of the social problems
which can be expected in a given society. The significance of these propositions lies in the implications of variable x. Basically, it is argued that
variable x operates in a position of dominant influence. While the other
variables also influence the development and use of information technology, the influence of variable x is the overriding factor.*

While it is fairly clear that there is a close correlation between the adoption of new forms of technology and the occurrence of new social problems, the process by which social problems evolve as a result of the
introduction of new technology is unknown. The objective of this article is
to offer a model which will facilitate the study of the process by which
information technology is adopted and new social problems emerge.

The first part of this article provides an outline of the various forms of information technology which have the greatest potential for creating social problems. In the second part, social problems which are most often associated with information societies are discussed. Finally, a model with six variables will be presented and each variable will be examined within the context of four different versions of the model.

Information Technology

Low-Power Television

A relatively new form of information technology is low power television. The FCC will eventually grant more than 7,000 low-power licenses to organizations wishing to enter the low-power television arena. Because these stations can be built for less than $150,000, many organizations which have never been in the broadcasting field will soon enter (Sears, newspapers, etc.). Already, more than fifteen newspapers have been granted permits by the FCC (*Presstime* 1983). This means increased cross-ownership of media and the possibility that newspapers could utilize low-power TV stations to transmit an electronic version of the newspaper.

Subscription Television

Subscription television is very similar to pay-TV, as offered by cable, except that the programming is broadcast, not cablecast. The medium uses an ultra-high frequency to send out movies and sports to those who have a special converter which will decode the scrambled signal. As of 1983 there were approximately twenty-seven stations operating and another sixteen being built. This represents rapid growth, though not a major threat to conventional television. In 1982, commercial TV industry revenues were close to $17 billion compared to $420 million for subscription television (Macy 1983). Subscription television is often cited as another form of information technology which will further fragment American audiences and contribute to the decline in homogeneity.

Cable Communication

From its humble beginnings in the early 1950s, cable has grown to more than 5,000 systems serving 32 million homes in the United States. More importantly, cable continues to add subscribers at the rate of 400,000 subscribers per month.

West European countries have only recently decided to wire urban areas with coaxial cable though cable growth could be rapid in certain European countries. France, as an example, hopes to have 1.4 million homes wired into one network by 1985. Among European countries, Belgium, Ireland,

FIGURE 21.1
Growth of Cable Systems

As of 1984 there were more than 34 million
cable subscribers and the industry was growing
at the rate of 400,000 per month.

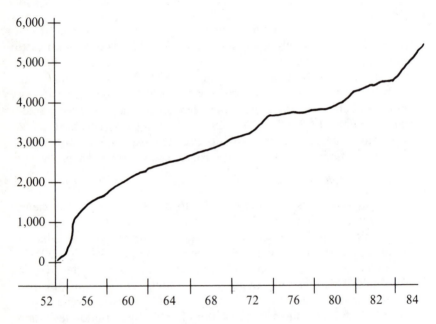

Source: U. S. Statistical Abstract, 1952-1984.

the Netherlands, Switzerland, and the United Kingdom all have at least 28
percent penetration rates (*Transnational Data Report* 1983).

Cable systems are of two types: one way and interactive. The number of
homes in the United States with interactive capability is approximately
350,000. While this number is expected to grow rapidly it is likely to
remain small compared to the total number of subscribers. Interactive
systems are capable of offering banking, shopping, courses, security, and
pay-per-view programming.

Cable TV is most often cited when the issue of privacy is raised due to
the system's ability to collect data on individual use of the media (Baldwin

and McVoy 1983). Other problems associated with cable communications include pornography, monopoly, and inequity.

Telephone Communication

Telephone and related technology are no longer the single domain of AT&T. In 1982, when AT&T, a corporate structure worth over $140 billion, was restructured the entire telecommunication picture in the United States changed. MCI, General Telephone, GTE, and others are now competing with seven Bell companies for telephone service. In addition, AT&T can now enter the lucrative computer area.

The importance of the telephone system is that it is the primary information infrastructure utilized for transmission of computer, voice, TV, and videotex data. In the United States, as an example, there are 180 million telephones compared to approximately 17 million personal computers.

The significance of the telecommunication industry in the U.S. economy is generally underestimated. AT&T estimates that American business spends nearly $700 billion a year on telecommunication, postage, written correspondence, travel, and meeting expenses. Currently, telecommunication systems account for a mere 10 percent of this total (*Business Week* 1982).

As will be seen later, the issue of information inequity is directly tied into the telephone system. This is due partially to the fact that the telephone system is the most pervasive form of information technology and partially because the telephone is more important than those forms of information technology which primarily provide entertainment.

Computer Communication

The following areas within the computer field have the greatest potential for altering social patterns in the next decade: microcomputers, supercomputers, artificial intelligent computers and large database firms. Estimates are that the personal computer business will be a $5 billion industry annually by 1990. In 1983 alone approximately $6 billion was spent on personal computers in the United States by 150 computer makers (Bukeley 1983). Adoption of personal computers is somewhat slower in other countries but there is little doubt that personal computers will be common throughout the developed world by the end of the decade.

The social significance of the rapid adoption of personal computers lies in the fact that personal computers facilitate networking among individuals and between individuals and institutions.

Mainframe computer technology is not receiving the headlines accorded to personal computers, but developments in this area are not only as dramatic but possibly more significant in terms of changing social patterns. In 1983, the world's fastest computers performed 80 million operations per second. Super computers, like Seymour Cray's new Cray X-MP, will perform 400 million operations per second, and in 1985 the Japanese firm, Nippon Electric Corporation, will make its SX-2 commercially available. The NEC computer will process 1.3 billion calculations a second. The Japanese hope to build a super personal computer which could perform 10 billion operations per second by 1990.

Artificial intelligence is also emerging as a significant industry with potentially profound social consequences perhaps altering employment patterns. Stanford University and Massachusetts Institute of Technology already have programs (DENDRAL and MACSYMA) which use heuristic problem-solving methods. AT&T's LISP program is able to correct complex grammatical errors such as split infinitives and redundancies. The objective of Japan's Fifth Generation Computer project is to design a computer with artificial intelligence capabilities. So far, the Japanese project has spurred twelve major American corporations to form a nonprofit joint venture to meet Japan's challenge. The cooperative effort is called Microelectronics and Computer Technology Corporation (MCC). The objective of MCC is to build a new class of intelligent supercomputers.

Computer communications encompass virtually all of the major social problems though privacy, misuse, social control, and monopoly are those most often associated.

Databases

Databases are growing worldwide with approximately 80 percent of all world databases located in the United States. At last count there were 1,350 databases in the United States representing a 40 percent increase over 1981, but the majority of these were business and scientific databases.

Access to the average database begins at $25 per hour and runs up to $300 per hour for more complicated searches. The potential for information inequity is considerable. Since 1975, the potential threat of databases, when used by the IRS and other government organizations and the corporate community, has come to the fore.

Fiber Optic Communication

Today, the bulk of data (computer), image (broadcast), and voice (telephone) is transmitted through copper wires in an analog form. Fiber optic, or lightwave, systems utilize the far more efficient digital form of transmission, which is sent through hair-thin glass fibers. Bell Labs invented the laser in the late 1950s and commercially tested lightwave technology in Chicago in 1977. The development of fiber optics began in 1970. By 1986, the telephone industry is expected to spend $400 million annually on these systems (U.S. Industrial Outlook 1982).

The cable industry is also expected to eventually switch from coaxial cable to fiber optic technology. United Cable Television, in Alameda, California, may be the first optical fiber cable system. Initially, it will provide 25,000 subscribers with 120 channels and such amenities as all-channel emergency announcement override, home security, FM voting, polling, parental controls, and pay-per-view.

Japan has made even more progress with fiber optics (Salvaggio 1982). In 1976, Japan began a $14 billion project named Hi-Ovis. Hi-Ovis, a small city, is wired with 350 kilometers of fiber (150 homes) providing two-way visual communication to every home. The success of Hi-Ovis has led to Japan's INS project which will entail the laying of fiber optics to all residential units in the next 15 years. The United Kingdom and West Germany have plans to wire their respective countries with fiber optics in the future.

Fiber optics systems, when combined with national telecommunication systems, make telebanking, teleshopping, telecommuting and telepolling possible on a national level. In a sense, fiber optics systems are the infrastructural basis for future information societies and all the social implications which go with information societies.

Satellite Communication

Since 1965, with the lauching of Early Bird, the world's first operational communication satellite, this form of information technology has grown in both sophistication and worldwide use. In 1986, INTELSAT will launch INTELSAT VI with 33,000 voice circuits or 120 TV circuits.

More significant is the potential of Direct Broadcast Satellite technology. Whereas conventional satellites need ground stations to pick-up their signals, direct-broadcast satellites only require a small antenna (3ft.) and a decoder. On September 23, 1982, the FCC issued a direct-broadcast satellite license to Satellite Television Corporation. Later that year the FCC issued seven more to applicants including RCA, American and Western Union Telegraph Company. International Resource Development, a market research firm, estimates that by 1990 there could be over 15 million rooftop earth stations (*Communications News* 1982). The potential social implications of these satellite systems are many, ranging from inequity to transborder information flow (see Le Duc 1983 and Smith 1976).

Teletext and Videotex

Videotex and teletext are electronic information retrieval systems allowing home subscribers or businesses to utilize commercial mainframe computers (see Martin 1982; Sigel 1983; Tydeman et al. 1982) by connecting a personal computer, or an especially designed converter, to one's telephone. Videotex is a technological service made possible by the convergence of the computer and the telephone. Individuals are able to send and receive information, shop, exchange messages, and receive electronic news in the home.

While there is no accurate count of how many homes subscribe to a videotex or teletext service in the United States, there are two major systems, CompuServe and The Source, and more than a dozen newer systems including Qube, BISON, Green Thumb, Dow Jones, EIS, and Viewtron, Knight-Ridder's $25 million venture in Coral Gables, Florida. Some are experiments, others are fully commercial.

Other countries which have videotex systems that are either commercially operational or are in the experimental stage include: Canada, Denmark, Finland, France, Holland, Japan, Spain, Sweden, United Kingdom, and West Germany (Tydeman, et al. 1982). Communist and Third World countries seem less interested in this form of information technology.

Estimates on how many households will subscribe to videotex vary from very few to 38.3 percent of all Americans by the year 2000. CompuServe, which is the largest system in terms of subscribers, had more than 80,000 subscribers in 1984 and was growing at the rate of 1,500 per week (CompuServe 1983).

Summary of Technological Use and Development

It is apparent that various forms of information technology are rapidly being adopted in the United States, Japan, and Europe and to a lesser extent in the Soviet Union and Third World countries. The majority of these systems are being used by both business and consumers. Taken as a whole, information technology is both bringing individuals closer together in the sense of McLuhan's "global village" and providing them with hundreds of specialized programs, creating a trend away from a homogeneous society. Information technology is also growing in speed and power allowing corporate and government agencies to better organize and analyze data on individuals.

Social Problems

Invasion of Privacy

Invasion of privacy has been a concern to civil libertarians since the advent of mainframe computers. Concern turned to virtual paranoia when computers and the telephone networks were interconnected to allow computers to talk to other computers. The birth of interactive cable and videotex has caused additional concern since both systems use the computer to collect and store information of a highly personal nature. The increasing use of electronic mail adds to the potential. As of September 1983, the U.S Postal Office alone handled 14.3 million messages on E-Mail, its computer-originated electronic mail service (Lublin 1983). In addition to the post office there are now at least seventy commercially available electronic mail services.

Scholarly studies on privacy in the information era have grown as well. Those studies which are especially relevant to information technology include: Burnham 1983; Dertouzos and Moses 1981; Donner 1981; Garfield 1979; Hiramatsu 1983; Linowes 1978; McLuhan and Powers 1981; Salvaggio 1983; and Westin 1972.

Information Misuse

Information is invisible, quickly transferred across great distances and easily destroyed. It is thus not surprising that the increase in information technology has given rise to increased apprehension in regard to information misuse (Allen 1982; Faber 1983; Goldhamer 1979; Krause and MacGahan 1979; Parker 1983; Salvaggio 1984).

Information technology is not only easy to misuse but the risk of prosecution is slight. Home video recorders are used to tape pay-tv movies. Writers can access literary and scientific databases, copy the material, and submit it for payment. Satellite dishes allow individuals to retrieve pay-programs free. But, most importantly, textual information such as medical, legal, financial, and personal information is easily accessed, manipulated, stored, and destroyed if one has a home computer and a telephone modem.

Social Control

The potential of controlling society through increased use of information technology has rarely been discussed in academic literature though it is worth considering here (Ellul 1983). Using new information technology, it may become possible to control movement. In Hong Kong, for instance, an electronic sensor which detects the number of times an individual's car passes a certain point is being tested. The sensor (on the bottom of a vehicle's chassis) sends a signal to another device buried in the road which then sends the signal to a main computer. The army is testing a "dog tag" which contains a micro-chip capable of keeping track of all soldiers in the field. Also being considered is an electronic ankle bracelet which will allow penal institutions to keep track of parolees. Each of these devices increases the potential of governments to monitor and control individual activity.

Information Inequity

Information is economic power and a commodity (Schiller 1978; Bell 1982). Information inequity operates at two levels: within societies and

FIGURE 21.2
Telephones Worldwide[1]

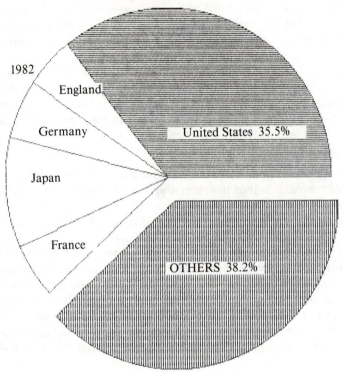

1982
England
Germany
Japan
France

United States 35.5%

OTHERS 38.2%

[1] Five Countries have 61.8% of all telephones
Source: U. S. Statistical Abstract, 1982.

among societies. The international problems are reflected in the formation of the New World Information Order; a consortium of Third World countries which have banded together to demand greater transfer of information technology.

Inequity is likely to be a serious problem to solve because it is not always in the best interest of those who can solve the problem to do so. Inequity often takes place as a result of geographic location. Cable firms, especially those with interactive capability, do not construct systems in rural areas or in nonaffluent areas due to the considerable cost of construction. Since cable is a private industry, it cannot be required by law to construct systems in these areas.

If new forms of technology, such as the computer and videotex, follow the pattern of international distribution set by the telephone, inequity will be the most serious problem of the information society.

Information Monopoly

The information industry is owned by hundreds of firms yet a handful of conglomerates provide a large percentage of the content. Such corporations as Time Inc., Cox Cable, Times Mirror, Gannett, Knight-Ridder, and IBM transcend single forms of telecommunications media and according to some scholars, control the bulk of information published (Bagdikian 1983). Other scholars point to the diversity of information offered through new forms of information technology (Compaine 1982). Yet, other scholars point to the trend toward large transnational corporations like Citibank, Exxon, and Chase Manhattan which are becoming increasingly powerful in the information sector (Schiller 1983).

Control

Control of information and information technology is both easier and more difficult in an information society. It is easier to acquire a greater variety of information because we can enter a large number of databases. It is becoming more difficult to get information crucial to the democratic process from the government due to the latter's paranoia concerning technological information leaving the country. The corporate sector is also making more information proprietary to guard against industrial espionage and sabotage (see Bell 1983; Forester 1982; Salvaggio 1983).

Aided by new technologies, an institution or a government can also more easily control the amount and kind of information which is released. The Reagan administration, for instance, is being accused of attempting to bring about an era of secrecy in the government due to the number of proposals President Reagan has submitted to limit the scope of the Freedom of Information Act (see Pear 1983).

Pornography

Traditionally, media programming has been regulated by the Federal Communications Commission (FCC). With the advent of subscription TV, pay TV, pay-per-view TV, video cassette recorders, video discs players and direct-broadcast satellites, the notion of regulating the content of electronic media is not only unnecessary but borders on the impossible. Various forms of information technology have thus taken advantage of the situation and are offering the public hardcore pornography. Sales of pornographic video cassettes are very high in the United States and the top seller in Japan. Cable franchises report that pay channels have great difficulty signing subscribers without the promise of R-rated films. Despite the fact that these newer forms of technology allow parents to control access to pornographic material, protest groups have been growing, and in some states laws have been enacted (see Smith 1983).

FIGURE 21.3
Variables Determining the Development of Information Technology

Ideology   —   Political Parties   —   Economic Systems

Technology   —   External Forces

Policy Making Organizations   —   Infrastructural Systems

Information Overload

Blumler (1980), Pelton (1981), and Klapp (1982) have argued that the sheer deluge of information in every field with which workers must keep up alienates individuals. Individuals suffering from information overload may devote less time to reading political, economic, environmental, and social issues due to the abundance of information which specialists must stay abreast of in an information society. Many believe that a consequence of specialization is that more individuals will leave the running of government to professionals thereby undermining the democratic process.

Model of Major Variables

Hypothesis one

Earlier, it was suggested that social problems evolve out of technological change through a complex process as opposed to a simple cause and effect relationship. It is suggested that there are a limited number of variables (Figure 21.3) which determine the manner and the extent to which information technology is developed and used in a particular society.

Four models, which loosely represent four different international systems, are examined here. Each model is a variation of Model One (Figure 21.4). Variable x is different in each model. Variable x represents that factor—outside of ideology—which exerts the greatest influence over the development and use of information technology. This will be discussed in greater detail later.

Scholars of international communication have traditionally divided world communication systems into three categories: Western societies, Communist societies and Third World societies. Here, a fourth category is added which distinguishes the American system, which has become increasingly private, from other Western societies. For purposes of simplicity the four models are referred to as the Competitive, the Public Utility, the Communist and the Third World. A much longer study would be required to place particular countries in each of these models. It is also likely that

FIGURE 21.4
Model One

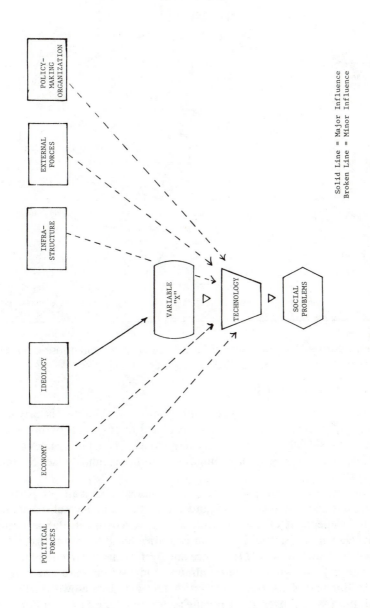

FIGURE 21.5
Competitive Model

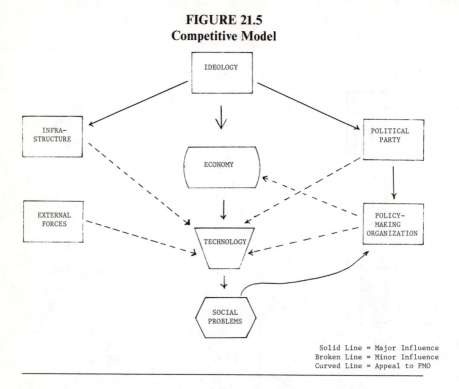

Solid Line = Major Influence
Broken Line = Minor Influence
Curved Line = Appeal to PMO

several additional models would be needed to represent those countries which do not fall into one of these categories.

Definition of Four Models

Very broadly, countries where information technology is privately owned and operated—and either unregulated or regulated by numerous organizations—fall into the Competitive Model. The United States clearly falls into this category. Canada might be another candidate (see Schmidt and Corbin 1983).

Those countries where information technology is private, or partially owned, but highly regulated and guided by a strong policymaking organization independent of the government are placed in the Public Utility Model. Japan and several European countries fall into this category (see Shiono 1978 and Wells 1974). France and W. Germany currently fall into this category, though their telecommunication systems are government owned. England plans to turn its British Telecom, a state-owned telecommunication system over to the private sector by selling 51 percent of Telecom to investors. This will move Britain closer to the Competitive Model

FIGURE 21.6
Public Utility Model

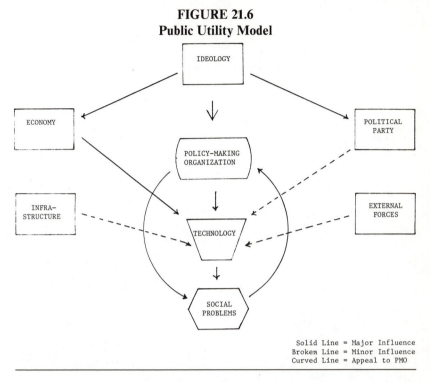

Solid Line = Major Influence
Broken Line = Minor Influence
Curved Line = Appeal to PMO

in some respects. A country like Norway, where telecommunications and broadcasting are state controlled, is problematic and may require a fifth model (Emery 1969).

Countries where information technology—computer, broadcasting, telephone, and satellite communication—is owned by the government and managed by a central communist party are represented in the Communist Model. Albania, Bulgaria, Czechoslovakia, Cuba, the People's Republic of China and the Soviet Union are included in this model.

The Third World Model is self-defined. In many respects Third World societies overlap with the above models because of the unique manner in which these countries adopt—or are not able to adopt—information technology. To this day, 90 percent of all telephones and television receivers are in 15 percent of the 157 nations (Pipe 1983). In both Third World countries and in Communist countries the government and /or the political party own or control all forms of information technology (Hollstein 1983).

Hypothesis two

It is hypothesized here that two variables dominate the development and use of information technology. The first variable is ideology. The second

FIGURE 21.7
Communist Model

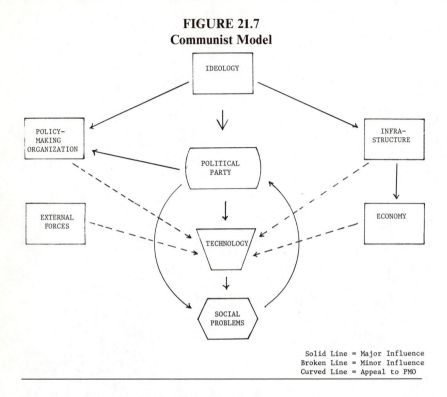

Solid Line = Major Influence
Broken Line = Minor Influence
Curved Line = Appeal to PMO

variable differs from society to society and shall be referred to as variable x. Variable x will be discussed in greater detail after discussion of the major variables in the context of the four different societies.

Ideology

The term *ideology* refers to the foundational doctrines which prevail in a society at a given time. While there is a sense in which one would not want to engage in a philosophical debate over what the ideology is in a given country on a given day, it is fairly safe to assume that most societies operate on a limited number of basic principles or set of ideas which are embedded in that society. These principles are sometimes formalized and written into a society's constitution. At other times religious, cultural, social, economic, and political ideas are only vaguely agreed upon by members of that society. And, in certain societies, two or more sets of ideals are in conflict.

There are a number of theoretical observations which shall be put forth here relative to ideology. First, because ideologies are so complex, involving every component of a society's value system, virtually every society will have a different ideology though several societies may share a large com-

FIGURE 21.8
Third World Model

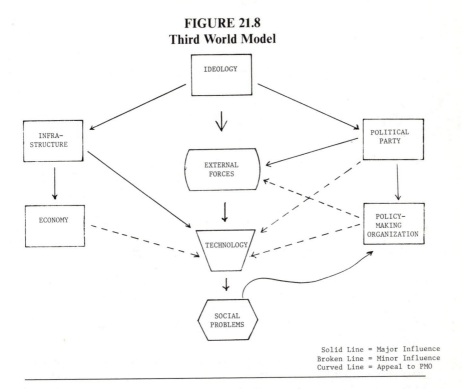

Solid Line = Major Influence
Broken Line = Minor Influence
Curved Line = Appeal to PMO

mon base. The manner in which ideologies overlap is represented in Figure 21.9 where the common base is the democratic system. Other components of ideology are the cultural, religious and social systems in which democracy can be embedded. In Japan, democracy is intermixed with religious aspects of Shintoism and Buddhism, the remnants of a monarchy and a socially homogeneous system. In the United States democracy is intermixed with Christian beliefs and a social system characterized by ethnic diversity.

Second, *ideology* may also refer to institutional influences which may not be stated in a political document, but nevertheless do prevail. In certain societies religion may become a major force guiding the development and use of information technology. This is especially true in Third World countries where Hindu or Islamic religions prevail. One example might be the case of Khomeni in Iran. And Vasquez (1983) points to the way in which radio was originally opposed on religious grounds in the Arabian Peninsula.

Third, it can be assumed that all ideologies are in a continual state of change, though the pace of change may vary from very slow in the United

FIGURE 21.9
Overlap of Ideologies

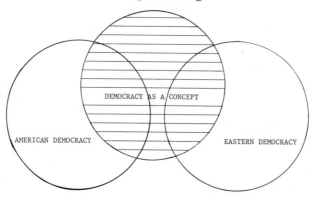

States to very fast in unstable societies. France's recent socialization of major industries is an example of an ideological shift in recent times. An important characteristic of Third World countries is that the dominant ideology changes, often before it's agreed upon by the various political forces.

The major point to be made here is that the root of all social problems associated with information technology is latent in a society's ideology. For this reason ideology is placed at the top of each model as a given constant and fundamental variable.

Political Party

As seen in the broken line from *political party* to *social problems,* the former is not a significant variable in the Competitive Model. Generally, the party in office is able to neither influence the development of technology nor create social problems of a lasting nature. An example is President Reagan's attempt to control information by passing a directive which would require government employees to undergo lifetime censorship of all publications and be subjected to a polygraph test. Congress (one of many American policymaking organizations has already delayed enforcement of the directive and several senators have indicated that the directive may be unconstitutional (see Roper 1983).

As a variable, political parties are most powerful in the Communist Model. In the Soviet Union, the communist party decides which form of information technology is important to economic and ideological objectives. The Soviet Union then develops and/or purchases those forms of technology. Other forms of information technology are viewed as detrimental to the ideology and are banned. Video discs and video cassette recorders and personal computers are examples of information technology which are not adopted on a mass scale because they are illegal.

Political parties also control information technology in Third World countries. Vasquez (1983) notes that the media's role is to promote national goals—a role which Third World societies share with Communist countries. It can be seen in Figure 21.8 that the political party is not the major influence over the development of information technology. It does, however, exercise a considerable influence over the use of information technology once external forces have transferred the technology (Ruggels 1976).

Economy

In the United States the economic system is based on a free marketplace. In the Competitive Model the marketplace or the economy is variable x. Certain forms of information technology are developed solely for military purposes regardless of the possible market, yet this is the exception rather than the rule.

In the Competitive Model, many forms of information technology are developed, but not adopted, because the market did not accept the technology. Bell's Picture Telephone is an example.

In societies which view information technology as a public utility, the economic system is a blend of government guidance and a free marketplace. Here many forms of information technology—videotex, fiber optics, mainframe computers, etc.—are developed on a national scale as they are perceived to be crucial to the economic system.

In both the Competitive Model and the Public Utility Model the economy is all important. However, the difference between the two systems, relative to information technology, is that the latter guides certain forms of technology while the former relies on the marketplace to guide the development of technology.

In the Communist Model, if information technology is perceived as beneficial to the economy (or to ideology), it is developed and adopted. The economic system itself, however, does not have a dominant influence over the development and use of information technology.

To a large degree Third World societies do not have viable enough economies to either develop information technology on a national scale or for their citizens to adopt technology which has been transferred from abroad (Vasquez 1983).

External Forces

All countries are influenced to some degree by external forces. As an example, the United States had not been committed to the development of supercomputers on a commercial basis prior to 1982. During that year it became obvious that Japan had made a firm national commitment to the development and commercialization of super computers and artificially intelligent computers. U.S. private industry reacted by setting up its own consortium (Microelectronics and Computer Technology Corp.) to compete.

External forces are also a powerful variable in the Public Utility Model and in the Communist Model. However, in the Third World Model external forces actually constitute variable x. Information technology is highly capital intensive and is beyond the means of Third World societies. If they are to utilize information technology it needs to be transferred from an external source. Thus, government use and public adoption of information technology in the Third World is dependent on an outside force. Only when a developed nation decides it is in its interest to transfer information technology to a Third World country will technology be adopted by that society.

Infrastructure

Daniel Bell (1983) has observed that every society is tied together by three different infrastructures: transportation, energy grids, and communication. An information infrastructure determines to a large extent what new forms of technology can be implemented. *Information infrastructure* refers to: (1) the physical side of technology—satellites, computer networks, telecommunications, telex, broadcast stations, and so on; and (2) the characteristics of ownership—private vs. government, concentrated vs. diversified—on the national level.

In the Competitive Model, the infrastructure has only minimal influence over technology. Computer, satellite, broadcast, and telephone communication are monopolized in their respective areas, but no company has a compelling interest in more than two areas (Compaine 1982). Innovative technological products have been developed and marketed by small firms and often by foreign firms. The microcomputer is an example of the former and the video cassette recorder as an example of the latter.

The infrastructure of those countries, represented in the Public Utility Model, is influential since the information industry is monopolized by a small group of large corporations. The number of firms which dominate the information industries of Western European countries is much smaller than found in the United States.

Countries represented in the Competitive and Public Utility Models have extensive information infrastructures in place. The fact that telephones are in more that 98 percent of all homes in the United States means that interactive information systems tied into all homes and industry is not only feasible but very likely to take place within the next fifteen years.

The Communist infrastructure is characterized by a large state owned telecommunication system utilizing broadcast, telephone, and satellite communication systems. Infrastructure would be placed in a more dominate position in Figure 21.7, except that the infrastructure is set-up and can be changed by the political party in office. Private information systems generally do not exist. The development of many of the services typical of information societies—video conferencing, electronic mail, telebanking, direct-broadcast satellite, computer conferencing, etc.—are not common due to the lack of a competitive atmosphere.

Few Third World countries have information infrastructures. Two-way home information systems are not within the realm of possibility for the next two decades since an extensive telephone infrastructure does not exist. In most cases Third World countries are not even in the position to accept certain forms of technological transfer as their information infrastructures are inadequate.

Policymaking Organization

Almost every country has at least one policymaking organization delegated with the responsibility of overseeing communications. Policymaking organizations appear similar on the surface though there are great variations. In almost every case the relative power, effectiveness, and accomplishments of policymaking organizations are an indirect outcome of that society's ideology.

In the Competitive Model, policymaking organizations are weak and decentralized because the idea of a powerful agency in charge of communication and information is antithetical to American ideology. As seen in Figure 21.5, these organizations are controlled by the other variables, rather than vice versa. The solid curved line in Figure 21.5 from *social problems* to the *policymaking organizations* and the broken straight line from the *policymaking organization* to the *economy* indicates that the policymaking organization is not in a position to influence variable x—the source of the problem.

The current phenomenon of computer crime and computer-hacking has called attention to the fact that there are no laws in effect which directly apply to computer tampering. This demonstrates that the numerous policy organizations (FCC, NTIA, Congress, the courts, and so on) are not in a position to solve a problem created by the marketplace. The broken straight line from the *policymaking organization* to *technology* points to these organizations' lack of control over the development and use of technology. Since the announcement of Japan's Fifth Generation Project there has been considerable discussion about meeting the Japanese challenge (Mitgang 1983; *New York Times* 1983), but the ideological constraints which delegate the power of the policymaking organization has effectively prevented them from taking any steps. As noted above, the only reaction to the Japanese challenge has come from the marketplace—not from a policy organization.

Policymaking organizations in the Public Utility Model—variable x— are very effective because these societies have a greater trust of government institutions (Homet 1979). The solid line in Figure 21.6 between *policymaking organization* and the *political party* points to the close working relationship between these two variables in the Public Utility Model. Such organizations in the Public Utility Model have considerable authority, greater centralization, and, generally, a sufficient budget. The organizations are, thus, in a position to guide the adoption of information technology— solid line from *policymaking organization* to *technology* — and are capable of creating social policy prior to the introduction of information technology—solid curved line from *policymaking organization* to *social problem.*

The fact that the *policy organizations* are in a position of strength does not necessarily mean that their long-range plans will be more successful than the laissez faire approach seen in the Competitive Model. The "Télématique" program launched by the French government in the late 1970s is an example of a policy organization and a political party working to develop information technology. So far the French government's attempt to take a leadership role in information technology has proven to be anything but a success (*Business Week* 1983).

A negative consequence which logically follows is that greater control by policymaking organizations over information technology means greater potential for government surveillance and control if the organizations are not able to remain independent of political parties. This scenario has already been recognized in Germany where the problem of privacy has come into conflict with government objectives. The Minister of Interior is working with the Bundeskriminalamt (federal intelligence bureau) in developing new guidelines for the Federal Data Protection Act. The Minister has been quoted as saying that "privacy" will be given a high priority as they develop public policy but that efficient "police work" and "national security" will also receive high priority (Gliss 1983).

Communist societies create policy organizations which are primarily puppets of government. Policy is created by the party in control and submitted to the policy organization as a directive. Technology is not so much guided by the policymaking organization as it is developed along mainline political objectives. Moreover, policy is to a large extent more concerned with information than with information systems (Paulu 1974). In some Communist countries the policy organization is basically a department of censorship.

In the Third World Model these organizations have as much power and effectiveness as resources and external forces permit. As seen in Figure 21.8, the policymaking organizations do not have a major influence (broken line) over the development of technology. Coups and external forces too often result in new political structures and new telecommunications policy.

Technology

The major question which will be considered here involves technological innovation and adoption by society. More specifically, how is technology driven—by what or by whom—in a particular society? If information technology is driven by the marketplace, as in the United States, the social consequences will be quite different than if information technology is driven by government policy.

Technology can be created in at least four primary ways. It can come about as a result of private enterprise (Competitive Model). The wide solid line in Figure 21.5 reflects the overwhelming influence economics has in developing technology and in the use of technology. Technology can also be produced as a result of government guidance (Public Utility Model); as a result of government production (Communist Model); or it can be imported from a foreign country, either as a foreign investment or as a form of technological transfer from a developed nation (Third World Model).

In the Communist Model virtually all technology is owned and operated by the government. Generally, there is little concern over market acceptance or diversity (Underwood 1983). Production of technology is a nonprofit process in the Communist Model. Information technology is looked upon as being crucial for communication and control. The relationship between the political party and technology is dichotomous. The party in office is generally able to control the use of information technology but is not able to develop technology to the degree desired. For example, telephones in the People's Republic of China average a little over 2.1 for every 100 persons (Hau 1983) and televisions are in fewer than 4 of every 1,000 homes.

Summary

The major variables in four models have been described. The point has been to show that different societies develop and use information technology in different ways. While all of the variables influence the development and use of information technology to some degree, two variables, ideology and variable x have the greatest influence.

The primary objectives have been to show that: (1) The relationship between information technology and social problems is not a simple cause and effect relationship but is a complex process; (2) for purposes of understanding the complexity of this process at least four different models representing four different types of societies can be studied; (3) the difference in each society is the manner in which information technology is developed and used; (4) the manner in which information technology is developed and used can be attributed to at least six variables; (5) the key variable in all six models is ideology; (6) the key variable which determines how information technology will be developed and used is variable x; and (7) ideology and variable x will largely determine the nature of the social problems which can be expected in a given society.

The significance of the above propositions lies in the implications of variable x. Basically, it has been argued that the variable x operates in a position of dominant influence. While the other variables also influence the development and use of information technology, the influence of variable x is the overriding factor. In all but one model—the Public Utility Model—variable x exercises a greater influence than the policymaking organization. Thus, when a social problem arises and is brought to the attention of the policymaking organization—the normal channel for complaints—the policy organization's ability to solve the problem is severely limited. In the United States's finely balanced system, the ability to control technology, and thus social problems, is traded off for the more attractive goal of a competitive marketplace. For example, the United States cannot simultaneously allow the whims of the marketplace to develop information technology and remain in control of its development and its use. In the Communist Model, a competitive atmosphere is traded off in favor of control over information technology and society. Thus, the Soviet Union cannot discourage the use of personal computers by its citizens while developing a national electronic mail system. The two are mutually exclusive.

It becomes clear, then, that the notion of "social problem" differs from culture to culture. Government control of information is viewed as a problem in Western societies while public access to information is viewed as a social problem in Communist Societies.

What has been offered here is a view of the process by which societies develop and use information technology. Additional research will need to be done on correlating each model with the dozen or more social problems discussed above.

Note

The author would like to acknowledge the National Geographic Society which made it possible to do field research in Nepal, Hong Kong and Japan.

References

Allen, B. Computer fraud: New findings, new insights. Paper presented at the 8th Annual Computer Security Conference, New York, N.Y., 1982

Bagdikian, B.H. *The media monopoly.* Boston: Beacon. 1983.

Baldwin, T.E. and D.S. McVoy. *Cable communications.* Englewood Cliffs, N.J.: Prentice-Hall, 1983.

Bell, D. The social framework of the information society. In T. Forester (ed.), *The microelectronics revolution.* Cambridge: MIT Press, 1980.

_____. Communication technology: For better or for worse? In J.L. Salvaggio (ed.), *Telecommunications: Issues and choices for society.* New York: Longman/An-nenberg, 1983.

Blumler, J.G. Information overload: Is there a problem? In E. Witte (ed.), *Human aspects of telecommunications.* New York: Springer-Verlag, 1980.

Broadcasting, March 14, 1983.

Bukeley, W. M. Microcomputers gaining primacy, forcing changes in the industry. *Wall Street Journal,* January 13, 1983.

Burnham, D. *The rise of the computer state.* New York: Random House, 1983.

Business Week. Telecommunications: Everybody's favorite growth business, the battle for a piece of the action. October 11, 1982.

_____. Télématique: A French export that doesn't travel well. July 4, 1983.

Cable took a beating in 81. *Broadcasting.* March 28, 1983.

Communications News. U.S. homes in 1990 to have over 15 million rooftop earth stations. November 14, 1982.

Communications News, 1983, 20.

Compaine, B. Shifting boundaries in the information marketplace. *Journal of Communication,* 1981, 31, 132-142.

_____. *Who owns the media?: Concentration of ownership in the mass communication industry.* White Plains, N.Y.: Knowledge Industry, 1982.

CompuServe, Telephone conversation, October 20, 1983.

Dave, N. W. The space race: Here comes Japan. *High Technology,* May 1983, 27-30.

Dertouzos, M. L. and J. Moses. *The computer age: A twenty-year view.* Cambridge, Mass.: MIT Press, 1981.

Donner, F. *The age of surveillance: The aims and methods of America's political intelligence system.* New York: Vintage, 1981.

Elliott, D. and R. Elliott. *The control of technology.* London: Wykeham, 1976.

Ellul, J. *The technological system.* New York: Continuum, 1983.

Emery, W. B. *National and international systems of broadcasting: Their history, operation and control.* East Lansing: Michigan State University Press, 1969.

Faber, J.P. Drug abuse in workplace costs employers billions. *Houston Chronicle,* July 25, 1983.

Forester, T., ed. *The microelectronics revolution.* Cambridge: MIT Press, 1982.

Garfield, E. 2001: An information society? *Journal of Information Science,* 1979, 1, 209-215.

Gliss, H. German privacy conflicts flare. *Transnational Data Report,* 1983, 6, 15-17.

Goldhamer, H. The social effects of communications technology. In H. D. Lasswell, D. Lerner, and H. Speier (eds.), *Propaganda and communication in world history.* Honolulu: East-West Center, 1980.

Hau, L. China Improving Communications Services. *Telecommunications,* October 1983, 155-157.

Hiramatsu, T. Privacy Law in Japan. *Transnational Data Report,* 1983, 6, 43-44.

Hollstein, M. Media Economics in Western Europe. In L. J. Martin and A. G. Chaudhary (eds.), *Comparative mass media systems.* New York: Longman, 1983.

Homet, R. *Politics, cultures and communication.* New York: Praeger, 1979.

Klapp, O. E. Meaning lag in the information society. *Journal of Communication,* 1982, 32, 56-66.

Krause, L. and E. MacGahan. *Computer fraud and countermeasures.* Englewood Cliffs, N.J.: Prentice-Hall, 1979.

Le Duc, D. R. Direct broadcast satellites: Parallel policy patterns in Europe and the United States. *Journal of Broadcasting,* 1983, 27, 99-118.

Logsdon, T. *Computers & social controversy.* Potomac, N.Y.: Computer Science Press, 1980.

Linowes, D. F. The privacy crises. *Newsweek,* June 26, 1978.

Lublin, J. S. Postal service finds little demand for new electronic-mail service. *Wall Street Journal,* September 16, 1983.

Macy, R. STV: Going, going, gone? *Broadcast Communication,* 1983, 28.

Martin, J. *Viewdata and the information society.* Englewood Cliffs, N.J.: Prentice-Hall, 1982.

Martin, L. J. and A. G. Chaudhary. eds. *Co mass media systems.* New York: Longman, 1983.

McLuhan, M. and B. Powers. Electronic banking and the death of privacy. *Journal of Communication,* 1981, 31, 164-169.

Mitgang, L. Many concerned U.S. may lose technology race. *The Houston Post,* March 6, 1983.

Multichannel News, June 13, 1983.

New York Times. Funds are asked for 'supercomputer.' April 2, 1983.

Parker, D. B. Computer abuse assessment. *Encyclopedia of computer science and technology,* Vol. 3, 1975.

_____. *Fighting computer crime.* New York: Charles Scribner's Sons, 1983.

Paulu, B. *Radio and television broadcasting in eastern Europe.* Minneapolis: University of Minnesota Press, 1974.

Pear, R. CIA easing request to exempt all its files from information act. *New York Times,* May 30, 1983.

Pelton, J. *Global talk.* Rockville, Md.: Sijthoff & Noordhoff, 1981.

Personal privacy in an information society. The Report of the Privacy Protection Study Commission. Washington, D.C.: U.S. Government Printing Office. 1977.

Pipe, R.G. WCY: Development of communications infrastructures. *Transnational Data Report,* 1983, 6, 7-9.

Presstime. Fifteen newspaper companies secure LPTV permits. February 1983.

Roper, J. E. Congress stalls Reagan's information control plans. *Editor & Publisher,* October 29, 1983.

Rose, E. Moral and ethical dilemmas inherent in an information society. In J. L. Salvaggio (ed), *Telecommunications: Issues and choices for society.* New York: Longman/Annenberg, 1983.

Ruggels, W. L. P. and D. Hall. Institutional limits on the introduction of communication technology. In G. Chu, S. Rahim, and D. L. Kincaid (eds.), *Institutional explorations in communication technology.* Communication Monographs No. 4. Honolulu: East-West Center, East-West Communication Institute, 1976.

Salvaggio, J. L. An assessment of Japan as an information society in the 1980's. In H. F. Didsbury (ed.), *Communications and the future,* Bethesda, Md.: World Future Society, 1982.

_____. Information societies and social problems: The Japanese and American experience. *Telecommunications Policy,* 1983, 7, 228-242.

_____. Information society research: An overview of the literature. In J. Bryant and Z. Dolf, eds. *Perspectives on mass communication research.* New York: Erlbaum, 1984.

Salvaggio, J. L. and S. Trettivik. Information inequity in an information society. Paper presented at the annual conference of the International Communication Association, Communication Technology Division, Minneapolis, May 1981.

Schiller, H. I. Computer systems: Power for whom and for what?" *Journal of Communication,* 1978, 28, 184-193.

_____. Information for what kind of society? In J.L. Salvaggio (ed.), *Telecommunications: Issues and choices for society.* New York: Longman/Annenberg, 1983.

Schmidt, J. S. and R. M. Corbin. Telecommunications in Canada: The regulatory crisis. *Telecommunications Policy,* 1983, 7, 215-227.

Sigel, E. *The future of videotex.* White Plains, N.Y.: Knowledge Industry, 1982.

Shiono, H. The development of technology and related laws in Japan. *Studies in Broadcasting,* 14, 1978.

Smith, B. S. Battle intensifying over explicit sex on cable TV. *New York Times,* October 3, 1983.

Smith, D. D. *Communication via satellite: A vision in retrospect.* Boston: A. W. Sijthoff, 1976.

Statistical abstract of the United States, 1982-1983. Washington, D.C.: U.S. Government Printing Office, 1982.

Transnational Data Report, 1983.

Tydeman, J., et al. *Teletext and videotex in the United States: Market potential technology public policy issues.* New York: McGraw-Hill, 1982.

Underwood, P. S. *Media entertainment in the Communist world.* In L. J. Martin and A. G. Chaudhary (eds.), *Comparative mass media systems.* New York: Longman, 1983.

U. S. industrial outlook. Washington, D.C.: U.S. Government Publishing Office, 1982.

U. S. News and World Report. Who is watching you?" July 12, 1982, 35-37.

Vasquez, F. J. Media economics in the Third World. In L. J. Martin and A. G. Chaudhary (eds.), *Comparative mass media systems.* New York: Longman, 1983.

Wells, A., ed. *Mass communication: A world view.* Palo Alto, Calif: Mayfield, 1974.

Westin, A. F. and M. A. Baker. *Databanks in a free society: Computers, record-keeping and privacy.* New York: Quadrangle/New York Times Book Company, 1972.

Weizenbaum, J. *Computer power and human reason: From judgment to calculation.* New York: W.H. Freeman, 1976.

22

Corporate Transborder Data Flow and National Policy in Developing Countries

Carrie L. Shipley, Dwayne L. Shipley, and Rolf T. Wigand

Developments in information technology have vastly expanded the capability of instantaneous, two-way communication and are creating major new service and equipment markets and industries. Transnational corporations are entering the information services business on a large scale internationally, and many believe developing nations will be adversely affected in the process. In this setting of progressing increases in transnational corporate control of local markets, the authors point out tendencies toward an increase in centralized decision making, loss of cultural autonomy and national sovereignty in developing nations,and inability of such nations to compete with transnational corporations through indigenous growth of their own informatics industries, among others. Finally, the possible methods of planning for transborder data flow in developing nations are discussed. Restrictions on such flows are considered and the issues of effectiveness and enforceability are explored. Three divergent alternatives for regulatory action on transborder data flow are discussed.

The appropriate use of information has become the basis for success in all transnational corporate activities. In this context, it has become evident that information constitutes both a resource and a commodity. It can help conserve other resources and enhance productivity. The right information is crucial to achieve corporate goals in a timely, effective, and efficient fashion. Sometimes such information can be bought and traded as a commodity just like other commodities on the world market, such as market research findings or economic data. In other situations information can serve as a resource comparable to land, energy, labor, or capital and is a crucial ingredient for the production or creation of other resources.

In the past decade, many new information technologies have been developed that can carry information to literally any place on earth, combining and integrating previously separate communication functions such as the transmission of data, voice, video, and facsimile communication (Wigand 1982). Previous distinctions and delineations among providers of information services, suppliers of information technologies, as well as users of information and technology are diminishing. The recent merging of two major information technologies, the computer and telecommunications, has had the greatest impact on international communication policies and on the way in which corporations can communicate across national borders.

The ability of corporations to transact business efficiently in international markets has always depended on fast, reliable transfers of information. Now, new technologies allow increased efficiency in such corporate activity. The term *informatics* has been coined to describe the combined use of telecommunications and computer technologies for handling vast quantities of information, i.e., for data processing or database access. The term *telematics* is sometimes used to mean telecommunication of computerized information. In our scenario, "transborder data flow" is said to occur when telematics operates across national boundaries.

This article discusses the impact of transborder data flow on developing countries, emphasizing the new communication technology and data processing and information access services used and offered by transnational corporations. Beginning with an overview of the value of information—specifically with regard to international flows—the importance of transborder data flow for communication within the transnational corporation and for the international sales of data processing, access, and storage services is considered. Next is an analysis of possible impacts of corporate transborder data flow on developing nations. Related issues of national sovereignty and cultural diversity are also investigated.

Finally, the issue of regulation is considered. The regulation of transborder data flow is being debated in many international settings, vehemently opposed by some and supported by others. There seems to be a definite trend toward an increase in regulation and control over such flows in many nations. These may greatly affect users of international data flows and, at the same time, may benefit economies and cultures of the regulating nations. Alternative controls are subsequently explored in this context.

Transborder Data Flow and International Business

For the purposes of international data transmission, networks of private and public communication linkages are connected via landlines, microwave systems, satellites, and other means. These are used by transnational corporations and others to receive and transmit data for a variety of purposes. Transnational corporations depend on such linkages for communication with subsidiaries and for access and sharing of data processing and database services, as well as for financial and administrative functions. Services include local and remote batch processing, interactive time-sharing and database management, inquiry services (bibliographic, market data, etc.), and educational or scientific services. The importance of such technologies to the transnational corporation is discussed by Vernon (1977, 1-2): "If scientists and engineers had not found a way to shrink international space over the past century or so, the odds are high that the multinational enterprise would be a rarity today. The international telephone, the computer, and the commercial aircraft have been indispensable to the growth of such enterprises."

Rapid development of informatics technology and its use for transborder data flow is helping to propel transnational corporations into a new era of increased advantage in developing nations. Vernon states that the new technologies offer "new opportunities to employ low-cost labor or to acquire low-cost materials," with the result that "those firms that do not seize the opportunities are exposed to the possiblility that their rivals might"(p. 5). Developing countries, thus, are confronted with new and increasing challenges to their ability to understand and, if necessary, to control complex economic factors affecting national interests. In this information age, developing nations need to be able to connect to international networks for communication and information access purposes. Yet they often lack the means to do so without transnational corporate help.

Table 22.1 illustrates the increasing proportion of the world's gross product that comprises information-related industries. One consequence of this trend is that more and more transnational corporations are getting into the information services business and, in the process, may be oblivious to the development needs and concerns of the nations involved. For example, a large amount of raw data is extracted from developing nations at a net expense in terms of potentially diminished sovereignty for the nation over its own resources and national information. As Lloyd (1980) indicates, returns of processed data are very expensive to access, in comparison, with some resulting negative effects possible on balance of payments (see Figure 22.1). Even without describing in detail the effects or benefits of the inter-

TABLE 22.1
Projected Revenues for the Total On-Line Database
Service Market, by Subject Area, 1979-1985
(in million dollars)

Subject Area	1979	1985	Average Annual Growth Rate (%)
Econometrics	67	200	20
Securities & commodities	140	300	13
Economics & finance	125	450	24
Natural resources	16	96	35
Demography	36	135	24
Industry	37	126	23
Law/accountancy	75	360	30
News	40	235	34
Marketing	105	510	30
Credit	240	710	20
Bibliography	62	225	27
Patents	22	95	28
Real estate	45	175	25
International	18	70	25
Other	140	560	26
Total	1,170	4,280	24

Source: *Transnational Data Report* 5 (7), 1982, p. 341. Originally published in INPUT, International Market Opportunities for On-line Database Services (Palo Alto, Calif., Sept. 1980).

national use of new communication technologies, it should be obvious that ramifications will be extensive:

> It is a historical fact that major changes in the way information is collected, stored, or distributed have had profound implications for the way societies function and institutions are structured. Earlier information breakthroughs, including the invention of language, printing, the telephone, mass media, and the computer itself, each, in its own time, revolutionized the structure and processes of human organization and interaction. There is no reason to expect that the impact of NIS [network information services] will be less far-reaching (Dordick, Bradley, and Nanus 1981, 3).

There are several areas that must be considered in an examination of the reasons for the developing nations' interest in the transborder data flow of multinational corporations. First, there is the use of transborder data flow for communication within transnational corporations—or internal data flows—which may be affecting locus of control over decision making within developing nations. Second, there are problems concerning the

FIGURE 22.1
Patterns of Transborder Data Flows

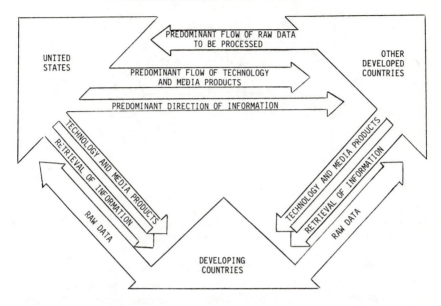

Source: Rein Turn, ed., *Transborder Data Flows: Concerns in Privacy Protection and Free Flow of Information*, vol. 1. Report of the AFIPS Panel on Transborder Data Flows (Washington: American Federation of Information Processing Societies, 1979), p. 5.

data-processing industry and corporate use of transborder data flows for external purposes. Finally, the manner and extent of impact on developing countries must be considered. In what ways does the recent convergence of communication and computer technologies benefit or hinder the development processes of developing nations?

Internal Data Flows

Obviously, centrally managed transnational corporations have a strategic advantage when utilizing the newest technologies to communicate internally, or with subsidiaries. The cost advantage alone should be enough to promote increasing transborder data flows, as lengthy intervals for message and data transmission and response time are cut to mere fractions of seconds (Wigand 1982, 1984). Long-distance telephone rates may be effectively limited, and companies are able to accelerate decision-making processes by accessing and utilizing timely information exactly where and by whom it is needed. In many cases, it is not only transmission of lengthy messages that is avoided, but transportation of bulky materials and even personnel may be reduced in certain instances. Dordick, Bradley, and

FIGURE 22.2
Motorola Administrative Message Service (MAMS)

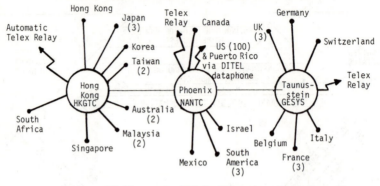

Delivers 175.9K messages (171.0 million characters) per month

Source: Telecommunications and Information Products and Services in International Trade,
U. S. Government Printing Office #97-59, 1981, p. 165.

Nanus (1981, 24) state, "Telecommunications systems provide an alternative to the physical transportation of people or paper and are used to the extent that they are cost-effective." Regarding the development of the new technologies, these authors go on to say:

> Most of the technological improvements in networking—such as digital transmission, message switching, and bandwidth conservation—have evolved in order to minimize the cost of communications. In fact, the development of these and other techniques has resulted, in part, in the cost of long-haul communications to drop dramatically over the past ten years, a trend which continues today (p. 24).

Utilization of these technologies has become so widespread for reasons of efficiency, that it is possible to say that many transnational corporations are entirely dependent upon them (Betz, McGowan, and Wigand 1984). Billions of bits of data move in mere seconds between subsidiaries and headquarters, making possible and facilitating the management of ever-expanding world enterprises. The new communication technology facilitates management by integrating dispersed subsidiaries into a cohesive system and by making possible immediate response to crisis or for feedback of information for planning purposes (see Figures 22.2 and 22.3).

Responsiveness to markets is augmented, as firms are able to "custom tailor" products and services to specification. Network information services help speed up corporate operations and are quickly becoming a necessary condition for international business management. Examples of uses for this technology include: data collection at remote sites, monitoring of inventories and trends, and maintenance of contact with customers or service groups (Dordick, Bradley, and Nanus 1981; Wigand 1982).

Data communication facilities may be located at sales and service offices, at manufacturing plants, and at administrative offices. Hewlett-Packard, for example, has 110 such facilities and 1,400 computers at over 200 locations worldwide (Van Rensselaer 1979, 90). Transborder data flows are the central nervous system of such corporations, with internal data flows used to monitor raw material stocks, production schedules, quality control, personnel records, tax and legal information, currency transactions, profit repatriation, and investment decisions internationally. Because of its intracorporate nature, the actual volume of data transmitted for such purposes is unknown. A cursory review of the scope of one such company's dependence on telecommunications will shed some light on the necessity of international data transmission for modern transnational corporate operations.

American Express utilizes transborder data flow extensively for the management of its insurance business, payment systems, assets, banking, and similar operations around the globe. Numbers of transactions on company on-line systems in 1981 were as follows (Transnational Data Report 1983):

FIGURE 22.3
Chase Manhattan Private Communications Network

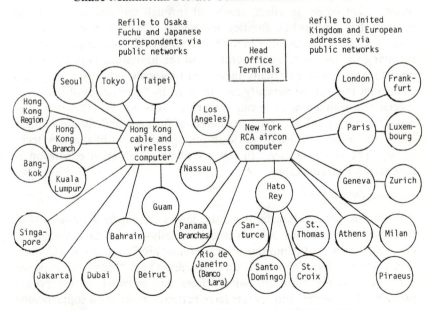

Source: Chase Manhattan, cited in *Transnational Data Report* 5 (1), 1982, p. 25.

- 310 million American Express Card and 360 million Visa and Master Card transactions processed;
- 250,000 of these transactions authorized daily with an average response time of five seconds;
- Over 350 million American Express Travellers Checks sold by over 100,000 banks and other outlets;
- 50 million insurance premium and claim transactions;
- About $10 billion/day automatically executed bank transactions internationally;
- 500,000 instantaneous responses daily to messages directing high-speed trading in securities, with 40,000 transactions easily sustained per day in one Shearson office.

American Express, it is evident, is heavily dependent on its telecommunication and data processing operations and could not function properly without speedy, reliable, worldwide communication.

Monitoring trends, collecting data, and maintaining customer relations are now more easily accomplished from a great distance. Unfortunately for the subsidiaries of transnational corporations, however, their dependence upon the headquarters will increase as the training, equipment, and software necessary for information processing are centrally controlled and allocated. It is evident that the tendency would be to increase centralized decision making, since home offices would have such ready access that top management could, in effect, reach out through the system to exercise more control. As Dordick, Bradley, and Nanus (1981, 23) predict, dependence on transborder data flow in the host country can also cause "problems involving the loss of resource control and the limitation imposed by the need to conform to externally mandated protocols." However, they also claim that this will be partially counteracted by "economies of scale and specialization inherent in resource sharing." It is possible that more decentralization could occur as parameters of distance and time no longer affect expansion because of world-shrinking influences of the new technology. The key issue here, though, is whether unlimited expansion is beneficial for all concerned.

External Data Flows

Regarding uses of transborder data flows external to a transnational corporation's operations, sales of internationally–offered data processing, database access, and communications services are increasing world-wide. Onstad (1979, 179) of Control Data Corporation said, "Many data processing service firms maintain sales offices and network access points throughout the world, and operate large batteries of the most sophisticated computer systems." Such uses range from elementary commercial and accounting operations to elaborate and sophisticated scientific, educational, and engineering applications.

A push-pull effect can be observed for potential data-processing markets. In the marketing tradition, "push" occurs when companies introduce their products or services into the marketing system, hoping to create consumer demand. The "pull" effect occurs when consumers are aware of the benefits of a product or service and demand that it be supplied to them. The "push" of international data services occurs because this type of service is a major growth industry with potential profits enticing companies into the field. One reason for this push also comes from the diminished viability of low-technology industries in the industrialized world today. To maintain a lead over up-and-coming high-technology manufacturing countries such as Japan, U.S. companies must develop new markets. According to a recent U.S. Senate telecommunications policy statement:

> Telecommunications and data processing services are . . . major growth industries in themselves. Mature countries such as the United States increasingly rely on these industries to offset the decline of low-technology sectors, and it has become virtually a requirement of U.S. economic health for such industries to expand abroad (1983, 109).

This development is not unique for the United States. Figure 22.4 demonstrates the revolutionary shifts in the labor force of just six industrialized nations. The "informatization" of society has taken place in many technologically advanced nations. Studies by the Organisation for Economic Cooperation and Development (1981), Porat (1978) and others put the average information labor force of leading industrialized nations at one-third of the total. In the United States the information labor force is approaching the 50 percent mark, while agriculture accounts for merely 3 percent, industry for about 20 percent, and the service sector with almost 30 percent. The industry labor force has been declining steadily since the 1950s and the agricultural labor force has been diminishing even more rapidly since the 1930s.

It has been recognized that if substantial gains in productivity are to be made, these gains will have to be made in the information and service sectors. Industrialized productivity growth is declining worldwide. Six industrialized nations, Canada, France, Germany, Japan, the United Kingdom, and the United States, showed an average growth in manufacturing output per man–hour of 5.5 percent during the 1960s and only 4 percent in the 1970s (Committee for Economic Development 1980). The figures look even more devastating when inspecting manufacturing trends in the entire economy:

FIGURE 22.4
Labor Force Distribution, by Sector

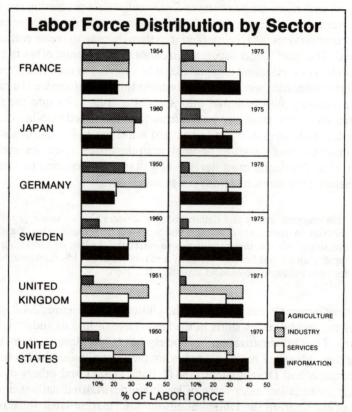

Source: Organization for Economic Cooperation and Development, *Information Activities, Electronics, and Telecommunications Technologies: Impact on Employment, Growth, and Trade,* p. 29. Cited in *The Futurist,* December 1982, p. 37.

From the end of World War II to 1964, U.S. labor productivity in the private business economy incresased at an annual rate of over 3 percent; but over the past several years, the annual growth rate has declined to less than 1 percent (Committee for Economic Development 1980, 2).

Although industrial productivity is slowing worldwide, substantial gains can be made in information productivity. With ever-increasing information output, increased automation, more and more substitution of telecommunication for transportation, and more and more individuals finding employment in the information and service sectors around the world, the role of information and communication technology in the pursuit of true productivity gains is of crucial importance. The resulting consequences for national economic policies can be observed in the multi-faceted activities of transborder data flow and related corporate activities for the production and distribution of associated technologies.

A "pull" from consumers and companies also exists within potential data processing markets. It arises for reasons similar to those stated previously with regard to internal transnational corporate uses of transborder data flow. For many industries at home and abroad, data processing has become a key factor of production (U.S. Senate 1983, 168). Access to these services can help reduce spiraling costs of operation and make it feasible to deal efficiently with the vast quantities of information that have become an integral part of the business world today. For data processing services to be useful beyond the information-access level, potential users need to be skillful in the collection and generation of data, and in sophisticated decision making (e.g., econometrics and financial simulation). It is not likely that such sophistication will come quickly or easily to many developing nations without training and other assistance. It will be worth considering, however, whether data processing services should be a high priority for development or are currently an "appropriate technology" for developing economies, a point to be addressed later.

Regarding the evolution of data-processing and information-access services internationally, it is possible to speculate that the size of the service entity is an important factor. It is indeed technically possible that many smaller firms may provide such services to other countries by leasing lines and selling their services without investing directly in other countries. But by far the major actors in the field will be the major transnational corporations already in existence and already possessing excess data-processing or information-storage capacity. While the cost of initiating a database service may be decreasing due to the development of lower cost computer technology, it is probable that large corporations already possessing the necessary facilities and offering on-line search services will enjoy a definite advantage.

The role of data-processing service growth in the international expansion and continued viability of many transnational corporations will soon be paramount. The larger corporation with a more advanced structure will tend to have a more global outlook, leading to transnational expansion. Hymer (1970) argues that size and administrative structure of transnational corporations play key roles in international corporate expansion. He stresses that effects of "bigness and fewness" of corporations include increases in "centrally planned world production" accompanied by a "decline in competition since the size of the market is limited by the size of the firm" (p. 443). This is perhaps the market control sought by the transnational corporation, as addressed by Müller (1973, 124): "Concentrated control of technology is one of the most effective means to establish oligopoly power over the market place, restricting the development of local competition and permitting . . . an astounding rate of profits, the greater majority of which leave the country."

While it is not known what the exact effects are of such direct investment, two opposing tendencies may be observed. First of all, improved communication is eliminating many barriers to trade and expanding the market available to most buyers. But on the other hand, as Hymer (1970, 443) notes, "direct foreign investment tends to reduce the number of alternatives facing sellers and to stay the forces of international competition."

The example that IBM provides is perhaps an appropriate one to illustrate the competitive advantage that size and structure gives to companies entering the international data-processing and information-handling market. IBM quickly achieved an annual sales average of over $1 billion soon after entering this market, excluding U.S. sales (Jacobson 1979, 151). It is reasonable to assume that because of its size, unique mobilization capabilities, and administrative structure, it was able to restrict the market-entry attempts of certain would-be competitors. In a joint venture with its partners Comsat and Aetna Life, IBM hopes to promote Satellite Business Systems (SBS) as an all-embracing communication network that "will eventually evolve into a 'gigantic service bureau,' providing computer power to users much the same way that utilities provide electricity today" (Lecht 1980, 1). Smaller, less competitive firms based in other countries are undoubtedly protesting their inability to enter the market or to compete in the face of such odds.

Other major corporations in or entering the data-processing and supply field include:

Citibank—offering its considerable in-house computer capacity to non-banking customers, as well as special services in credit analysis, on-line financial and securities databases, investment and economic planning, software, etc.;
Exxon—expected to have subsidiaries in this field producing $10 to $15 billion in revenues by the end of the 1980s;
American Telephone & Telegraph—data processor and distributor; enormous resources being mobilized under "de-regulation" (Schiller 1981, 39).

While these and other services being offered by major corporations may indeed threaten competition, almost 80 percent of network usage will be by the Fortune 500 and "about one-third of the medium-size industries in the nation (U.S.)." The remainder "will still find the network marketplace too expensive" (Dordick, Bradley, and Nanus 1980, 1). Thus, even if they were prepared to utilize such systems, most smaller businesses of the developing world are unlikely to be able to afford such participation in the near future. The resulting inability of national corporations to compete with transnational corporations (due to inferior information-handling capacities) will mean an increasing trend toward governmental limitation of access to transborder data flow in some nations.

Information and Development

Many complex issues confront the developing world as it decides to what extent it should allow unlimited transborder data flow or access to what it considers sovereign information. To understand these issues, it is first necessary to examine specific considerations that contribute to the recent developments in transborder data flow regulation. Among these are *infrastructure, appropriate technology, growth of national level firms, and protectionism.*

A dichotomy seems to exist in developing countries. Their delegates to international conferences seem to agree that new communication technologies should be utilized, and yet one finds very little existing infrastructure that would make such use possible. There are few developing-world businesses ready to utilize the benefits of communication technologies on a scale large enough to warrant such major investment. Similarly, governmental or social organizations who may need these technologies lack trained personnel and other resources to make such investment worthwhile. Bortnick (1981, 21) relates a first-hand experience in Nigeria:

> As I traveled through the countryside I noticed innumerable microwave towers, receiver antennas, and telephones were available virtually everywhere; however, nothing worked. I was unable to complete a domestic phone call at any place in which I was located because there were no technical people to service the equipment. Equipment problems and power failures were constant.

Bortnick indicated that although several developing nations, among them Indonesia and India, made "significant progress in establishing domestic telecommunications networks," most of them "face substantial problems in acquiring, installing, and maintaining modern communications capabilities." Often, however, the problem may not lie with the developing country, but with the "inappropriate and expensive equipment which they could neither use nor service" (p. 29).

Industrialized countries have provided demonstrations of the efficiency and usefulness of such technologies, at least for themselves. Subsequently, through "trickle-down marketing" and trade and investment in developing nations, the technologies have been unreasonably promoted and adopted in some cases. But advantages of appropriate technology cannot be denied (such as access to scientific information in data banks) that could aid these nations in developing and using their resources more efficiently. Thus, at the Intergovernmental Bureau of Informatics Conference in 1980, the majority of developing nation delegates seemed anxious to promote policies that would further the indigenous growth or regulated transfer of new communication technologies in their respective countries (*Transnational Data Report* 1980).

So the question we must concern ourselves with is how such growth may occur. As already discussed, competition from large, well-structured corporations of the industrialized world can effectively curtail the efforts of certain firms (public or private) in developing nations to enter information age markets. Transnational corporations are able to do this when allowed market entry because of competitive advantages related to economies of scale, size, and structure. According to Biersteker (1978, 10), these also include: (1) Economies of scale, finance, marketing; (2) vertical integration; (3) capital intensity of production; and (4) attractive wage scales. This results, first, in the "displacement of indigenous production," and second, among other things, in "dependency on the multinational corporation" and reduced "autonomy in decision making." Again, this latter problem is exacerbated as transnational corporations use information derived from less developed countries to make decisions on development of resources; for example, decisions that are thus not made by members of the nations involved.

In spite of all this, Vernon (1973-4, 117) states that most developing nations "are still engaged in trying to woo the undecided foreign investor, often by blandishing promises of tax exemptions, duty-free imports, and even capital grants." What is needed in the case of informatics industries, however, is often the reverse. Leeway is necessary for competition to be possible in this high-technology sector, and some sort of protection may be needed to obtain this in certain circumstances. Hymer (1970, 444) states that this is "an important rationale for the infant entrepreneur argument supporting protection. Temporary protection of a weak firm from a stronger firm can improve the competitive structure of the industry in future periods by maintaining numbers." The increase in competition may be costly for the negotiating country, but "benefits would accrue to the world as a whole."

From the developing world's market viewpoint then, it is apparent that some form of protection may be beneficial for information-access and data-processing industries in developing nations. Indeed, most European

countries were afforded this advantage by the United States in the manufacturing sector for years after World War II. Specific methods that can
accomplish this protection without severe "protectionism" or hindrance to
the free flow of information are already enacted by some countries or are
being discussed in international fora by others. These methods deserve to
be considered. First, however, one must ask why such protection may be
necessary from a broader perspective that reaches beyond marketplace
considerations.

National Sovereignty, Cultural Integrity, and Transborder Data Flow

A broader potential social concern of corporate transborder data flow
lies within a context of national, cultural, economic, and political considerations. An example of a national sovereignty issue is that satellites can, to
a certain extent, be controlled by the company that made them, and national telecommunications may thus be interruptible. Supposedly, Hughes
or the U.S. Department of Defense can "turn off" Indonesia's Palapa satellite (Jacobson 1981). Less obvious, but perhaps more controversial, is the
use of satellites for "remote sensing," which allows corporations to "secure
a comprehensive inventory of the physical features of any nation, regardless of that state's willingness to have such a geographical profile developed" (Schiller 1981, 119).

Transnational corporations utilize such knowledge to their economic
advantage: e.g., the location and quantities of mineral deposits in developing nations. Here is one area in which potential development decisions can
be made outside of the host country. In addition, national governments
tend to lose negotiating power when information on their country's resources is available to others. This becomes especially important when a
government talks about turning over its remote sensing technology (e.g.,
the U.S.'s Landsat satellites) to transnational corporations for private sector
ownership and operation (Frank 1981). Ideally, some argue, remote sensing
should be used to serve the scientific and development interests of the
entire world.

National sovereignty is also at stake when countries rely on foreign data-
processing and information-access services. Such reliance implies a certain
vulnerability when nations do not maintain data banks (scientific and
technical) within their own borders. Of course, were it possible for nations
to be more trusting of each other, and trustworthy, with regard to guaranteed access, such concern would not be necessary. But it is partly for
sovereignty reasons that the EURONET database and communication network was set up by the European Economic Community (Wigand, Shipley,
and Shipley 1983). According to Joinet (1979, 118):

Information is power and economic information is economic power. Information has an economic value, and the ability to store and process certain types of data may well give one country political and technical advantage over other countries. This, in turn, leads to a loss of national sovereignty through supranational data flows.

Sociocultural issues also influence the trend toward regulation of transborder data flow in less industrialized nations. First is the danger to personal privacy that occurs when large amounts of personal data are collected and transmitted between countries. More data flow regulation has occured for the ostensible protection of information on citizens than any other cause (Wigand, Shipley, and Shipley 1983). Next is the issue of impact on labor and the work process. As transnational corporations move production facilities where labor is cheapest, "sweatshops" reminiscent of the early industrial revolution may be formed, even in high-technology or data-processing applications. Pollack (1982, 18E) states:

Dull, menial data entry work . . . has been done offshore for more than fifteen years. Magazines, for example, often use foreign facilities to enter subscription forms into computerized mailing lists. Large companies like American Telephone and Telegraph use them to convert massive amounts of paper data into computerized form. At least a dozen companies offer such services using facilities in the Caribbean, Ireland, India, Korea, Taiwan and the Phillippines. Pacific Data Services, a Dallas company, recently opened up a data entry business in China, which will have ninety-six computer terminals operated for three shifts a day.

He continues with the prediction that satellite technology will result in the increase of data sweatshop activity in developing nations by allowing even smaller, tight-deadline data entry jobs to be done overseas. The nature of the work that employees are engaged in changes greatly, in some cases becoming more abstract, in others more repetitive.

Another primary concern arises in many countries when the reverse occurs, and nations discover that they are losing thousands of jobs to the countries where they send their data to be processed. That this is an area of major economic concern is made evident by policy enactment in some countries requiring certain types of data to be processed within the country. This relates to the information-as-commodity issue, since along with the international transmission of data goes a transfer of jobs (Sanger 1983; Wigand, Shipley, and Shipley 1983).

This leads to the need to consider the cultural impact of transborder data flow. Transnational corporations are the most influential agents "in determining the mix of production, the character of consumption, the social values attendant on both, and the informational messages circulating worldwide" (Schiller 1981, 21). The effect of a worldwide oligopoly in the information-access and processing trade is that limited styles or methods of organizing information are imposed on cultures that may prefer to operate under different premises had they the economic leeway or capability to develop suitable alternatives. The rule of thumb for database development, for example, is that *costs* determine *users* determine *content* and the classification system. The result is that those who can most readily afford access determine the cultural slant of database content. Ease of use and information retrieval for those most able to pay is, of course, the primary consideration of system developers.

Nations hoping to "leapfrog" over the industrial age into the information age will require a great deal of training as well as material assistance. Unfortunately, those who supply these needs bring with them the values and professional methods of their culture. All of this must be balanced, however, with the social benefit of free access to information. The new information infrastructure should be designed to not only provide cultural diversity in its composition but ensure that the individual's rights are protected. If information flows only to those who can afford to pay the high price of transborder data flow, then "large companies and wealthy individuals would be able to increase their advantage over small business and the less affluent" (Dordick, Bradley, and Nanus 1981, 4).

Nanus (1981, 148) states that the social benefit of information access is that it can "lead to the transformation and expansion of human awareness and social intelligence." But such benefits to society as a whole are unfortunately counterbalanced by the increasing gaps between rich and poor, between elites and masses, gaps that are created in part by disproportionate access to information and information technologies. If trends continue in the current diffusion of informatics access, then it appears that such gaps, much as those in other areas, will continue to widen. National elites are strengthened in their entrepreneurship, as stronger links are forged with foreign sources of services and capital, through information networks. The widening of the elite-mass gap is a direct consequence of this encouragement of "external orientation among elites," says Biersteker (1978, 24).

Both transnational corporations and national industries, then, share the burden for the lack of concern for social benefits possible through equitable information access. According to Vernon (1973-4, 104), "despite the fact that the leaders in each major market were nationals, however, there was no evidence of a greater concern for social goals. Whether the social goals of their home countries or the social goals of other countries." This attitude, it may be seen, worsens problems of true national sovereignty, as it relates to protection of the wider good of society rather than protection of elite interests only. Because of inefficiency, however, state-owned enterprises are no better at serving the social good. "What is at issue," Vernon continues, "is usually a struggle among competing elites." Thus, there are no easy solutions to transnational corporate control and domination of international information networks and services.

The threat to national sovereignty and cultural integrity is a very real one, and solutions must be found if worldwide information inequities are to be alleviated. Smith (1981, 176) states that "indeed, the threat to independence in the late twentieth century from the new electronics could be greater than was colonialism itself." How then can this problem be effectively addressed? Some of the solutions that have been promoted in the case of developing countries are not easily implemented. An attempt to answer this question is next, with particular concern for effectiveness, alternatives, and enforceability.

Restrictions, Barriers, and Planning for Transborder Data Flow

As of 1979, over 60 countries had instituted official information and communication policies. While there is a degree of variety in the forms this may take, the trend for most is to set up administrative authorities "to direct or coordinate the procurement of data processing equipment and training of operators, and to set priorities in its use and application" (Pipe 1979). However, when contrasted with the immense organizational powers of transnational corporations, it seems that these authorities have little actual power to promote controlled utilizations of the new technologies according to developmental planning efforts.

Effectiveness

Part of the reason for a lack of success in controlling the international corporate informatics movement lies with a lack of coordinated negotiating power among developing nations. Hymer (1970, 44) states that "the less developed the country, the greater its disadvantage in the bargaining process." The remedy he proposes is increased use of resources to develop controllable, competitive firms within developing nations to help create a "stronger bargaining position." Unfortunately, due to lack of resources and planning capability, this process can be a very difficult if not even a hopeless one.

In spite of this manifest inability of developing nations to significantly counteract the transnational corporate informatics oligopoly, corporate executives of many industrialized nations are clamoring about informatics protectionism. In recent testimony before a U.S. congressional subcommittee, for example, they claimed that many barriers to information flow and access are being erected that hinder transnational corporate expansion in some global markets. Barriers include those initiated for political and social reasons, and for the protection of national and cultural sovereignty and personal privacy. It was also claimed that major economic restrictions were affecting U.S. corporate action abroad. These include tariffs and discriminatory pricing, inconsistent technical standards, monitoring of information and data flow (often for enforcement of privacy legislation), and restriction or denial of market entry (U.S. House of Representatives 1980). Most of these restrictions have been enacted in more developed, or "second tier" countries, however, with little control existing as yet in less-industrialized nations.

Alternatives

What can developing nations do to effectively establish a place for themselves in this era of advanced communication technology? The question is entangled in practical and philosophical issues of whether such technology, or ability to gain access to databases and processing services, is appropriate in the unique social and economic positions in which each nation finds itself. Many alternatives have been suggested or implemented in developing nations. Three of these are presented here.

Hamelink (1983) recommends a policy of "dissociation," an alternative which some may find extreme but others believe would turn out to be helpful. In his view, nations should disengage themselves, as necessary, from the informatics adoption and diffusion process to gain needed time to make rational, well-considered decisions about national development. This "breathing space" would "allow indigenous needs to be formulated before embarking on development paths that carry unclear but powerful alternatives" (Schiller 1981, 167).

A second type of recommendation, perhaps more feasible, is that of Ripper and Wanderley (1981). They suggest that in countries where the amount of transborder data flow is still minimal, regulation on a "case-by-case" basis would be a suitable way to approach the problem. Two benefits of this approach were noted. One is that implementing the policy would be fairly simple. The second is that understanding the problem and developing criteria for judging its impact could be fostered gradually. This would include, for example, the identification of types of uses and users of data flows (Kuitenbrouwer 1980).

A third method involves use of trade barriers and other policies in a broader regulatory application. This is the sort of tactic that is most unacceptable to U.S. corporate executives. It may be most desireable for developing nations to select from these policies those regulations that would most benefit their individual situations without unduly antagonizing those international parties on whom they often depend for economic and political assistance in developmental activities.

Enforcement

A major problem with use of regulatory or trade barriers to control informatics access it that the new technologies make it fairly easy to circumvent such barriers. The Satellite Business System has fully integrated four broad areas of communication: data, voice, facsimile, and audio/video teleconferencing. This integrated service is offered on an on-line basis via satellite without having to depend on existing telephone networks. The resulting high-speed service is so fast that in one minute a database can be transmitted that required an entire hour just three years ago. Similarly, a one-billion byte file, or eight billion bits, could be distributed to one or many locations simultaneously in just thirty minutes (Wigand 1982). The Satellite Business Systems network can handle 6.3 million bits of information per second. AT&T's proposed satellite network will handle 600 million bits. One American businessman (Nilson 1980, 13) said, "Soon we will be dealing with data transmission in terms of a few tenths of seconds here, a few tenths of seconds there. How are you going to police this activity?" Two other examples illustrating this difficulty are:

- United States regulations prohibited data transmission by ordinary telephone lines, but businessmen from all over the world (Africa, Asia, Latin America) just did it. Chance of discovery: 100 to 1 (Kuitenbrouwer 1980, 13)
- Last December, AT&T LongLines began operating the world's longest high-capacity digital coaxial system . . . a 2,300-mile route stretching from Plano, Illinois to Sacramento, California. The $26 million all-digital system can handle 140 million bits of information per second. At that rate, it would take one-tenth of a second to transmit the contents of a typical 76-page newspaper (Wiley 1983, 94)

Viability of national postal, telephone, and telegraph systems are especially susceptible to the influences of the new technology. Corporations offering instantaneous satellite transmission capabilities may seriously threaten economic viability of such systems by making inroads into first-class postal and prime-time telephone revenue generation.

Ability to enforce data flow regulations will depend first of all on a nation's resources for detection or noncompliance. If it is feasible to monitor compliance, then, in addition, there will need to be penalties severe enough to help ensure conformance with the law. Compliance can also be encouraged by making sure that regulations are clear and well-understood by those whose use of transborder data flow is restricted (Sabatier and Mazmanian 1980). But also of great importance is the ability of governmental agencies to deal effectively with transnational corporate executives. Feld (1976, 1503-4) writes:

> The overwhelming economic power of MNEs [multi-national enterprises] has made it difficult for many Third World governments to come up with effective regulations for either the establishment of a new subsidiary or the continuation of an MNE's operation. Although national governments have the political and legal power to pass and enforce laws limiting the operations of MNEs in accordance with national policies, they are often hesitant to exercise these powers because of the real or perceived costs entailed.

Anxieties tend to be overplayed, he continues, with a resulting loss of developmental support by transnational corporations because of "misunderstandings and tensions." It would be much easier for nations to implement regulatory programs that transnational executives could perceive as being legitimate and not overly injurious to business interests.

Among developing nations, Brazil has become a leader in planning for the control of national informatics development. Brazil pioneered the case-by-case method of corporate transborder data flow control by authorizing on an individual basis the linkage of corporations to international informatics systems. Approval by the Special Secretariat of Informatics is for a maximum of three years, but is it "renewable, with consideration of the preservation of the internal market, the protection of national enterprises and individual privacy" (de Oliviera Brízida 1981, 19). This includes control over the linkage of transnational corporations with subsidiaries, access to foreign data banks, use of foreign computers for data processing, and importation of software. Applications are denied when national alternatives are available (de Oliviera Brízida 1981).

These types of policies can at least give developing nations a better understanding of possibilities for control of transborder data flow, even if they do not make it logistically feasible on a large scale. In the long run, implementation of truly effective controls in most developing nations will require the negotiating power of a larger, regional or international entity or commission. Vernon (1973-4, 117) states that "it would take some fairly complex international agreements to keep the competition in bounds, but the effort could easily be worthwhile." International assistance could also be useful in the explorations of such ideas as imposition of an international fee charged to transnational corporations to help finance regional information access and processing centers in the developing world. The MacBride Commission suggested "a tax on TNCs [transnational corporations] that exploit raw materials in LDCs [less-developed nations], an international tax on communication transnationals, as well as duty on the use of the electromagnetic spectrum and the geostationary orbit" (Jussawalla 1982, 117).

International solutions, however, can run into major political problems that usually prevent successful enactment. Ability of developing nations to agree on similar matters has been limited at best in the past, and this is not likely to change with worsening global economic conditions. Another concern is the recently stated position of the United States (U.S. Senate 1983, 195):

> The U.S. position on this matter . . . would be to strongly oppose any actions that would interfere with the ability of producers and users to make optimum use of information as a productive resource. This will lead to a more efficient utilization of resources. It will also lead to greater revenues for both private entities and, ultimately, for taxing authorities.

The obvious question generated by this statement is: For whose benefit? Will such actions lead to a more efficient utilization of resources? Ideally, users in less developed nations should also be able "to make optimum use of information as a productive resource." This is difficult when access to databases, which may even contain information derived from the developing nation, is controlled by transnational corporations. "Private entities" who profit from free flow are primarily transnational corporations deriving "greater revenue" from use of information and telecommunications within developing countries.

It is difficult to disagree with the U.S. position that opposes transborder data flow regulation because "surveillance of specific kinds of information," in order to determine taxable value or enforce privacy legislation, "could too easily become surveillance of all information for political reasons" (U.S. Senate 1983, 195). This is just one aspect of regulatory enforcement that must be studied for practical alternatives that both ensure freedom of expression and allow developing nations to have greater control over their own futures. Hymer (1970, 445) states that "it does not appear to be socially efficient to allow corporations to monopolize information on new possibilities created by science." This is undoubtedly true of the new communication and data processing technologies, as well as of the uses to which they are put. Nations should also be able to determine more of their own parameters of transborder data flow in order to promote cultural integrity and diversity.

Conclusion

Third World regulation of transborder data flow requires balance among many factors ranging from the economics and cultures of nations to governmental ability to negotiate effectively with each other and with corporations. The flow of computerized information benefits international business, and also in various ways, it can help developing nations. It is to the advantage of less developed nations to have full access to foreign data banks, especially scientific, economic, and technical, and to data-processing services when their use augments the development process. But when the unhindered flow of international information services jeopardizes a nation's plans for improving its economy or safeguarding its culture, or when corporate transborder data flow strengthens a corporate position to the detriment of national interests, then regulation is often promoted.

This regulation may take the form of a fee charged on transnational corporate data flows in order to finance national or regional information services. Or it may temporarily establish barriers to the flow of such trade in order to promote development of national level industries. It may even require a temporary moratorium on all transborder data flows in order for needed time to be gained for reasoned planning of a nation's informatics development.

But in any case, the pragmatics of national negotiation with transnational corporations dictate an approach that must make allowances for the interests of all concerned. The vast majority of transnational corporate executives are probably interested in the welfare of nations in which they conduct business, even if simply for economic or market development reasons. Thus, it may be worthwhile for developing nations to cultivate the participation of these executives in the planning process. A division of responsibility for aspects of the regulatory process might even be beneficial by inviting corporations to jointly provide staff and funding for regional research and development centers. Allocation of responsibilities would have to be based on such factors as the capabilities and interests of the parties involved.

Regulations should, ideally, demonstrate as much flexibility as possible, allowing for individual, case-by-case analysis of corporate transborder data flow usage, as well as for problems that might arise in the public or national sectors. If international industry proves itself capable and worthy of self-regulation, it should be able to do so. Should nations desire to cultivate the good will of transnational corporations, it is also important that transborder data flow regulation not be implemented in arbitrary ways that may be unresponsive to justifiable needs. It may be worthwhile for nations to provide incentives for corporate compliance with, or assistance in, achieving national goals. Economic incentives, such as preferential taxation, could be utilized with cooperating corporations. Such preference could be justified if enough benefit was derived by the nation to compensate for foregone tax income.

While it is not possible to predict that any of these methods will be practicable, or even that such cooperative approaches are feasible given human nature and hidden factors of political interest, they are certainly worth exploring in lieu of any better alternatives. Since governments and/ or international agencies are the most likely entities to protect the cultural diversity and strength of economies in developing nations, it is their responsibility to investigate possible solutions such as those given here. Davis (1983) points out, "What is at stake is the liberal ideology that more information access serves everbody's best interests, and does so in a relatively equitable manner." He also notes, however, that the central issue, ultimately, is a question of power—power for the corporation versus for the individual nations involved. Many factors must be weighed in the balance before workable, enforceable solutions, which offer the most benefit to society as a whole, are achieved.

References

Betz, M.J., P.J. McGowan and R.T. Wigand *Appropriate technology: Choice and Development.* Durham, N.C.: Duke University Press, 1984.

Biersteker, T.J. *Distortion or development? Contending perspectives on the multinational corporation.* Cambridge, Mass.: MIT Press, 1978.

Bortnick, J. *Telecommunications and information products and services in international trade.* Washington, D.C.: U.S. Government Printing Office, Serial No. 97-59, 1981. Statement to Committee on Energy and Commerce, House of Representatives.

Committee for Economic Development. *Stimulating technological progress: A statement by the research and Policy Committee of the Committee for Economic Development.* New York: Committee for Economic Development, 1980.

Davis, T.E. Personal communication, November 15, 1983.

De Oliviera Brízida, J. The Brazilian transborder data flow policy. *Transnational Data Report,* 1981, 4 (3), 19-21.

Dordick, H.S., H.G. Bradley, and B. Nanus. Information inequality. *Computerworld,* June 16, 1980, p.1.

_____. *The emerging network marketplace.* Norwood, N.J.: Ablex, 1981.

Feld, W.J. UN supervision over MNCs: Realistic expectations or exercise in futility. *Orbis,* 1976, 19(4), 1503-1504.

Frank, R.E. Quoted in H.I. Schiller, *Who knows: Information in the age of the Fortune 500.* Norwood, N.J.: Ablex, 1981.

Giuliano, V.E. 'Rear view' vision limits growth. *Information World,* 1980, 2, 11.

Hamelink, C. *Cultural autonomy in global communications.* New York: Longman, 1983.

Hymer, S. The efficiency (contradictions) of multinational corporations. *American Economic Review,* 1970, 60(2), 443-444.

Jacobson, R.E. The hidden issues: What kind of order? *Journal of Communication,* 1979, 29(3), 151.

_____. Quoted in H.I. Schiller, *Who knows: Information in the age of the Fortune 500.* Norwood, N.J.: Ablex, 1981.

Joinet, L. Quoted in G. Russell Pipe, National policies, international debate. *Journal of Communication,* 1979, 29(3), 118.

Jussawalla, M. International trade theory and communications. In M. Jussawalla and D.M. Lamberton (eds.), *Communication economics and development.* New York: Pergamon, 1982, pp. 92-99.

Kuitenbrouwer, F. Legal and regulatory issues a main focus of the TDF Conference. *Transnational Data Report,* 1980, 3 (3/4), 10-13.

Lecht, C.P. Quoted in J. Beeler, IBM may become service bureau. *Computerworld,* March 10, 1980, p. 1.

Lloyd, A. Data flow may lead to domination. *Transnational Data Report,* 1980, 3 (3/4), 9.

Müller, R. The multinational corporation and underdevelopment of the Third World. In C.K. Wilber (ed.), *The political economy of development and underdevelopment.* New York: Random House, 1973, pp. 124-148.

Nanus, B. Cited in a H.S. Dordick, H.G. Bradley, and B. Nanus, *The emerging network marketplace.* Norwood, N.J.: Ablex, 1981, p. 148.

Nilson, M.C. Quoted in F. Kuitenbrouwer, Legal and regulatory issues a main focus of the TDF Conference. *Transnational Data Report,* 1980, 3 (3/4), 13.

Onstad, P.C. Data processing services and transborder data flows. *Transborder data flows and the protection of privacy.* Paris: OECD, 1979. Proceedings of the OECD Symposium, Vienna, Austria, September 20-23, 1977.

Organisation for Economic Cooperation and Development. *Information activities, electronics, and telecommunications technologies: Impact on employment, growth, and trade.* Paris: OECD, 1981, p. 29.

Pipe, G.R. National policies, international debate. *Journal of Communication,* 1979, 29(3), 114-123.

Pollack, A. Latest technology may spawn the electronic sweatshop. *The New York Times,* October 3, 1982, 18E.

Porat, M.U. Communications policy in an information society. In G. Robinson, (ed.), *Communication for tomorrow.* New York: Praeger, 1978, pp. 3-60.

Ripper, M.D. and J.L.C. Wanderley. The Brazilian computer and communications regulatory environment and transborder data flow policy. Paper presented at the IBI World Conference on Transborder Data Flow Policies, Rome, June 23-27, 1980.

Sabatier, P. and D. Mazmanian. The implementation of public policy: A framework of analysis. *Policy Studies Journal,* 1980, 8(4), Special Issue #2, 538.

Sanger, D.E. Waging a trade war over data. *The New York Times,* March 13, 1983, 1F.

Schiller, H.I. *Who knows: Information in the age of the Fortune 500.* Norwood, N.J.: Ablex, 1981.

Smith, A. Quoted in H.I. Schiller, *Who knows: Information in the age of the Fortune 500.* Norwood, N.J.: Ablex, 1981.

Transnational data report. National independence dominates IBI World Data Flow Conference, 1980, 3 (3/4), 1-2.

Transnational data report. Why American Express depends on telecommunications, 1983, 6(1), 18.

U.S. House of Representatives, Committee on Government Operations. *International information flow: Forging a new framework,* Thirty-second Report with Additional Views. Washington, D.C.: U.S. Government Printing Office, H.Report 96-1535, 1980.

U.S. Senate, Committee on Commerce, Science, and Transportation. *Long-range goals in international telecommunications and information.* Washington, D.C.: U.S. Government Printing Office, S. Prt. 98-22, 1983.

Van Rensselaer, C. Centralize? Decentralize? Distribute? *Datamation,* 1979, 25 (4), 90.

Vernon, R. (More) on multinationals: Does society also profit? *Foreign Policy,* 1973-4, 13, 113.

———. The multinational enterprise as symbol. *Storm Over the Multinationals.* Cambridge: Harvard University Press, 1977.

Wigand, R.T. Direct satellite broadcasting: Selected social implications. In M. Burgoon (ed.), *Communication Yearbook 6.* Beverly Hills, Calif.: Sage, 1982, pp. 250-288.

———. Direct satellite broadcasting: Definitions and prospects. In G. Gerbner and M. Siefert (eds.), *World communications: A handbook.* New York: Longman, 1984, pp. 246-253.

———, C. Shipley, and D. Shipley. Transborder data flow, informatics and national policies: A comparison among 22 nations. Paper presented at the International Communication Association Annual Convention, Dallas, May 28, 1983.

Wiley, D. The new AT&T. *Communication News,* 1983, 20(9), 88-108.

About the Contributors

James D. Anderson is professor of Library and Information Studies and associate dean of the School of Communication, Information, and Library Studies at Rutgers University. He is author of numerous publications on indexing. In 1979-80, he headed the design team of the Modern Language Association's Bibliography Revision Project, and developed the Contextual Indexing and Faceted Taxonomic Access System (CIFT), which is now used to produce both contextual indexes and classified arrays for the MLA's print and online *International Bibliography.*

Jerome Aumente is professor and director of the Journalism Resources Institute, Rutgers University, and was founding chairperson of the Department of Journalism and Mass Media at Rutgers. He is a former journalist and Nieman Fellow at Harvard University and is currently completing a book to be published by Pergamon Press entitled *Words on Paper, Words on the Screen: Publishing and the New Electronic Pathways.*

David Blomquist is supervising editor, Prototype Development, of CBS Venture One, the CBS videotex research group. During the CBS/AT&T videotex trials in 1982 and 1983, he edited features and opinion sections of the experimental service, wrote about national politics, and directed part of the project's market research program. Educated at the University of Michigan and Harvard, he wrote for the *Detroit News* and taught at Harvard before joining CBS. He is a member of the Education Committee of the American Political Science Association.

Robert Cathcart is professor in the Department of Communication Arts and Sciences at Queens College of the City University of New York. He coedited, with Gary Gumpert, *Inter/Media: Interpersonal Communication in a Media World,* and is the coauthor, with Gary Gumpert, of numerous articles, including "Mediated Interpersonal Communication: Toward a New Communication Typology," which appeared in the *Quarterly Journal of Speech;* and "Stereotyping: Images of the Foreigner," in *Intercultural Communication: A Reader.*

James W. Chesebro is associate professor in the Department of Communication Arts and Sciences at Queens College of the City University of New York. In the past he was president of the Eastern Communication Association.

John Cook is visiting assistant professor of speech communication at Texas Tech University in Lubbock, Texas. His research and teaching interests include interpersonal, instructional, and organizational communication and persuasion.

Frank E. X. Dance is professor of speech communication at the University of Denver, and is president of Frank E. X. Dance Associates, Inc., his own firm of consultants in communication. Dance is author and editor of eight books (two of which have been translated into other languages) and over fifty scholarly articles. He has served as editor of the *Journal of Communication* (1962-1964), editor of *Communication Education* (1970-1972), president of the International Communication Association (1967), and president of the Speech Communication Association (1982). His most recent book is *Human Communication Theory: Comparative Essays.*

Stanley Deetz is associate professor in the Department of Communication, School of Communication, Information, and Library Studies at Rutgers University. He is author of numerous articles and book chapters on theory and philosophy of interpersonal and organizational communication, and is currently chairperson of the Philosophy of Communication Interest Group of the International Communication Association. His books include *Phenomenological Research in Rhetoric and Communication.*

Roger Jon Desmond is associate professor in the Department of Communication at the University of Hartford. His primary research interests are the development of media comprehension in young children, and the relationships between media use and cognition. His articles have appeared in *Human Communication Research; Communication Research; Communication Monographs;* and *Policy Sciences.*

Herbert S. Dordick is a member of the faculty of the Annenberg School of Communications at the University of Southern California and a fellow at the Communications Institute of the East-West Center in Hawaii. He is the coauthor, with Burt Nanus and Helen Bradley, of *The Emerging Network Marketplace; The Executives's Guide to Information Technology*, with Frederick Williams, (1983); and the forthcoming *Modern Telecommunications: A Non-Technical Guide.*

Gary Gumpert is professor in the Department of Communication Arts and Sciences at Queens College of the City University of New York. He coedited, with Robert Cathcart, *Inter/Media: Interpersonal Communication in a Media World*, and is the coauthor, with Robert Cathcart, of numerous articles, including "Mediated Interpersonal Communication: Toward a New Communication Typology," which appeared in the *Quarterly Journal of Speech;* and "Stereotyping: Images of the Foreigner," in *Intercultural Communication: A Reader.*

James D. Halloran is director of the Centre for Mass Communication Research at the University of Leicester, England, a position which he has occupied since 1966. He has been president of the International Association for Mass Communication Research since 1972, and has served as a consultant to UNESCO and the MacBride Commission. Halloran is author of numerous books and articles on sociology and mass communication, including "A Case for Critical Eclecticism," *Journal of Communication*, 1983; "The Context of Mass Communication Research," in *Communication and Social Structures: Critical Studies of Mass Media Research,* (1981); *Attitude Formation and Change,* 1970; and *Demonstrations and Communication: A Case Study,* 1970.

Richard F. Hixson is a professor in the Department of Journalism and Mass Media, School of Communication, Information, and Library Studies, Rutgers University. His primary areas of research and teaching are the history of American news media and communication law. He is the author of *Isaac Collins: A Quaker Printer in 18th-Century America,* 1968; *Mass Media: A Casebook,* 1973; and *The Press in Revolutionary New Jersey,* 1975. He is also the author of numerous articles and of 30 entries on American journalism pioneers in *The Academic American Encyclopedia,* 1980.

Irving Louis Horowitz is Hannah Arendt Distinguished Professor of Sociology and Political Science at Rutgers University. He is author or editor of over twenty books, and editor-in-chief of Transaction/*Society*, the nation's largest independent social science private complex of publications. He has increasingly become concerned with closely inspecting the interrelationships between the two sides of his work, political sociology and scholarly publishing, by examining the implications for democratic societies of the new information technology, especially those elements which have a direct linkage to publication and information processing.

Harvey Jassem is assistant professor in the Department of Communication at the University of Hartford. He has written in the areas of new media, futures studies, and media regulation, policy, and economics. His publications appear in such journals as *Journal of Broadcasting; Policy Sciences; Communications and the Law;* and *Journalism Quarterly.*

Young Yun Kim is professor of communication at Governors State University in University Park, Illinois. Her teaching and research interests include communication theory, research methods, intercultural communication, cultural adaptation, nonverbal communication, and communication training. She has conducted surveys in several Asian and Hispanic communities in Illinois, and has published her work in a number of journals, including *Human Communication Research; Communication Yearbook*; and *International Journal of Intercultural Relations.* She is coauthor of *Communicating with Strangers: An Approach to Intercultural Communication,* and coeditor of *Methods for Intercultural Communication Research.* She also serves as the editor-elect of the *International and Intercultural Communication Annual.*

Elizabeth Loftus is professor of psychology at the University of Washington, Seattle, where she has taught since 1973. She is author of more than 10 books and over 100 scientific articles. Her fourth book, *Eyewitness Testimony,* won a National Media Award in 1980 from the American Psychological Foundation. Her other books include *Memory,* 1980; *Essence of Statistics,* 1981; and *Mind at Play,* 1983—the last two coauthored with Geoffrey Loftus.

Denis McQuail is professor of Mass Communication at the University of Amsterdam, Netherlands. Prior to that he taught in the Sociology Department of Southampton University, England. His main research interests have been in political communication, audience research, and media evaluation research for policy. He worked as academic adviser to the British Royal Commission on the Press 1974-1977, and as adviser to the Scientific Council for Government Policy in the Netherlands (1981-1982). Professor McQuail is currently engaged in a cross-national study of new electronic media policymaking in Europe. His publications include *Televisison and the Political Image* (with J. Trenaman), 1961;*Television in Politics* (with J. Blumler),1968; *Towards a Sociology of Mass Communication,* 1969; *Communication,* 1975; *Communication Models* (with S. Windahl), 1982; *Mass Communication Theory,* 1983.

Gordon Miller is a doctoral candidate at the School of Communication, Information and Library Studies of Rutgers University. His research deals with the concept of information in modern physics and psychology in relation to theories of perception.

Dennis Mumby is an assistant professor in the Department of Speech Communication at St. Cloud State University. He has written on metaphors in television and business organizations.

Dan Nimmo is professor of communication at the College of Communications and in the Department of Political Science at the University of Tennessee, Knoxville. His research interests are in political communication, especially government-news media relations, political campaigns, and the role of mass communication in political behavior. Nimmo is author of numerous books and articles, including *Newsgathering in Washington; The Political Persuaders; Political Communication and Public Opinion in America; Government and the News Media*; and *Mediated Political Realities*. He is the editor of *Communication Yearbook 3* and *4*.

Sheizaf Rafaeli is a research associate and doctoral candidate in the Institute for Communication Research at Stanford University. His areas of interest and previous work are the new technologies of interactive communication, as well as political communication.

Ronald E. Rice is assistant professor at the Annenberg School of the University of Southern California. His research interests include human communication networks, social impacts of telecommunications, diffusion of innovations and public communication campaigns. He is coeditor of *Public Communication Campaigns* with William Paisley, and editor of *The New Media: Communication, Research and Technology.*

Everett Rogers is Janet M. Peck Professor of International Communication in the Institute for Communication Research at Stanford University. He is a former president of the International Communication Association, and author or coauthor of numerous books and articles including *Communication of Innovations; Communication Networks; Communication in Organizations; Diffusion of Innovations*; and his most recent—*Silicon Valley Fever.*

Brent D. Ruben is professor and director of the Ph.D. Program, School of Communication, Information, and Library Studies, Rutgers University. He is the author of *Communication and Human Behavior,* 1984; and the author/editor of *Interdisciplinary Approaches to Human Communication,* 1979; *Beyond Media,* 1979; *General Systems Theory and Human Communication,* 1975; and other books and articles. He was also founding editor of the International Communication Association, *Communication Yearbook* series.

Jerry Salvaggio is associate professor and director of the International Telecommunications Research Institute at the University of Houston, University Park. He is the author of *Telecommunications: Issues and Choices for Society,* 1983, and of more than a dozen articles and book chapters on communication technology and social issues. His most recent book, *Broadcast Communications Technology,* will be published in 1985.

Herbert I. Schiller is professor of communication, University of California, San Diego. He is vice-president of the International Association for Mass Communication Research and a trustee of the International Institute of Communication, London. Schiller is author of *Who Knows: Information in the Age of the Fortune 500,* 1981; *The Mind Managers,* 1976; *Mass Communications and American Empire,* 1969; and the forthcoming *Information and the Crisis Economy,* in which his present article will also appear.

Jonathan W. Schooler is a doctoral student at the University of Washington. He did his undergraduate work at Hamilton College.

Carrie L. Shipley is a master's degree candidate at the Department of Communication Arts at Cornell University. Concurrently, she is an assistant instructor of communication at George Mason University in Northern Virginia. Her recent article, coauthored with Rolf T. Wigand and Dwayne L. Shipley, appeared in the *Journal of Communication,* 1984.

Dwayne L. Shipley has been a consultant in management and development. He is currently a researcher for the Homer Hoyt Institute in Washington, D.C., and is pursuing graduate studies in urban development at American University.

Alfred G. Smith is professor of anthropology in the College of Communication, University of Texas at Austin, where he was director of the Center for Communication Research from 1973 to 1978. He was president of the International Communication Association in 1973-74. Smith specialized for many years in Malayo-Polynesian linguistics. This led to studies of language in culture, and that to studies of national and international systems of communication. His publications range from *Gamwoelhaelhi Ishilh Weleeya* and *Ki Luwn Specl Kosray* to *Communication and Culture,Communication and Status,* and *Cognitive Styles in Law Schools.*

Lawrence Wheeless is professor of speech communication at Texas Tech University. He is author of two volumes, including *Introduction to Human Communication* (with James McCroskey), 1976, and nearly 50 articles and book chapters dealing with communication theory and research methods. Professor Wheeless has served as editor of *Communication Quarterly* and associate editor of *Human Communication Research,* and has served on the governing boards of the Eastern Communication Association and the International Communication Association.

Rolf T. Wigand is a professor of communication and public affairs at Arizona State University. He has published nationally and internationally in the areas of organizational communication and behavior, social and policy issues of satellite broadcasting, telecommunication issues, and the impacts of office automation on the work environment. He is chairperson for the Information Systems Division of the International Communication Association.

Frederick Williams is professor of communication in the Annenberg School of Communications at the University of Southern California, where he served as founding dean from 1973 to 1980. He is author of *The Communications Revolution,* 1983; coauthor, with Herbert Dordick, of *The Executive's Guide To Information Technology,* 1983; and coauthor, with Victoria Williams, of *Microcomputers in Elementary Education,* 1984.

Subject/Name Index

James D. Anderson and Gordon Miller

Citation Index

Gordon Miller and James D. Anderson

(A list of references appears at the end of each chapter)

Adler, M. (1967), 72
Aitchison, I. (1977), 310
Allen, B. (1982), 435
Allport, G.W. (1955), xxi
Almond, G. (1960), 353
Althusser, L. (1971), 373, 375, 376
Alwitt, L.F., et al. (1980), 264
American Law Institute (1967), 83
American Newspaper Publishers Association (1983), 199
Andersen, P.A., Garrison, J.P., and Anderson J.F. (1979), 271
Anderson, D.R., et al. (1980), 274; (1981), 264
Anderson, J.D. (1979), 314; (1982), 318; (1983), 317
Anderson, J.R. (1976), 231, 236; (1977), 246; (1983), 231, 232, 236, 239, 240
Anderson, J.R., and Ross, B.H. (1980), 232
Apfelbaum, M.J. (1983), 90
Aronson, E. (1968), 254
Aronson, S. (1971), 116
Artandi, S. (1973), 18; (1979), 8
Ashby, W.R. (1964), xxii
Atkin, C.K. (1972), 260; (1973), 260; (1981), 361
Atkinson, R.C., and Shiffrin, R.M. (1968), 241, 269
Atkinson, R.C., and Wickens, T.D. (1971), 267
Atwood, A.R. et al. (1982), 261
Austin, D. (1974), 314
Axley, S. (1983), 381

Bachrach, P., and Baratz, M.S. (1962), 353
Baddeley, A.D. (1982), 239
Baddeley, A.D., et al. (1975), 234
Baddeley, A.D. and Hitch, G.J. (1974), 269, 271, 272
Baddeley, A.D., Thomson, N., and Buchanan, M. (1975), 269
Bagdikian, B.H. (1983), 437
Baggett, P. (1979), 274
Baldwin, T.E., and McVoy, D.S. (1983), 431
Bates, M. (1976), 292
Becker, L. (1982), 42
Becker, L.B., and Krendl, K.A. (1983), 261
Becker, T. (1981), 413
Bekerian, D.A., and Bowers, J.M. (1983), 234
Belkin, N.J., and Robertson S. E. (1976), 18
Bell, D. (1976), 3, 172; (1982), 435; (1983), 10, 14, 15, 437, 446
Benjamin, G. (1982), 413
Benjamin R. (1980), 152
Bennett, T. (1982), 45
Bennett, W.L. (1981), 360, 361, 362, 363, 365; (1983), 348
Bennis, W., and Slater, P. (1968), 171
Benson, J. (1977), 384
Berelson, B., Lazarsfeld, P., and McPhee, W. (1954), 345, 352, 354
Berger, C.B., and Calabrese, R.J. (1975), 266
Berger, P. L., and Luckmann, T. (1966), xxii, 16, 372
Bergstrom, B., Gillberg, M., and Arnberg, P. (1973), 268

Berlo, D. (1960), 209
Berlyne, D.E. (1960), 264
Bernard, H. and Killworth, P. (1977), 170
Berndt, T.J., and Berndt, E.G. (1975), 274
Bertalanffy, L. von (1968), xxii
Betz, M.J., McGowan, P.J., and Wigand, R.T. (1984), 460
Bezzini, J., and Desmond, R. (1982), 143
Biersteker, T.J. (1978), 468, 471
Bindra, D. (1969), 267
Birch (1981), 3
Blasko, L. (1982), 187, 195, 423
Blomquist, D. (1981), 414
Blumer, H. (1969), xxii, 16, 365
Blumler, J.G. (1980), 438
Blumler, J.G., and Katz, E. (1974), 261
Boffey, P.M. (1983), 395
Bohlin, G. (1973), 268
Boorstin, D.J. (1961), 19
Bormann, E.G. (1972), 348; (1982), 348
Bortnick, J. (1981), 467
Boulding, K. (1956), xxi, 19
Bower, G.H. (1981), 234
Bowers, B. (1978), 172
Boyd-Barrett, J.D. (1982), 42
Bransford, J., Barclay, J., and Franks, J. (1972), 233
Brittain, J.M. (1984), 159
Brizida, J. de Oliviera. See de Oliviera Brizida, J.
Broad, W.J. (1983), 14
Broadbeck, M. (1956), 258
Broadbent, D.E. (1958), 253, 270; (1971), 267, 270; (1983), 240
Brock, T.C., Albert, S.M., and Becker, L.A. (1970), 259, 264, 265
Brock, T.C., and Balloun, J.L. (1967), 259
Brody, E.B. (1969), 329
Brookes, B.C. (1980), 320
Brown, A.S. (1979), 236
Brown, J. (1958), 230
Brown, R. (1965), 257
Bruner, J.S. (1973), xxi, 19
Budd, R.W. and Ruben B.D. (1979), xxii, 16
Buhler, K. (1934), 65
Bukeley, W.M. (1983), 431
Burnham, D. (1983), 435
Burrell, G., and Morgan, G. (1979), 49
Business Week (1982), 431; (1983), 395, 448
Butterfield, E.C.: Lachman, Lachman, and Butterfield (1979), 295, 299, 306, 310

Butterfield, F. (1982), 395; (1983), 403

Califano, J.A. (1983), 399
Cambron, J. (1983), 204
Campbell, A., Converse, Miller, and Stokes (1960), 350, 358
Campbell, A., et al. (1966), 350, 358, 359
Cannon, D.L., and Luecke, G. (1980), xxii
Cannon, W.B. (1932), xxi
Canon, L.K. (1964), 258, 259
Capron, H.L., and Williams, B.K. (1982), 203, 215
Carey, J.W. (1982), 122
Carey, J. (1969), 46; (1975), 48; (1982), 408, 422, 423
Carey, J., and Kreiling, A.L. (1974), 42
Carey, J., and Siegeltuch, M. (1982), 196
Carpenter, T. (1983), 205, 215, 216
Cassirer, E. (1944), 67
Cathcart, R. and Gumpert, G. (1983), 113, 114, 206
Cermak, L.S. (1972), 238
Chaffee, S.H. (1972), 260; (1975), 353, 354
Chaffee, S.H., and Miyo, Y. (1983), 253, 254, 255
Chaffee, Z. (1947), 88
Chalkey, M. (1982), 65
Cherry, C. (1953), 263; (1966), 18; (1971), xxii, 10
Cherry, R.S. (1981), 265
Chesbro, J.W. (1985), 11
Ciano, J.M., and Carne, E.B. (1982), 10, 11
Clark, J.W. (1909), 289
Clegg, S., and Dunkerley, D. (1980), 372
Cleveland, H. (1982), 15
Coelho, G. (1958), 336
Coen, R.J. (1980), 398
Cohen, A.R., and Latane, B. (1962), 255
Collins, A.M., and Loftus, E.F. (1975), 235
Collins, A.M., and Quillian, M.R. (1972), 235
Collins, W.A. (1978), 274; (1982), 264, 265, 274
Commission on Freedom of the Press (1947), 420
Committee for Economic Development (1980), 463, 464
Communications News (1982), 433
Compaine, B. (1982), 437, 446
CompuServe (1983), 434

Lachman, R., Shaffer, J.P., and Hennrikus, D. (1974), 306

Lakoff, G., and Johnson, M. (1980a), 372, 377, 378, 382; (1980b), 372, 378, 382

Lane, D.M. (1980), 265, 272

Lane, D.M., and Pearson, D.A. (1982), 272

Lang, G.E., and Lang, K. (1981), 42

Langer, S.K. (1967), 67, 72; (1972), 67, 72; (1982), 67, 72

Lasswell, H. (1948), 66-67; (1971), 420

Lasswell, H., and Kaplan, A. (1950), 362

Lawson, C.A. (1963), xxi

Lazar, E.D. (1980), 80, 81

Lazarsfeld, P. (1941), 46

Lazarsfeld, P., Berelson, B., and Gaudet, H. (1948), 129, 356

Lazarsfeld, P., and McPhee, W. (1954), 356

Lazarsfeld, P., and Merton, R.K. (1971 [1948]), 406

Le Duc, D.R. (1983), 433

Lecht, C.P. (1980), 466

Leonard, L.E. (1977), 295

Levin, S.R., and Anderson, D.R. (1976), 264

Levy, S. (1983), 114

Lewin, T. (1982), 76

Library of Congress, Automated Systems Office (1980), 315

Library of Congress, Subject Cat. Div. (1980), 306

Lichty, L. (1982), 347

Linder, D.E., Cooper, J., and Jones, E.E. (1967), 255

Lindsay, P.H., and Norman, D.A. (1977), 362

Linowes, D.F. (1978), 435

Lipetz, B. (1970), 18

Lippmann, W. (1922), 348

Littlejohn, S. (1983), 130

LJ/SLJ Hotline (1982), 397

Lloyd, A. (1980), 457

Loftus, E.F., (1973), 235; (1975), 233; (1980), 292, 305

Loftus, E.F., Miller, D.G., and Burns, H.J. (1978), 233

Loftus, G.R. (1972), 228

Loftus, G.R., and Loftus, E.F. (1976), xxi, 8, 19, 226, 291, 294, 318

Logan, G.D. (1979), 270

Long, L., and Boertlein (1976), 171

Lowe, D.G., and Mitterer, J.Q. (1982), 263

Lowe, R.H., and Steiner, I.D. (1968), 259

Lowery, S., and DeFleur, M. (1983), 128, 129

Lowin, A. (1967), 258; (1969), 258

Lublin, J.S. (1983), 435

Luhmann (1978), 330

MacCannell, D. (1976), 172

Machlup, F. (1962), xxii

MacKay, D.M. (1952a), 19

Mackworth, T.F. (1969), 263

Macy, R. (1983), 429

Malbin, M. (1982), 413, 424

Malmo, R.B., and Surwillo, W.W. (1960), 268

Mandler, G. (1967), 230

Marcel, A.J. (1980), 242; (1983), 228, 242

Marks, K. (1982), 81, 82

Martin, J. (1982), 433

Massad, C.M., Hubbard, M., and Newtson, D. (1979), 266

Masteron, J.T., Beebe, S.A., and Watson, N.H. (1983), 69

Masuda, Y. (1980), 9, 16

Maxwell, M. (1982), 383

McCormack, T. (1961), 47

McCroskey, J.C., and McCain, T.A. (1974), 207

McCroskey, J.C., and Wheeless, L.R. (1976), 252, 264, 266

McDermott, V. (1975), 129

McGaugh, J.L., and Herz, M.J. (1971), 238

McGee, M. (1985), 349

McGrath, J.E., and McGrath, M.F. (1962), 362

McGraw, K.O., and McCullers, J.C. (1979), 268

McGuire, W.J. (1966), 257; (1968), 258; (1969), 255

McLuhan, M. (1964), xxii, 19, 43, 407; (1974), 132, 144

McLuhan, M., and Powers, B. (1981), 435

McQuail, D. (1969), xxii; (1983), 43

McSpadden, M., and Loftus, E.F. (1983), 234

Mehrabian, A. (1981), 208

Meiklejohn, A. (1960), 88

Merleau-Ponty, M. (1962), 369

Metz, R. (1979), 395, 402

Meyer, D.E., and Schvaneveldt, R.W. (1971), 235

Pollack, A. (1982), 470; (1983a), 205; (1983b), 205

Pondy, L. (1983), 372

Pool, Ithiel de Sola (1977), 174; (1982), 154; (1983), 9, 12, 13, 16

Popper, K.R. and Eccles, J.C. (1978), 291, 292

Porat, M. (1977), 3

Porat, M.U. (1978), 463

Posner, M.I. (1978), 269, 270; (1982), 271

Posner, M.I., and Snyder, C.R.R. (1975), 263, 269, 270

Postman, L., and Phillips, L.W. (1965), 237

Postman, N. (1982), 176, 177

Presstime (1983), 429

Prothro, J.W., and Grigg, C.M. (1960), 353

Publishers Weekly (1982), 401

Rabbitt, P. (1977), 274

Rapoport, A. (1973), xxi; (1979), 170

Rappaport, R.A. (1971), 329

Read, W.H. (1983), 12

Reddy, M. (1979), 380

Reed, S.K. (1972), 229

Reid, A.L. (1977), 172

Reitman, J.S. (1974), 269, 272

Remington, R.W. (1980), 263

Renshon, S. (1977), 361

Revelle, M.S., Humphreys, L.S., and Gilland, K. (1980), 267

Rice, M.L., Huston, A.C., and Wright, J.C. (1982), 265

Rice, R.E. (1980), 167, 168; (1982), 168, 170, 171

Rice, R.E. and associates (1984), 162, 164, 168, 169

Rice, R.E., and Paisley, W. (1982), 162

Rice, R.E., and Richards, W. (1984), 168

Richmond, V.P. (1977), 261

Ripper, M.D., and Wanderley, J.L.C. (1981), 473

RMH Research Associates, Inc. (1982), 187, 195, 196

Robbins, L. (1945), 173

Robinson, J., and Converse, P. (1972), 174

Robinson, J., and Levy, M. (1983), 348

Roediger, H.L. (1980), 225, 226

Roediger, H.L., Neely, J.H., and Blaxton, T.A. (1983), 236

Rogers, C. (1980), 338

Rogers, E.M. (1962), 353; (1973), 134; (1983), 162

Rogers, E.M., and Adhikarya, R. (1979), 260, 261

Rogers, E.M., and Agarwala-Rogers, R. (1976), 167

Rogers, E.M., and Cavalcanti, C.P.B. (1980), 355

Rogers, E.M., Daley, H.M., and Wu, T.D. (1982), 162

Rogers, E.M., and Kincaid, D. (1981), 95, 110, 111, 168, 356

Rogers, E.M., and Larsen, J.K. (1984), 100

Rogers, E.M., and Shoemaker, F.F. (1971), 131, 260, 261, 353

Roper, J.E. (1983), 444

Rose, E.D. (1983), 13, 14

Rosenfeld, H.N. (1980), 159

Rosengren, K.E. (1980), 43, 45; (1983), 49

Routheir, M.E. (1979), 69

Ruben, B.D. (1972), xxi; (1975), 327; (1979), xxii; (1983), 326; (1984), xxii

Ruben, B.D., Holz, J.R., and Hanson, J.K. (1982), 11

Ruben, B.D., and Kealey, D.J. (1979), 335

Ruben, B.D., and Kim, J.Y. (1975), xxii

Ruben, B.D., and Kwasnik, B. (1985), 10

Ruben, R.B., and Ruben, A.M. (1982), 260

Ruchinskas, J. (1982), 164

Ruchinskas, J., and Svenning, L. (1981), 164

Ruesch, J. (1968), 332

Ruesch, J., and Bateson, G. (1951), xxi

Ruggels, W.L.P. (1976), 445

Rytina, S., and Morgan, D. (1982), 171

Sabatier, P., and Mazmanian, D. (1980), 475

Salamon, G. (1979), 274

Salton, G., and McGill, M.J. (1983), 294, 312, 319

Salvaggio, J.L. (1982), 433; (1983), 4, 14, 435, 437; (1984), 435

Samson, E.E. (1981), 245

Sanger, D.E. (1983), 470

Savage, R.L. (1981), 353

Schacter, D.L., and Tulving, E. (1982), 232

Schiller, A. (1981), 396

Schiller, A., and Schiller, H.I. (1982), 396

Schiller, H.I. (1978), 435; (1981), 466, 469, 471, 473; (1983), 13, 437; (1985), 13

Schmidt, J.S., and Corbin, R.M. (1983), 440